On Concepts, Modules, and Language

On Concepts, Modules, and Language

Cognitive Science at Its Core

EDITED BY ROBERTO G. de ALMEIDA

AND

LILA R. GLEITMAN

OXFORD
UNIVERSITY PRESS

OXFORD
UNIVERSITY PRESS

Oxford University Press is a department of the University of Oxford. It furthers the University's objective of excellence in research, scholarship, and education by publishing worldwide. Oxford is a registered trade mark of Oxford University Press in the UK and certain other countries.

Published in the United States of America by Oxford University Press
198 Madison Avenue, New York, NY 10016, United States of America.

Library of Congress Cataloging-in-Publication Data
Names: de Almeida, Roberto G., editor. | Gleitman, Lila R., editor.
Title: On concepts, modules, and language : cognitive science at its core /
edited by Roberto G. de Almeida & Lila R. Gleitman.
Description: 1 Edition. | New York : Oxford University Press, [2018]
Identifiers: LCCN 2017010025 (print) | LCCN 2017031978 (ebook) | ISBN 9780190464790 (UPDF) |
ISBN 9780190667528 (EPUB) | ISBN 9780190464783 (hardcover : alk. paper)
Subjects: LCSH: Cognition. | Cognitive science. | Semantics (Philosophy) | Language acquisition.
Classification: LCC BF311 (ebook) | LCC BF311 .O485 2017 (print) | DDC 153—dc23
LC record available at https://lccn.loc.gov/2017010025

For Jerry Fodor

CONTENTS

Preface ix
Contributors xi

Introduction: A Fodor's Guide to Cognitive Science 1
Roberto G. de Almeida

PART I Language and the Modularity of Mind

1. Two Notions of Modularity 25
   Noam Chomsky
2. Exploring the Limits of Modularity 41
   Merrill F. Garrett
3. The Modularity of Sentence Processing Reconsidered 63
   Fernanda Ferreira and James Nye
4. The Unity of Consciousness and the Consciousness of Unity 87
   Thomas G. Bever
5. Semantics for a Module 113
   Roberto G. de Almeida and Ernie Lepore
6. Center-Embedded Sentences: What's Pronounceable Is Comprehensible 139
   Janet Dean Fodor, Stefanie Nickels, and Esther Schott
7. Getting to the Root of the Matter: Acquisition of Turkish Morphology 169
   Natalie Batmanian and Karin Stromswold
8. Scientific Theories and Fodorian Exceptionalism 191
   Zenon W. Pylyshyn

PART II Concepts and the Language of Thought

9. Fodor and the Innateness of All (Basic) Concepts 211
   Massimo Piattelli-Palmarini
10. The Immediacy of Conceptual Processing 239
    Mary C. Potter
11. On Language and Thought: A Question of Formats 249
    David J. Lobina and José E. García-Albea
12. The Neurobiological Bases for the Computational Theory of Mind 275
    C. Randy Gallistel

Index 297

PREFACE

Far from enjoying any sort of consensus, cognitive science has always been boiling with disputes over its foundational assumptions—including the degree of its adherence to functionalism and nativism, the role of computations in cognitive processes, the very nature of mental representations, the nature of concepts, and the constraints on the architecture for cognition. And for about 60 years now, Jerry Fodor has been at the center of these disputes. The causes he has championed—carrying the ensign of the cognitive revolution—have led to major advances in how we conceive of the mind's internal representations and how these representations are put to use in mental processes.

The present volume epitomizes the excitement and controversies brought about by the ideas that have been the object of Fodor's characteristic analytic treatment (and occasional experimental investigation). The volume brings together newly commissioned contributions from some of the most influential cognitive scientists—some of whom are also central figures of the cognitive revolution—representing the wide spectrum of research within the field, including linguistics, psycholinguistics, visual attention, philosophy, and neuroscience.

The broad intellectual scope of the book is on the foundations of cognitive science and on one of its most important and prolific exponents. And true to the centrality of Fodor's ideas, two main topics emerge as common threads: the nature of concepts—thus the elements that are constituent of the "language of thought"—and modularity, in particular the modularity of language and vision, with implications for the architecture of the mind more generally. Both topics come loaded with hypotheses and empirical work, which are sure to promote yet further intellectual debate and experimental investigation, thus fueling advances in the field.

For convenience, we have organized the chapters into two major sections representing those two main topics, although there are lots of interactions between the modularity issue that permeates chapters in the first section, and issues on concepts and the language of thought that permeate those in the second. The view of conceptual tokening that Fodor has argued for is essentially (and perhaps unsurprisingly) modular and atomistic: to wit, the objects one perceives and attends to are causally linked to token symbols that stand for mental representations of those objects. And because Fodor assumes perception is modular, the causal links between things in the world and their representations are independent of any beliefs the perceiver may hold in stock.

Finally, for countless years, both editors have benefited enormously from Fodor's friendship, besides his intellectual brilliance. We are thankful to him for all. We also want to thank the contributors to this volume for *not* turning this into a feast, holding on to the premise of the volume, which was to be critical of Fodor as a sort of perverse homage, but a homage that all—including him—would find more fruitful for advancing our understanding of the mind.

Roberto G. de Almeida and Lila R. Gleitman
Montreal and Philadelphia

CONTRIBUTORS

Natalie Batmanian
Department of Psychology
Rutgers University
Piscataway, NJ, US

Thomas G. Bever
Departments of Psychology and Linguistics
University of Arizona
Tucson, AZ, US

Noam Chomsky
Department of Linguistics and Philosophy
Massachusetts Institute of Technology
Cambridge, MA, US

Roberto G. de Almeida
Department of Psychology
Concordia University
Montreal, Quebec Canada

Fernanda Ferreira
Department of Psychology
University of California
Davis, CA, US

Janet Dean Fodor
Program in Linguistics
The Graduate Center
City University of New York
New York, NY, US

C. Randy Gallistel
Department of Psychology
Center for Cognitive Science
Rutgers University
Piscataway, NJ, US

José E. García-Albea
Department of Psychology
Universitat Rovira i Virgili
Campus Sescelades
Tarragona, Spain

Merrill F. Garrett
Department of Psychology
University of Arizona
Tucson, AZ, US

Lila R. Gleitman
Department of Psychology
University of Pennsylvania
Philadelphia, PA, US

Ernie Lepore
Department of Philosophy
Center for Cognitive Science
Rutgers University
Piscataway, NJ, US

David J. Lobina
Faculty of Philosophy
University of Oxford
Oxford, United Kingdom

Stefanie Nickels
Department of Psychiatry
Harvard Medical School
Center for Depression, Anxiety, and Stress Research
McLean Hospital
Belmont, MA, US

James Nye
Department of Psychology
University of South Carolina
Columbia, SC, US

Massimo Piattelli-Palmarini
Department of Psychology
University of Arizona
Tucson, AZ, US

Mary C. Potter
Department of Brain and Cognitive Sciences
Massachusetts Institute of Technology
Cambridge, MA, US

Zenon W. Pylyshyn
Center for Cognitive Science
Rutgers University
Piscataway, NJ, US

Esther Schott
Department of Psychology
Concordia University
Montreal, Quebec, Canada

Karin Stromswold
Department of Psychology
Center for Cognitive Science
Rutgers University
Piscataway, NJ, US

Introduction

A Fodor's Guide to Cognitive Science

ROBERTO G. DE ALMEIDA

The so-called cognitive revolution—the second, by some accounts, after Descartes'—began taking shape over 60 years ago. Intellectual revolutions, as you probably know, are almost never the labor of a single mind (perhaps, again, *sauf* Descartes'). They are usually the result of scientific and philosophical discontent with modes of explanation and with the very nature of the explanan dum. And they lead inexorably to changes in theory and empirical object. Or so they should. The "second" revolution on the workings of the mind brought forth a torrent of new guiding assumptions in linguistics, psychology, computer science, and philosophy, among other disciplines. In this revolution, heads did not roll: they turned. The history of these disciplines and how they came to be together under the big tent of cognitive science cannot be reduced to just a few names, even if they are the names of true pioneers. But history *is* unfair and sometimes reduction is the only feasible way to convey the transformations a field might go through: the proper names become metonymic for the ideas, the ideas become standard (or, as it happens, generate classical controversies), and the history of the revolution is largely told by the names of those who are taken to push it forward. The short history I want to tell is like that. No matter how one maps it out, it has Alan Turing as an early influence, even if his influence was felt only later. And, of course, Noam Chomsky and modern—*Chomskyan*—linguistics are mainstays. Along the same lineage, the cognitive revolution owes Jerry Fodor some of its most fundamental ideas. Perhaps this lineage should be traced back to Plato, Ockham, Descartes, Locke, Hume, and a few others, with no clear discontinuity—certainly passing by Frege, the early Russell, and the early Wittgenstein. But in contemporary work, Fodor's name is metonymic with a kind of cognitive science—possibly *the cognitive science*—that many of us care about doing. One could refer to it as *Fodorian* cognitive science.

The chapters collected in this volume are a celebration of that kind of cognitive science, of its most fundamental ideas, and, in particular, of Fodor's contributions to psycholinguistics, to the theory of concepts (thus, to a theory of the elements of the *language of thought*), and to cognitive architecture, more broadly. I won't really call it *Fodorian* cognitive science because Fodor's contributions are so entrenched—and so inspiring and, because of that, at times so controversial—that I like to call it simply *cognitive science*. But this volume does not constitute the kind of celebration you might expect, the homage he would refuse. In line with his polemical style, the goal was to bring to fore a critical evaluation of the foundations of many of these ideas; we[1] wanted to put them to test, but also to move them forward (or, if it's the case, move away from them). We wanted, in sum, to examine the status of these ideas and how they might set the agenda for what is to come.

Now, here is some background on how we got thus far. Fodor's main contributions to cognitive science gather around language and thought and, of course, the nature of the language *of* thought, its elements—concepts—and how they connect to the world. This is not to say that all his philosophy of mind and philosophy of language—not to mention his many incursions into experimental work—can be *reduced* to language and thought, but these are the key terms of a deep and wide theorizing about the nature of the human mind.

Fodor entered the scene around 1959 when, coming out of doctoral work with Hilary Putnam in Princeton, he arrived at MIT. The "second" cognitive revolution, then, was "in the air." Chomsky was there, himself surrounded by behaviorist territory—and at striking distance. George Miller was then infiltrated in that territory, at Harvard, starting a program of research that was full-blown cognitivist, having among his aims the marriage between the new linguistics and a psychology that was increasingly soaking up computational metaphors.[2] The story is long and its plot reads somewhat like a thriller, as far as intellectual thrillers go. Chomsky (1959) had just famously exposed the limits—and the explanatory inadequacies—of behaviorism: to put it simply, there had to be a mechanism underlying both language attainment and language use, and that mechanism was far more complex than simply pairings of stimuli and overt or covert responses. The plot thickens, for much of this revolution was also taking place elsewhere—in computer science and in psychology, in domains such as in memory and attention. But here the focus will be mostly on the bits of how psycholinguistics and the core of cognitive science became what it is today (or what it is supposed to be).

At MIT, in the early 1960s, Fodor (re)encounters other young co-conspirators, including Tom Bever and Merrill Garrett, both of whom had been infiltrated in behaviorist territory before: Bever at Harvard, Garrett at the University of Illinois. Fodor had been a visitor at Illinois, where he exchanged ideas on the nature of psychological explanation with Charles Osgood, one of the leading behaviorists then. I put this very politely, because this visit—and those exchanges—only deepened the canyon that separated the two then competing worldviews, particularly on what pertains to language and cognitive processes.[3] Osgood later remarked that science can be "faddish,"[4] with which I agree, but for reasons that will soon become clear.

Meeting his younger co-conspirators was instrumental in advancing the cognitivist brand of psycholinguistics into enemy territory. It was then and thence that, with the little camouflage that Jerry's old *Austin Healey* provided while crossing Cambridge, MA, or in the trenches of their improvised lab, Jerry, Tom, and Merrill plotted about changing psycholinguistics; or perhaps plotted about new ways of testing linguistic postulates deploying experimental methods. The psycholinguistics of the early 1960s was still dominated by what we can call "psychology of language," mostly destitute of its linguistic core. There were, of course, very notable exceptions: Miller was then the main driving force behind a new linguistically informed experimental psycholinguistics. What was in the air—or in a few of those minds—helped establish the materialistic mentalism that was rejected by the dominant "behavioral science." Crucial to these advances was *Chomskyan* linguistics (the adjective stands for what was then already a small legion), which was beginning to thrive, thus providing the impoverished psychology of language with the algorithms it was supposed to run. To be clear, that's not the beginning of psycholinguistics, for since the early 1950s the term was already being thrown around, labeling other forms of contact between the structuralist linguistics of thence, experimental psychology, and theory of communication. And certainly that was not the beginning of experimental psychology of language, which can be traced back to James Cattell and Wilhelm Wundt. But it was the pinnacle of theoretical work on the formal properties of the apparatus that yields a language—and, by hypothesis, the mechanisms for its use—combined with the experimental paraphernalia of incipient cognitive psychology, both heavily guarded by philosophical functionalism about psychological explanation. Those were the beginnings of Cartesian Psycholinguistics.

A small portrait of Turing could have been hanging there somewhere, perhaps in the improvised lab, as a reminder of the agenda for cognitive science (nobody called it that, then): the prospects for a theory of language hinged on understanding the nature of its underlying rules and representations. And so did the prospects for a theory of mind (at least some of its main faculties or *modules*, as we'll see). Many experiments ensued and many techniques were developed, beginning with the "clicks" that perceptually (and illusorily) marked the boundaries between clauses within sentences. We were then beginning to "see" what the mind does when a sentence comes its way. The black box was cracked open; rats and pigeons were spotted in the unemployment line. The results of this collaboration appeared in many experimental and theoretical papers spanning over 10 years, with its apotheosis being Fodor, Bever, and Garrett's classic *The Psychology of Language: An Introduction to Psycholinguistics and Generative Grammar* (1974).[5] The one-liner would read like this: the computations involved in understanding a sentence do not go *pari passu* with the transformations that grammatical principles determine for its structural analysis—from its surface form to its kernel—nor are the computations effected by some analysis by synthesis in which the grammar provides a set of possibilities, a search space. The computations involved in sentence comprehension rather rely on heuristic procedures, perceptual strategies for analyzing sentence constituents based primarily on the detection of clause boundaries together with analysis of constituent structure within those clauses.

There were at least two main arguments for not tying the computations involved in sentence perception to actual sequences of syntactic transformations; and these arguments are as valid today as when they were first put forth, even if both grammatical theory and parsing models have moved away from transformations. One is that the putative linear perception of a sentence allows, at every moment, for myriad possible structures compatible with the input, yielding a "search space" that is less than practical, perhaps close to impossible, for quick online structuring and interpretation. Another is that different transformations (when we take movement of constituents into account) lead to partial analyses that can be deemed incompatible with surface input—even ungrammatical—raising the need for almost constant course corrections in parsing and interpretation.

Phrase structure grammar, it should be clear, was said to underlie sentence structure: it was "psychologically real," as people used to say back then. That was not under dispute within the cognitivist camp. But the process by which phrase structure is perceptually computed was said to rely on independent phrasal packaging mechanisms. This view later evolved into different parsing models, from the "sausage machine," proposed by Lyn Frazier and Janet Dean Fodor, all the way to the "good enough" approach that Fernanda Ferreira has championed more recently.[6] Parsing models have since swung between these views—with linguistic principles either actively engaged in the moment-by-moment analysis or operating on the product of other processing routines that are taken to be more or less dedicated to processing language (I will leave aside those who believe there are no linguistic principles at all).

Even if the empirical data did not fully support the clause as the perceptual unit, the 1960s and 1970s psycholinguistics chapters of the cognitive revolution became paradigmatic of what cognitive science came to be (or was supposed to be): largely interdisciplinary, a collaborative enterprise without boundaries between established disciplines or departments (well, now I may be daydreaming). It is not the case though that all rebels were speaking in unison but their voices were then shouting in a similar direction. The beauty of rationalist revolutions is that no blood spills on the streets.

A few years before, in his *Psychological Explanation* (1968), Fodor had focused on the metatheory for cognitive science—or at least on one of its main philosophical approaches: *functionalism*. Along the lines of Chomsky's (1959) attack on B. F. Skinner's *Verbal Behavior*, in *Psychological Explanation*, comes a more detailed plan for attacking the philosophical foundations of behaviorism and its positivistic roots—but chiefly the anti-mentalism represented mainly by G. Ryle. The emphasis was on understanding what constitutes an explanation in a "special science" like psychology. The plan had two fronts. One was the idea that an explanation of "behavior" could not dispense with the underlying causes of which overt behavior is only a consequence (and limited at that). Another was the rejection of reductionism for psychology. Vienna gave us great music (as Fodor knows it well) and great philosophy (though he might question its consequences), but the logical-positivistic thesis that eventually all sciences could be reduced to physics did not bode well for psychology, not at least for *cognitive* psychology.

Both in *Psychological Explanation* and in papers collected later in *Representations: Philosophical Essays on the Foundations of Cognitive Science* (1981), Fodor argued for the special status of functionalistic explanations. If you are in doubt, here is a clarification: functionalism in philosophy of mind is materialism, but the kind of materialism that does not appeal to the nuts and bolts of the machine (or neurotransmitters and neuroanatomy, for that matter): it takes functional properties to suffice at a certain level of explanation. This level is something Alan Newell, Zenon Pylyshyn, and other bona fide cognitive scientists have called "symbolic" or what David Marr called the "representation and algorithm" level. Whichever label one chooses for it, or whichever way one partitions the analysis, it is at the symbolic/algorithmic level that a cognitivist explanation about rule-governed processes is best conceived. And it is also perhaps at that level where we should begin to approach the so-called knowledge-based processes, the intentional kinds. (Not to be forgotten: cognitivist/functionalist explanations also appeal to folk-psychological mental states or attitudes: it is because I planned to write these very words—following a long chain of desires, beliefs, hopes, doubts, and (in)actions—that I actually came to do it.) And no matter how one conceives of this relation—between the symbols and what they represent, between the rules and their following in the course of mental processes—no particular status is given to the "other" level, the biological or implementational one. The issue is about explanation and *qua* explanatory level, appealing *just* to biology won't do. Of course, one should not be in any way discouraged from actually pursuing an investigation of the biological level, quite the opposite. Functionalism is materialism, one must insist: It is assumed that functional processes supervene upon physical ones, as Fodor keeps saying. But before the functional magnetic resonance imaging (fMRI) machine is plugged in, one needs to make sure to have a good working theory at the symbolic/algorithmic level.[7] Revolutions are more effective when transformations take place at the infrastructure; and, in the cognitive revolution, the infrastructure is not in the biological bases of behavior but in the functional mechanisms that the biological substrate executes. There is, however, a very active minority of cognitive *neuro*scientists (the prefix is more fashionable, these days) who are realists about representations—who follow something like a methodological law: neuroscientists who are cognitive scientists *have* to postulate representations.[8]

There is for sure a direct connection between the battles fought in psycholinguistics and in philosophy of mind. The cognitive revolution needed to have its guiding assumptions about the object of investigation clear: the internal states of the organism. But at the same time it also needed to show that its theoretical and empirical accounts of these internal states had validity. This amounted to both, postulating the nature of the internal system of representations and processes underlying those states *and* providing empirical evidence for their workings. Psycholinguistics was at the center of this program of research because it had all the ingredients necessary to build a theory of mind: it had the symbols and algorithms for linguistic computations, and a mechanism for yielding internal representations. And in true cognitive-science fashion, no type of evidence was ruled out. Just as distributional arguments and crosslinguistic evidence were important for advancing *linguistic* arguments, they were also important for advancing arguments about the nature of the mind's *internal code*. Experimental evidence—coming from all corners of booming cognitive psychology—was also instrumental in pushing forward the agenda. Fodor has been committed to the science of the mind *and* to its philosophical foundations, and very few—in the last couple of centuries—have been able to keep these commitments as prolifically as he has; and few have transited between the science and the philosophy of mind with the same ease.

The Language of Thought (*LoT*, 1975) is typical of this attitude, a landmark for psycholinguistics and for the view that the mind is best conceived as a computational device. It is in *LoT* where we see three of Fodor's main threads coming together like in no work before: philosophy of mind/science, psycholinguistics, and the roots of his Computational/Representational Theory of Mind (henceforth, C/RTM). It is in fact his view of cognitive architecture that begins to emerge, with implications for several of his lines of work. *LoT* was much more than the "speculative psychology" he claimed it was. It detailed what a commitment to C/RTM entails:[9] first there ought to be representations, if explanations appeal to anything other than simple overt behavior, and representations are the medium for processes, which are carried out as computations. *LoT* had it moreover that many mental states were relations to propositions, that to believe or to desire *P* was to be in a relation—a *computational* relation—to a representation of *P*, which was couched in the vocabulary of the internal code. Computations leading to mental states were taken to be sequences of events akin to derivations (e.g., the sequences of syntactic operations; or the sequences from premises to conclusion in syllogistic reasoning). This is in essence what constitutes the common operations of putative cognitive processes. And, by hypothesis, the language of thought bears many of the properties of natural language: it is recursive, productive, compositional, and it is a typical computational system, for its processes too are computations over (symbolic) representations. There is a caveat, though: as Fodor warns us in the last chapter of *LoT*, quite possibly a few ("more than none") cognitive processes behave that way, but most likely not all do. Cognition is to a large extent holistic, context-sensitive (think about, e.g., decision making). And there might be lots of propositional attitudes that are not computationally derived—for example, those whose causes are not psychological.[10] But if we were to have a (cognitive) psychology, a good way to start was to devise a theory of the internal representations and how these representations were manipulated in mental processes.[11]

The plan for cognitive science taking over all (relevant) psychological accounts of typically *cognitive* processes was not complete, of course. First, because there was no detailed plan to follow: cognitive science from its inception has been anarchic, and it was then barely holding on to a few postulates on what constitutes the proper level of analysis. And second, because the conception of the mind that was then emerging begged many questions: What was the nature of the code? Or how many codes were there? Which processes were supposed to be computational and which ones were not? As more specific hypotheses about the nature of representations and processes were ironed out, yet deeper questions internal to the program were raised. One of Fodor's key concerns was mental content—roughly how symbols get to represent what they do and how they enter into putative intentional processes. This appears early on in *LoT* and in the original essays of *Representations*. In fact, accounting for the nature of the units of representation—let's call them *concepts*—became one of Fodor's main missions, spanning over 50 years of hard labor. And not surprisingly, this is perhaps *the* central issue in cognitive science, for it underlies many others, from the nature of visual processes of object recognition, to language comprehension and production, and certainly to many "high-level" processes we can call *thinking*. If concepts are the building blocks of the representations manipulated in all these processes, if they are the building blocks of all processes that employ anything having to do with content (all that's relevant about perception and cognition, as far as I can tell), then how are they represented, and how are they developed in the organism?

Fodor once said that every Monday morning there was a meeting at MIT to decide what would be innate that week; whoever had the most outrageous proposal would chair the works. I don't think this is entirely a joke, as it is clear that nativism of some sort is the only route to the postulation that internal states develop and change partly in response to environmental causes. Poverty of stimulus arguments stand not only for language but for concepts too. It was in this context—perhaps in one of those Monday morning meetings—that conceptual nativism became central to Fodor's work. In his early treatment of conceptual nativism, he showed that the process of concept attainment couldn't be anything nearly what many cognitivists and practically all empiricists were postulating it was: a process of learning. More than an assertion, there was an argument, a puzzling one. Fodor suggested that what was being shown by Jerome Bruner, Jean Piaget, and others, under the rubric of "concept learning," was what he called "belief fixation." Roughly, decisions about the extension of a given word/concept—say, *wyz*—presuppose the existence of the criteria (features or properties such as *ROUND* and *GREEN*) upon which those decisions are based. Thus, what the organism has at its disposal are the very premises for inductively fixating the belief or hypothesis that the referent is a *WYZ* (and all this requires, of course, a vocabulary of representations, a language of thought). This kind of argument made strong waves in the canals of the Abbaye de Royaumont, near Paris, where, in 1975, Massimo Piattelli-Palmarini brought together Chomsky, Fodor, and other nativists for an epic debate with Piaget and his constructivist colleagues. Legend has it that some of the best arguments pro nativism still echo in the Cloister.[12]

Nativism about concepts, contrary to the popular joke—that we know the likes of *AIRPLANE* and *ELECTRON* from birth—assumes that the conceptual stock must be primitive. The problem is that, on pain of committing to analyticity (see later discussion) or, worse, to the idea that concepts are structured (the problems do overlap), the conceptual stock has to be vast, having more than just the sensory primitives of classical empiricism. Even the classical empiricists—Locke and Hume—were committed to some form of nativism, except that their commitment was to the sensory basis or to the conditions for picking out the sensory basis. But the sensory apparatus—or what the sensory apparatus, by hypothesis, yields—underdetermines the bases upon which belief fixation relies. Hence, the only way out of this morass is to assume that indeed the *conditions* for fixating *AIRPLANE* and *ELECTRON* are innate. It's the structure of the mind that allows for the triggering of concepts by experience. And because all concepts are acquired like that or because most concepts are *triggered* like that, they ought to be considered all primitive, *atomic*, not molecular.

As the reader surely noticed, despite all denials, Fodor flirts with empiricist postulates, but not with the kind of empiricism that is radically anti-nativist. In fact, he denounces a strict dichotomy between empiricism and nativism. Fodor is empiricist with regard to the primacy of the perceptual input in causally determining—or triggering—the conceptual stock. Since all lexical concepts are primitive, or all lexical concepts arise from primitive functions, the main worry is how the organism works on triggering or fixating its supposedly vast stock. He assumes that it is probably the basic level—*DOG*, not *POODLE* or *ANIMAL*—what is first triggered by the environment, and one works out different levels of generalization or specificity along the way. Notice that, contrary to what one would suppose—if classical empiricism were to be enforced—it is not *RED* and *LINE* that the child picks up, but putative links with referents that are possibly at the basic level of abstraction. And even in the case of *RED* and *LINE*, what determines their primitive status is not that they are sensory, it is that they bear *properties*.

The arguments in *Representations* and *LoT* surely raise lots of questions. The work of understanding how concepts get linked to their referents requires fine-tuning a cognitive architecture that affords these links. Enters, then, *The Modularity of Mind* (1983). Though *Modularity* does not appear to be "causally connected" to the early work on concepts, it plays an important role in Fodor's program. It is the centerpiece of much of his work linking the psychology of perception, C/RTM, and the idea that higher cognitive processes involve large doses of belief fixation. It is *via perception* in fact that belief fixation begins to take place, with the triggering or matching of concepts by referents. Perception—at least the classical empiricism way—was a process of matching a thing to an Idea, a process that was atomistic, as Fodor noticed. It is somewhere here that the modularity of perception and atomism about meaning meet: roughly, seeing a cow triggers *COW* (even if you think there are features, seeing a spot or a horn triggers *SPOT* or *HORN*).

The story about atomism in conceptual representation and the story about the modularity of perception, then, are complementary: if you believe, as Fodor does, that much of conceptual tokening is "brute-force" linking between referents and their representations, you are somewhat committed, as he is, to the modularity of perception. What you see—to stick to vision—is independent of what you believe; and the concept you token is too independent of other sorts of beliefs you might have. What Fodor proposed in *Modularity*, more specifically, is that the perceptual analysis process is highly constrained. In his version of modularity, Fodor takes perceptual analysis to be encapsulated from the rest of cognition, with modules, notably vision and language, separate from each other and from other systems. The modules have their own rules and have access to their own representations, mostly the ones that are causally connected to the analysis of input post-transduction—they are causally connected, in sum, to the kinds of stimuli that are the modules' natural kinds. Crucially, modules, in their task of producing perceptual analyses, are not influenced by the beliefs that the organism has at its disposal. It is here where Fodor traces the line between perceptual computations and what he called the holistic, Quinean, central system, where all outputs of modules eventually meet. There is an epistemological thesis here as well: observation and inference ought to be kept apart, just like perceptual computations and beliefs ought too.

I won't say much more about modularity because several of the chapters in the present volume assess the modularity hypothesis, what became of it, and even how it can be reframed in current cognitive science.[13] But I want to call attention to *Modularity*'s sizeable impact in the psychology of perception, where it set the agenda, the guiding hypotheses on how language (in particular, but not exclusively) might be perceived. Fodor's formulation assumes that the module for language is dedicated to input analysis (though here he spars with Chomsky),[14] producing what ought to be minimally some form of syntactic or perhaps something like a logical representation of the linguistic input. The general idea of modularity was as well in the air when Fodor wrote his influential monograph, but he refined the hypotheses and marked the boundaries between two main psycholinguistic camps: those who assume some level of autonomy for language perception (and its internal computations) and those who assume perception to be, in the term coined by Pylyshyn, "cognitively penetrable." Most parsing models from the early 1980s were predicated on how much or at what point in time they allowed for non-linguistic information (non-sentential context, beliefs, expectations) to influence structural decisions. This issue has never really been settled. And although I am not keen on appealing to arguments from philosophy (or sociology) of science to legislate on matters in need of theoretical and empirical treatment, it is worth emphasizing that, as Feyerabend (1975) once put it,

> No idea is ever examined in all its ramifications and no view is ever given all the chances it deserves. Theories are abandoned and superseded by more fashionable accounts long before they have had an opportunity to show their virtues. (p. 35)

Maybe the modularity hypothesis is not at that stage yet—it has neither been abandoned nor superseded, despite the enormous amount of research conducted on behalf of its constituent postulates. But it is clear that fashions change—and research grants go with them.

We have to admit, then, just like in Osgood's reaction to the new psycholinguistics in the 1960s and 1970s, that science can be faddish. But it seems that, in its current stage, cognitive science does not have many viable alternatives other than to assume—as a working hypothesis—that some of its main systems might be encapsulated and, moreover, that some or perhaps most of its representations and processes are symbolic and computational. One might think, of course, of scores of alternatives to the architecture that C/RTM breeds. Think for instance of connectionism, which was trumpeted, when it came out in the late-1970s, as a revolution (within *the* revolution, I suppose). Connectionism was supposed to provide cognitive science with what it appeared to lack: some strong neurological plausibility; it was supposed to rescue physicalism while holding on to the idea that representational states (the activated nodes) are entertained in the course of cognitive processes. Moreover, what gave connectionism its most plausible selling point was the idea that representations were causally connected as if they were (actual) neuronal networks—with their activation and inhibition functions as well as learning capabilities operating as massively parallel, interconnected units. But soon—perhaps not soon enough—it became clear that connectionism failed to account for many of the key properties that C/RTM took to be front-and-center. Fodor's move to Rutgers University, in the late 1980s, afforded a closer collaboration with Pylyshyn, at the Rutgers Center for Cognitive Science, which they founded (not to be discounted were also the strategic proximity with the opera at the Met and the sailing on the Hudson). They were then engaged in dismantling the tenets of connectionism as an explanatory model for the mind. Many of the tools for that job were already out, in Pylyshyn's (e.g., 1984) and Fodor's (e.g., 1987) own work.

In a seminal paper, Fodor and Pylyshyn (1988) argued that connectionist representations and processes failed to account for some of the key properties of cognitive systems: that they are compositional, productive, and systematic. Crucially, complex representations have constituent structure, which activated nodes in connectionist networks lack. Fodor and Pylyshyn's position on the nature of cognitive architecture has wide consequences for the nature of cognitive representations and processes and, more broadly, for how work on cognitive science ought to progress. Productivity here is key, for if complex representations (thoughts, sentences, perhaps the output of visual processes) do not have constituent structure, are not systematic and, ultimately, if complex expressions are not compositional, then cognitive processes can't be productive. And if cognitive systems aren't productive, how do we manage to say, understand, and think expressions we never said, understood, or thought before? To put it even more dramatically, it seems that the only way to conceive of a mind with an infinite capacity out of its finite resources is to assume that its elementary representations enter into complex structures that are systematic, compositional (and recursive), and thus productive.

It is healthy for any science to have competing paradigms, except that alternatives to symbolic cognitive architecture clearly aren't up to the task. Connectionism cannot account for recursivity, so it appeals to the likes of recurrent networks, which merely mimic recursion. And, as Fodor and Pylyshyn put it, connectionist representations are not compositional: contrary to symbolic expressions, which actually contain their constituent representations, higher nodes that stand for more complex representations do not contain the lower token simplex nodes/representations to which the higher ones dynamically respond. Conversely, a node that stands for a complex representation does not really entail the simplex nodes that are supposed to stand for its constituents. In fact, there is nothing lawful in an association between nodes to the point that a node that stands for *P&Q* can be associated with *P* but not with *Q*. Overall, connectionism cannot give an account of the productivity and systematicity of complex representations: because they are not compositional and do not allow for hierarchical structures and recursion, the only way connectionism can mimic productivity and systematicity is by creating new nodes. But it is not only connectionism that fails to account for the productivity of mental representations: a variety of frameworks (e.g., usage-based language representation, embodied cognition) do too. The main point about Fodor and Pylyshyn's view of the architecture of cognition is that the finite elementary symbols/representations ought to yield for an infinite capacity and the only way known to humankind that this can be achieved is by assuming that cognitive capacities are truly productive (and compositional and systematic), which thus far—circa 2017—only symbolic architectures do.

In his work on cognitive architecture Fodor[15] has emphasized the role of compositionality in complex representations (sentences, thoughts). Compositionality became, in fact, the ensign in the crusade—a "nonnegotiable assumption" in Fodor's take on thought and language. One might suppose that the very idea that the meaning of a sentence/thought should be compositional borders triviality; but it is often the seemingly trivial ideas the ones that make the most noise in cognitive science (take commonsense psychological explanations as a twin example). Compositionality is satisfied, to be clear, when the meaning of a complex expression (sentence/thought) is obtained from the meaning of its constituents (say, morphemes or concepts) and how they are syntactically arranged. As trivial as this might be, opposition to this general principle is the rule rather than the exception. The vast majority of positions in philosophy of language, linguistics, and cognitive psychology, to name the main parties in this dispute, take the meaning of an expression to be rather a function of "semantic features" of the expression's constituents, or to be images, or to be statistical averages (viz., prototypes), or stipulations, or inferential roles, or activation patterns, or to be contextually determined, or something else (the list is vast—and all "or's" are inclusive). How is then Fodor (and colleagues) supposed to take offenders to task? In philosophy of language and mind, Fodor and Ernie Lepore mounted a scathing review of the main positions out there in the market, starting with their *Holism: A Shopper's Guide* (1992). The intricacies of their analyses are way beyond the few words I can write here, but the message is clear: holism is the antithesis of compositionality and thus holism has to be false unless one gives up on the idea that sentences and thoughts are productive and systematic. The crux of the problem goes back to Quine's position on the analytic/synthetic distinction. Since as far as I know nobody has ever came up with the principles for sorting out content-constitutive from contingent properties of a complex representation, the only way to account for lexical-conceptual content while preserving compositionality is to appeal to atomism (of course, contrary to Quine's solution).

Fodor and Lepore's attack on analyticity (of the lexical-content kind) did not stop there: in a series of articles published in the collection *The Compositionality Papers* (2002), they turned their analytical wrath against other offenders. They argued for a position that preserves the "classical" compositionality principle and worked on the details of their approach in typical fashion: showing that a variety of proposals for combining concepts would not work for being committed one way or another to the analytic/synthetic distinction. The solution Fodor and Lepore propose is to assume that lexical concepts are atomic—that is, denotations of token lexical items. Complex representations are obtained only via syntactic/logical form operations introduced by particular types of lexical items. Under their approach a lexical item is complex only in the sense that it specifies, beyond its denotation, a rule for its composition—namely, something akin to an argument structure or a rule for determining the logical form of the expression it partakes. To put it lightly, it's not the content of a token item that is complex, it is its structural/compositional properties—namely, syntax. This view has far-reaching consequences for the nature of semantic/conceptual representations, for the nature of compositionality and, of course, for how language maps onto meaning. With no solutions in sight for the analytic/synthetic distinction, one's choices besides atomism are harsh: either committing to the distinction or abandoning it and adopting some form of holism. Even though these two options lead to a dead end for semantics, a common methodological strategy in the lexical-semantics literature is to shove the problem under the rug and to embark on an empiricist approach to finding the ultimate constituents, the primitives of all lexical concepts.

In several works, notably in *Psychosemantics: The Problem of Meaning in Philosophy of Mind* (1987), *A Theory of Content and Other Essays* (1990), and

The Elm and the Expert: Mentalese and Its Semantics (1994), Fodor addresses key issues on the nature of content, in particular, on the link between tokens and the properties that concepts express, while mounting a defense of C/RTM for commonsense belief/desire psychology. But it is in *Concepts: Where Cognitive Science Went Wrong* (1998a), where many of these problems are brought to fore in the context of psychological theories. *Concepts* is perhaps Fodor's most developed work on the nature of concepts, and in particular on the metatheory of conceptual representation and development. The book picks up where several other works left off, chiefly *The present status of the innateness controversy*, one of the original chapters of *Representations* (1981). But to get into *Concepts* we need to take a small detour and revisit the early days of lexical semantics.

Fodor's first incursion into the field of lexical semantics (or concepts) was a collaboration with Jerrold Katz, starting when they met in Princeton in the late 1950s and, again, at MIT, in the 1960s. Together they worked on some of the principles of what later became Katz' much more developed semantic theory. In their early work, Katz and Fodor (1963) were strongly committed to a form of lexical-semantic representation that was entirely built on constituent features or "semantic markers." Semantics, for them, was supposed to constitute an autonomous component of linguistic analysis—one that would take the output of structural descriptions provided by syntax and produce a semantic description of token items, based on their constituent features and how they combined. But there was no account of analyticity then, that is, there were no principles governing the selection of semantic markers as constituents of lexical content. And Fodor, soon after, jumped ship.

It is ironic that lexical atomism was borne out of lexical decomposition but that is what happened when Fodor entered into what became known as the "linguistic wars"—though waging a war of his own, one that was not necessarily on the side of the "interpretive semantic" establishment, much less on the side of the opposing "generative semantics." One of the main battles of the "wars" was on the very nature of the division of labor between syntax and semantics: the "generative semantics" movement then assumed that a linguistic description ought to include both syntactic and semantic variables—that putative semantic properties such as causality would constitute part of the grammatical/semantic "deep" constituents that linguistic analyses would yield. The generative-semantics' view, then, was that syntax was not autonomous and that structural analyses of sentences ought to include predicates that were effectively deep-structure representations of surface verbs and their syntactic relations. To put it in other words: the translation of a sentence into its semantic representation required, among its operations, decomposing morphologically simplex verbs into predicate structures containing primitive, morphologically covert predicates (the likes of *CAUSE*) and their syntactic relations to other sentence constituents. Fodor's (1970) paper, *Three reasons for not deriving "kill" from "cause to die"* effectively showed that sentences containing the periphrastic *cause to die* were not synonymous with those containing *kill*. For instance, we can have "John caused Mary to die on Friday by poisoning her food on Thursday," but not "John killed Mary on Friday by poisoning her food on Thursday." Unless *cause to die* does not mean *CAUSE TO DIE*, which would be shocking, we should expect the simplex verb and its periphrastic pair to hold their distributional properties—that is, to "behave" the same way—or at a minimum to yield the same semantic representation. But they didn't, so, Fodor concluded, "kill" couldn't possibly mean *CAUSE TO DIE*. It was the end of Fodor's fleeting commitment to semantic decomposition and the beginning of a life-long crusade against it.

The papers that followed, with Merrill Garrett and Janet Fodor,[16] among others, included empirical—namely, psycholinguistic, experimental—investigations of the *kill/cause-to-die* asymmetry and related cases, showing that semantic decomposition does not seem to be at play when we understand sentences. If what we do when we understand sentences is indeed to recover their semantic/conceptual representations (what else?), we should expect processing complexity effects to arise when simplex verbs by hypothesis turn into complex structures at the semantic or conceptual level. Recall that C/RTM is in effect and more complex computations ought to yield something like greater processing time or some other complexity effect compared to simplex ones. But complexity effects were not obtained in the majority of experiments investigating the semantic complexity of verbs, in experiments that have employed a variety of methods and sentence types.[17]

It is never the case that theoretical advances—or choice between alternatives—are solely determined by empirical data. Arguments do carry the heaviest load. In the case of lexical concepts, linguistic and philosophical arguments against decomposition allied to the virtual lack of experimental support for decomposition could be taken as the triumph of the alternative—atomism. Fodor takes up the task of developing atomism more prominently in both *A Theory of Content* and in *Concepts*. In this later work, in particular, he looks deep into current (then and now) theories of concepts taking the "nonnegotiable assumption" of compositionality to be the yardstick for measuring the goodness of a concept theory, on the assumption that concepts are the elements of thoughts and that thoughts are compositional. In Fodor's analysis, all decompositional views get similar diagnoses. Concepts can't be *definitions* a la Katz and Fodor or a la Ray Jackendoff and others; definitions *can* be compositional but, remember, having definitions entail a commitment to the infamous analytic/synthetic distinction, which does not exist at press time. Besides definitions, Fodor's analysis centers on the prototype theory and its kith and kin: concepts can't be prototypes because prototypes do not compose when they enter into complex expressions—that is, they do not contribute their content (their prototypes) to complex concepts, which, by hypothesis have their own prototypes. Think about the *PET FISH* problem: *PET FISH* should have its own prototype, which does not have among its constituents the prototypes of *PET* and *FISH*. And finally, Fodor shows that if compositionality is to be taken seriously, concepts can't be *theories* either; obviously, theories do not compose and they are at the extreme end of the holism continuum if such a continuum exists. Strictly speaking, holism can't be true because, among a constellation of problems, if our concepts were dependent on all our beliefs, at a minimum this would violate the publicity of concepts and no two people would ever be talking about the same thing. Moreover, nobody would ever be able to entertain the same thought twice, for the constituents of thoughts would be constantly and forever changing. Neither the publicity nor the stability arguments, of course, deter the proliferation of holistic theories as the current popularity of, say, "embodied" cognition can attest.

Then, if the arguments against holism are right, and if we hold on to the compositionality yardstick, we are left with atomism yet again. It is the only view of conceptual representation that is both compositional and not committed to an analytic/synthetic distinction; the only view of conceptual representation that is compatible with C/RTM. The story seems coherent and well knit, but I am not showing all its knots. The general point is, as Fodor wrote in *Representations*,

> If we *are* going to have a cognitive science, we are going to have to learn to learn from our mistakes. When you keep putting questions to Nature and Nature keeps saying "no", it is not unreasonable to suppose that somewhere among the things you believe there is something that isn't true. (p. 316)

The question of decomposition is one for which Nature keeps saying "no." The case against conceptual decomposition—or, conversely, the case *for* atomism—is one in which arguments and much of the experimental evidence point in the same direction. But the last I checked, most concept theories in psychology and lexical-semantic theories in linguistics haven't addressed the key issues that Fodor raised in *Concepts* and in many of the papers that appeared in his *In Critical Condition: Polemical Essays on Cognitive Science and the Philosophy of Mind* (1998b): instead most theories opted for vexingly ignoring arguments against holism, for the impossibility of an analytic/synthetic distinction, and for the central architectural postulate of compositionality. There are sociological arguments for this neglect, but I won't descend to that.

Much of Fodor's subsequent work, including *Hume Variations* (2003), and *LOT2: The Language of Thought Revisited* (2008), was dedicated to advancing the cause of C/RTM and making the case for atomism. I say "advancing" but, true to his work, theoretical reflection often involves long and healthy therapeutic sessions (often in group, often in the Insolvent, with Granny, or Aunty, or Snark, or Mr. James, or simply beloved Greycat). The challenges are great, but not insurmountable. For instance, assume that atomism is indeed the only theory compatible with C/RTM and that what C/RTM postulates is that higher cognitive states are essentially relations to propositional attitudes. If concepts are atoms and if atoms are elements of mentally represented propositions—thus, elements of thoughts and their causal relations—how can holism be avoided? In other words, if it is postulated that higher cognitive mechanisms are predicated on the causal relations between beliefs and desires expressed as propositions, on what basis do conceptual/propositional relations obtain? As an admittedly simplified example, consider again the case of *kill/cause to die*. How can the inference *x kill y→y die* be obtained unless *kill* is something like *cause to die*? Causally determined inferential relations are what functionalism takes to be central to cognitive processes, but the conditions under which inferences are to be obtained appear to be incompatible with atomism, and are surely in conflict with rule-governed, Turing-like computations. Early on, there was an appeal for meaning postulates—a la Carnap—to take care of inferences that appear to be content constitutive. But, in *Concepts*, Fodor all but abandoned that solution on grounds that meaning postulates that are simply inferences holding between lexical concepts without being *necessary* (viz., encoding empirical knowledge) are, to put it mildly, too weak an alternative.[18] Besides the problems that one faces trying to put together the idea that concepts are atomic with the idea that psychology is intentional *and* computational, there are problems in the architecture front. As Fodor argues in *The Mind Doesn't Work That Way* (2000), C/RTM (or just CTM) is in trouble for it does not seem to work with abductive inferences, which constitute much of the workings of higher cognitive processes. This is a problem for the architecture of cognition *tout court*—"higher" cognition that is—but not so much for processes that are modular. Something's got to give.

Fodor's perennial existential crisis is the existential crisis of cognitive science—it's ours to own. His latest book, with Pylyshyn, *Minds without Meanings: An Essay on the Content of Concepts* (2015), tackles the nature of the connection between the referents—the things out there in the world—and their symbolic representations. Fodor and Pylyshyn take primitive visual attentional mechanisms, the kinds that lock into properties of the world, to establish the causal links between distal stimuli in the "perceptual circle" and their atomic mental representations. Pylyshyn[19] has demonstrated that we attend to and track multiple objects simultaneously and that the connections that are established between the token referents—the things tracked—and their representations are initially "preconceptual." That is, the link serves simply as an individuating mechanism, a form of deixis, as if the visual-attentional system could put its "fingers" on the things tracked or point at them.

Now, let's see what's "inside" the system that affords those links: to begin with, nothing like a "meaning" or an intension (with "s"). In fact, they say it is a "mistake," one that has plagued semantics for about a century (again: *they* say), to identify meaning with intension, following Frege.[20] Here is how they frame the problem: Assume expressions or concepts *JT* (say, Justin Trudeau) and *CPM* (Canadian Prime Minister) both refer to *that* individual *R*. One would imagine that *JT* and *CPM* both hold the same intensional content such that the extension *R* is determined by that content. But as Frege (1892) had shown, the system breaks down in propositional attitude expressions: that the supposed coextension of *JT* and *CPM* does not hold, for an individual can at the same time believe that *JT* refers to *R* while not believing that *CPM* refers to *R*. Fodor and Pylyshyn assume that there is an alternative to the Fregean appeal to intension: since nobody knows what intension is, let alone what a *naturalistic* account of meaning/intension amounts to, it has to go. The alternative is that RTM takes concepts to be "individuated by their extensions *together with their vehicles*" (p. 74).

In other, very rough words, the concept/symbol does not actually *contain* any intensional property, for conceptual individuation is simply a link with its referent. "Meaning is a myth," they proclaim. I'm confident they are not interested in eliminating semantics as a career option, but by claiming that all there is is reference, they are also saying that a lot of the semantic vocabulary—synonymy, paraphrase, translation, and so on—is on its way out.

Turning to the nature of referential links, a key issue is what happens to concepts that are not and never been within the perceptual circle. Those are cases in which Fodor and Pylyshyn take to be the result of long chains of referential connections, cases in which actual referents somehow *were* within the perceptual circle of somebody some time ago. Thinking about Moses, in that sense, implies having a symbol that stands for Moses assuming Moses had been somehow referred to directly sometime, somewhere. Even if we let that pass, for proper names have their own peculiarities, reference to things and events past follow similar chains. Forget "brute force" here: this is more like the case of Plato's "earlier souls" which first triggered the concepts that we now refer by inheritance.

We can't fully evaluate these proposals just yet, not here. While reference within the perceptual circle is well anchored in perceptual and attentional (hence, naturalistic) links, much needs to be said about the representations beyond the "circle," about many concept types and, yet again, about the purported relations between concepts that give rise to categories and other types of inferential processes bearing on the content of propositions. (Quick question: If they don't run on intensions, what do they run on?) But if Fodor and Pylyshyn are at a minimum half-right, the cognitive science of concepts will be required to do some work on its foundations, much like their missing epigraph would have recommended,

> If you slip . . .
> Pick yourself up
> Dust yourself off
> And start all over again
>
> (Jerome Kern & Dorothy Fields)

For long our belief boxes have been holding a symbolic expression meaning that Fodor has been the most prominent figure in some of the most important battles leading to cognitive science's current stage, to its autonomy from behaviorism and physicalism, and for its focus on the nature of mental representations and processes. He has set the agenda for some of the most important debates shaping the core of the field—from the nature of cognitive architecture to the nature of concepts. One certainly can't tell what would have been of cognitive science, of its second revolution, without some of the metonymic names fighting its most important battles against behaviorism (then and now) and against the reductionism that physicalism (then and now) promotes. And one doesn't know in particular what would have been of all this without Fodor. But there is no doubt about what happened to the field when he came into play.

Sometimes battles are fought alone, sometimes under quixotic delusions, as the knight in the well-known story put:

> Fortune is arranging matters for us better than we could have shaped our desires ourselves, for look there, friend Sancho Panza, where thirty or more monstrous giants present themselves, all of whom I mean to engage in battle and slay... (M. de Cervantes, *Don Quixote*).

It just so happens that sometimes windmills *are* indeed giants worth slaying. In Fodor's case, there were giants, the targets of his unique analytic mind, some of whom still linger despite being inflicted with mortal arguments. Nobody knows what will be of Fodor's work 300 years from now (assuming exceptions are made, I shall update this guide). Descartes' contributions to philosophy are still at the forefront of the debates on how the mind works. Hume's work was, on his own account, initially "overlooked and neglected,"[21] but look at him now. We do know that Fodor's impact has been immediate upon entering the cognitive science scene—and that he has been engaging and slaying giants ever since. In the process, anarchic as it has been, the cognitive revolution achieved many of its goals. Old Granny does not visit anymore, though her psychographic messages keep recurring in connectionist writings. History—fairly or unfairly—will hold Fodor as a metonym for the kind of cognitive science that was, is, or ought to be.

AUTHOR'S NOTE

I plead guilty to false advertisement, for I do not—and cannot—provide anything near a *complete* guide to all the many attractions. What is provided here is a very rough map to some of the issues that have occupied Fodor's mind and have helped set the agenda for cognitive science. I also limit the scope of the discussion to the topics that have occupied the minds of the editors and contributors to this volume, undeniably under Jerry Fodor's spell. Even the title of this introduction is, of course, inspired in the title of one of Fodor's papers ("Fodor's Guide to Mental Representations," 1985; millennials are supposed to Google "Fodor's guide" to get the joke). We—Lila R. Gleitman and I—are certainly most grateful to Jerry for all. I am also indebted to Caitlyn Antal, Tom Bever, Noam Chomsky, Lila R. Gleitman, and Ernie Lepore for comments on earlier versions of this chapter, and to the National Sciences and Engineering Research Council of Canada (NSERC), for support.

NOTES

1. I occasionally use "we" to refer to both editors of this volume or as a generic pronoun.
2. See Miller, Galanter, & Pribram (1960) and their interest in exploring "cybernetic ideas" in psychology—specially "plans" as cognitive programs. These "cybernetic

ideas" were well under development in the 1950s (see, e.g., Newell, Shaw, & Simon, 1958; and the papers in Feigenbaum & Feldman, 1963).

3. A product of this visit was Fodor (1965), an analysis of behaviorists' account of meaning as "mediating" responses.
4. In Rieber (1980, p. 80).
5. See chapters by Bever and Garrett in this volume.
6. See, in this volume, chapters by J. D. Fodor, Nickels, & Schott and by Ferreira & Nye.
7. We could just as well take the symbolic level to be part of biology. Here I yield to convention and treat them as separate levels of analysis.
8. See, for instance, Gallistel's chapter in this volume.
9. For ease of exposition, I will collapse two theses, RTM and CTM. You can be committed to the idea that there are representations of some sort without being committed to the idea that processes over those representations are computational, Turing-like. If you are committed to the latter, you have to be committed to the former, and that commitment in turn restricts the nature of representations (viz., to those that are computable). For the most part, Fodor is committed to both, but see his *The Mind Doesn't Work That Way: the Scope and Limits of Computational Psychology* (2000), where he discusses varieties of CTM, and why he assumes that CTM, strictly speaking, only holds for modular processes typical of input systems—not holistic ones, typical of central-system processes. I return to this later in the discussion on modularity. See also, de Almeida & Lepore (this volume).
10. If a mosquito bites you, most likely the cause of your desire to scratch the itch you got—and ultimately whether or not you actually scratch yourself—is not computationally derived, not in any sense that, say, the conclusion in a *modus ponens* is.
11. See, in this volume, the chapter by Lobina & Garcia-Albea, on the relation between LoT and the faculty of language.
12. See chapter by Piattelli-Palmarini in this volume. See also Fodor & Piattelli-Palmarini's *What Darwin Got Wrong* (2010), where Darwin's natural selection theory is taken to be analogous to behaviorism's learning theory, presupposing nothing in terms of the organism's internal states in the process driving evolution.
13. See chapters in this volume by Chomsky, Garrett, Ferreira & Nye, de Almeida & Lepore, Pylyshyn, and Potter.
14. See, in particular, Fodor (1983, 2000) and Chomsky's chapter in this volume. For an early treatment of Chomsky's notion of modularity, see Chomsky (1980).
15. Besides Fodor & Pylyshyn (1988), see also Fodor & McLaughlin (1990).
16. See J. D. Fodor, Fodor, & Garrett (1975) and Fodor, Garrett, Walker, & Parkes (1980).
17. I say "majority" because there have been a few experiments claiming to support verb-semantic decomposition, all of which face some harsh problems. A recent review of these appears in de Almeida and Manouilidou (2015).
18. The idea that there are non-content-constitutive meaning postulates is not necessarily a weak, unconstrained alternative; it might be simply the best one can get out of rule-like processes in an otherwise holistic environment, thus at least preserving a weak version of CTM without being committed to "inferential role semantics." But this cannot be worked on here (see de Almeida, 1999, for an early attempt).
19. See Pylyshyn's chapter in this volume.

20. The reader might want to brush up on the so-called Frege cases (viz., "the morning star" and "the evening star" as both referring to Venus; and the problem posed by the use of these expressions in propositional attitude statements) and, on the way back, to look at Putnam's case (the Twin Earth argument). Both types of cases have been subject to Fodor's scrutiny (see, e.g., Fodor, 1987, 1994, and 2008). It should be noted that neither Frege nor Putnam takes meaning to be "in the head." Fodor's reading is that at least in Frege's case expressions or concepts are token mental representations—that, e.g., *THE MORNING STAR* is a concept, in fact a different one from *THE EVENING STAR* even though both refer to Venus.
21. This refers to the reception of his *Enquiry Concerning Human Understanding*. See Hume's (1777/2009) *My Own Life*.

REFERENCES

Chomsky, N. (1959). A review of BF Skinner's Verbal Behavior. *Language*, *35*(1), 26–58.

Chomsky, N. (1980). *Rules and representations*. New York, NY: Columbia University Press.

de Almeida, R. G. (1999). What do category-specific semantic deficits tell us about the representation of lexical concepts. *Brain and Language*, *68*, 241–248.

de Almeida, R. G., & Manouilidou, C. (2015). The study of verbs in cognitive science. In R. G. de Almeida & C. Manouilidou (Eds.), *Cognitive science perspectives on verb representation and processing* (pp. 3–39). New York, NY: Springer.

Feigenbaum, E. A., & Feldman, J. (1963). *Computers and thought*. New York, NY: McGraw-Hill.

Feyerabend, P. (1975). *Against method: Outline of an anarchistic theory of knowledge*. New York, NY: Verso.

Fodor, J. A. (1965). Could meaning be an rm? *Journal of Verbal Learning and Verbal Behavior*, *4*(2), 73–81.

Fodor, J. A. (1968). *Psychological explanation: An introduction to the philosophy of psychology*. New York, NY: Random House.

Fodor, J. A. (1970). Three reasons for not deriving "kill" from "cause to die." *Linguistic Inquiry*, *1*(4), 429–438.

Fodor, J. A. (1975). *The language of thought*. New York, NY: Thomas Y. Crowell.

Fodor, J. A. (1981). *Representations: Philosophical essays on the foundations of cognitive Science*. Cambridge, MA: Bradford Books/MIT Press.

Fodor, J. A. (1983). *The modularity of mind: An essay on faculty psychology*. Cambridge, MA: Bradford Books/MIT Press.

Fodor, J. A. (1987). *Psychosemantics: The problem of meaning in the philosophy of mind*. Cambridge, MA: Bradford Books/MIT Press.

Fodor, J. A. (1990). *A theory of content and other essays*. Cambridge, MA: MIT Press.

Fodor, J. A. (1994). *The elm and the expert: Mentalese and its semantics*. Cambridge, MA: MIT Press.

Fodor, J. A. (1998a). *Concepts: Where cognitive science went wrong*. New York and Oxford, England: Oxford University Press.

Fodor, J. A. (1998b). *In critical condition: Polemical essays on cognitive science and the philosophy of mind*. Cambridge, MA: MIT Press.

Fodor, J. A. (2000). *The mind doesn't work that way: The scope and limits of computational psychology.* Cambridge, MA: Bradford Books/MIT Press.

Fodor, J. A. (2003). *Hume variations.* Oxford: Clarendon Press/Oxford University Press.

Fodor, J. A. (2008). *LOT 2: The language of thought revisited.* Oxford, England and New York, NY: Oxford University Press.

Fodor, J. A., Bever, T. G., & Garrett, M. F. (1974). *The psychology of language: An introduction to psycholinguistics and generative grammar.* New York, NY: McGraw-Hill.

Fodor, J. A., Garrett, M. F., Walker, E.C. T., & Parkes, C. H. (1980). Against definitions. *Cognition, 8*(3), 263–367.

Fodor, J. A., Lepore, E. (1992). *Holism: A shopper's guide.* New York: Wiley-Blackwell.

Fodor, J. A., & Lepore, E. (2002). *The compositionality papers.* Oxford, England: Oxford University Press.

Fodor, J. A., & McLaughlin, B. (1990). Connectionism and the problem of systematicity: Why Smolensky's solution doesn't work. *Cognition, 35*(2), 183–204.

Fodor, J., & Piattelli-Palmarini, M. (2010). *What Darwin got wrong.* New York, NY: Farrar, Straus and Giroux.

Fodor, J. A., & Pylyshyn, Z. (1988). Connectionism and cognitive architecture: A critical analysis. *Cognition, 28*(1–2), 3–71.

Fodor, J. A., & Pylyshyn, Z. W. (2015). *Minds without meanings: An essay on the content of concepts.* Cambridge, MA: MIT Press.

Fodor, J. D., Fodor, J. A., & Garrett, M. F. (1975). The psychological unreality of semantic representations. *Linguistic Inquiry, 6*, 515–531.

Frege, G. (1892). On sense and reference. In P. Geach & M. Black (Eds.), *Translations from the philosophical writings of Gottlob Frege*, 2nd ed. (pp. 56–78). Oxford, England: Basil Blackwell.

Hume, D. (1777/2009). My own life. In D. F. Norton & J. Taylor (Eds.)(2009). *The Cambridge companion to Hume*, 2nd ed. (pp. 522–529). Cambridge, England: Cambridge University Press.

Katz, J. J., & Fodor, J. A. (1963). The structure of a semantic theory. *Language, 39*(2), 170–210.

Miller, G. A., Galanter, E., & Pribram, K. H. (1960). *Plans and the structure of behavior.* New York, NY: Holt, Rinehard & Winston.

Newell, A., Shaw, J. C., & Simon, H. A. (1958). Elements of a theory of human problem solving. *Psychological Review, 65*(3), 151–166.

Pylyshyn, Z. W. (1984). *Computation and cognition: Toward a foundation for cognitive science.* Cambridge, MA: MIT Press.

Rieber, R. W. (Ed.) (1980). *Dialogues on the psychology of language and thought.* New York, NY: Plenum Press.

PART I

Language and the Modularity of Mind

1

Two Notions of Modularity

NOAM CHOMSKY

Jerry Fodor opens his deservedly influential monograph on modularity (Fodor, 1983) by recalling that the butterflies were fluttering in a joint seminar that we taught in 1980. There were actually two highlights of the seminar: Fodor's ideas about modularity, which grew into the caterpillars presented in the monograph, and his early thoughts about the intriguing conceptual atomism that he was developing at the same time. I benefitted from lively discussions with him about both—with a mixture of puzzlement, accord, disagreement. These remain, but can be sharpened in the light of progress since. I will keep here to the issue of modularity.

The monograph was inspired by an observation about perception of speech by Merrill Garrett: that parsing is "basically . . . a reflex." Correspondingly, the focus is on input systems and fixation of belief in central systems. The major example is parsing in language, with some observations about vision and other input systems.

Fodor briefly alludes to a different process that is "basically . . . a reflex" (p. 100), acquisition of language: "the neural mechanisms subserving input analysis develop according to specific, endogenously determined patterns under the impact of environmental releasers." With one important qualification, that has been a guiding thesis of the study of generative grammar within what has come to be called "the biolinguistic framework" and "the generative enterprise" since its origins in the 1950s.

The qualification is that these neural mechanisms are not limited to "subserving input analysis" (parsing). They also subserve production: the normal use of language to express thoughts either externally or in "internal dialogue."[1] In this crucial respect, language is quite different from vision and the other input systems that Fodor discusses. Correspondingly, it is not clear that language falls within Fodor's framework, let alone that it can serve as the major illustration.

It is perhaps worth noting that it is the production of expressions, not parsing, that has been the central concern of scientific-philosophical inquiry into language since the early days of the modern scientific revolution. Galileo and the Port Royal logicians and grammarians were awed by the "marvelous invention" of a means to construct "from 25 or 30 sounds that infinity of expressions, which bear no resemblance to what takes place in our minds, yet enable us to reveal [to others] everything that we think, and all the various movements of our soul." Galileo regarded that as an achievement "surpassing all stupendous inventions," even those of "a Michelangelo, a Raphael, or a Titian." For Descartes, this was a primary difference between humans and any beast-machine and provided a basis for his mind-body dualism. Wilhelm von Humboldt conceived language to be "a productive activity" that makes "articulated sound capable of expressing thought"—"audible signs for thought," in the words of William Dwight Whitney. For the last great representative of this tradition, Otto Jespersen, the central question of the study of language is how its structures "come into existence in the mind of a speaker" on the basis of finite experience, yielding a "notion of structure" that is "definite enough to guide him in framing sentences of his own," crucially "free expressions" that are typically new to speaker and hearer.

In contrast, the input aspect of language use does not seem to have been a major concern.

Evidently, the input and output (production) systems are linked. No one understands only Japanese and speaks only Swahili. The natural assumption—which, to be clear, I've always assumed to be correct—is that language is a module of a "central system," which is accessed in the many kinds of use of language, including input analysis and externalization in production.

Like Fodor, I will keep here to speech. Externalization, when it takes place, along with parsing, appears to be independent of sensory modality, a matter of some importance that I will put aside, though it bears directly on a question that I shall address: How significant are input modules for inquiry into the nature of human language?

The tradition does not distinguish production from generation, the latter akin to the process whereby an axiom system generates proofs and theorems, more generally the way a finite program can determine an infinite array of symbolic objects. That distinction, which is crucial, had become quite clear by mid-twentieth century, thanks to the work of Gödel, Turing, Church and others, making it possible to capture more clearly and to pursue intensively some of the ideas that animated the tradition. The postulated acquisition-based module of the central system is a generative process accessed for production and input analysis. It is what has come to be called an I-language (internal, individual, intensional), in earlier years called "a grammar" in one of the uses of this systematically ambiguous term.[2]

It is also worth emphasis that in much of the tradition, the aspect of production that was salient was its free creative character, the constantly innovative use of "free expressions" in ways that are appropriate to circumstances but apparently not caused by them, and that elicits thoughts in the hearers that they can then formulate themselves. That aspect of human action, for Descartes most vividly revealed in language use, remains at the border of scientific inquiry, or beyond, a fact recognized in the most sophisticated studies of voluntary action. As the point is put ("fancifully") by Emilio Bizzi and Robert Ajemian, "we have some idea as to the intricate design of the puppet and the puppet strings, but we lack insight into the mind of the puppeteer."[3] Similarly in the case of language, in terms of Humboldt's now often-quoted aphorism that language involves "infinite use of finite means," we now have some idea as to the nature of the means that are used, but the mind of the user remains a mystery. The problem reaches beyond Fodor's "First Law of the Non-existence of Cognitive Science" (107).

The two types of modules suggest two different ways of approaching the nature of language: as a parsing system (with production and the nature of the linkage left to the side) or as an internal cognitive system accessed by various uses of language. The former approach is sometimes held to be the only one possible, with human language taken to be "by definition an experience-dependent mapping of auditory and visual stimuli onto meaningful objects, actions, and concepts."[4] The traditional perspective, and the development of certain aspects of it in the biolinguistic-generative framework, is, however, at least an alternative, and one that I think is more revealing of the nature of language, for reasons to which I will return.

Central-system modularity may appear to be inconsistent with Fodor's rejection of modularity for central systems, but it actually is not. Let us put that problem aside for the moment, just assuming consistency.

One major contribution to the biolinguistic program was Eric Lenneberg's fundamental work (1967), which founded modern biology of language and along with much else, formulated the basic issues of evolution of language with clarity and insight that remain unsurpassed, and also provided important evidence about dissociation of language from other cognitive faculties. The latter topic has been extensively pursued since, yielding the conclusion that the language faculty is "a distinct module of the mind/brain, based on domain-specific organizing principles,"[5] and accordingly lending support to the thesis that acquisition-based modularity of central systems is a real phenomenon. It is this concept of central-system modularity that is developed, for example, in Chomsky (1975, chap. 1) and many other publications before and since. This "modular view of learning" is "the norm in neuroscience" today, Randy Gallistel observes, referring in particular to the "module for learning language." It is appropriately called an "organ," he continues, because "the specialization of structure and function that we see in organs is an appropriate way to think about the specialization of structure and function we see in the various learning organs." In general, learning is based on specialized mechanisms, "instincts to learn" in specific ways, yielding modules within the brain that perform specific kinds of computation such as in the remarkable navigational feats and communication capacities of insects. Apart from "extremely hostile environments," these modules develop and change states under the triggering and shaping effect of external factors, more or less reflexively, and in accordance with internal design. That is the "process of learning,"[6] though "growth" might be an appropriate term.

In this respect, language acquisition falls together with vision and other input systems of the mind-brain—though language crucially differs from them in that it provides not only an "input system" but also an output system and, I presume, a central generative system that both access.

Uncontroversially, the systems involved in navigation, vision, and other subcomponents of the organism ("modules") are in substantial part genetically determined. For language, the theory of the genetic component has been called "universal grammar (UG)" in contemporary work, adapting a traditional term to a new framework.

Curiously, though adopted without serious question elsewhere, the assumption for language is considered highly contentious, if not refuted by "field linguists."[7] In one formulation, "Universal Grammar is dead."[8] The only coherent interpretation of this thesis is that language is acquired by other cognitive capacities that are somehow unique to humans. The suggestion faces two problems: one is the failure to deal with even the simplest cases, such as those to be discussed. The other is the radical (double) dissociations that have been found since Lenneberg's pioneering work in the 1950s. I will put these beliefs aside here.[9]

The central systems incorporate principles, which enter into behavior. For bee communication, for example, internal principles enable calculation of the changing angle between the sun's azimuth and the direction of the food. We assume these to be neurally coded in some manner, though how is apparently not well understood. For vision and language we find such principles as (1), (2), respectively:

(1) The Rigidity Rule
(2) The Rule of Structure-dependence[10]

In these and other cases, the crucial question is why the principles hold.

The Rigidity Rule, as defined by Donald Hoffman (1998) in his study of visual intelligence, holds that when other rules permit, image projections are interpreted "as projections of rigid motions in three dimensions," even with highly impoverished stimuli. That seems initially problematic. The environment throughout the evolution of the visual system rarely contained rigid objects, and the experimental work on the principle shows that presentations are perceived falsely. Questions thus arise about the internal nature of the visual system, the factors in its development in the individual, and its evolution. Related questions were raised by Descartes in his work on the visual system, for example, when he speculates (plausibly) that presented with the drawing of a triangle, a child will not take it to be the "composite figure of the triangle drawn on paper . . . but rather the true triangle," because "the idea of the true triangle was already in us," as an innate concept. In Ralph Cudworth's formulation, the intelligible idea of an object is not "stamped or impressed upon the soul from without, but upon occasion of the sensible idea excited and exerted from the inward active and comprehensive power of the intellect itself," based on its innate structure, a version of the idea that experience conforms to the modes of cognition.

One of many illustrations of case (2), the principle of structure-dependence, is given by (3)–(6):

(3) Birds that fly instinctively swim
(4) The desire to fly instinctively appeals to children
(5) Instinctively, birds that fly swim
(6) Instinctively, the desire to fly appeals to children

The structures of (5) and (6) are, roughly, as indicated by bracketing in (5′) and (6′) respectively:

(5′) Instinctively, [birds [that fly]] [swim]]
(6′) Instinctively, [[the desire [to fly]] [appeals [to children]]]

The structural descriptions of (5′) and (6′) reveal clearly the difference between linear and structural proximity. In both cases, "fly" is the closest verb to "instinctively" in linear distance, but the more remote in structural distance.

Examples (3) and (4) are ambiguous ("fly instinctively," "instinctively swim/appeal"), but in (5) and (6) the adverb is construed only with the remote verb, raising immediate questions: why does the ambiguity disappear, and more puzzling, why is it resolved in terms of the computationally complex operation of locating the *structurally* closest verb rather than the much simpler operation of locating the *linearly* closest verb?[11]

The principle of structure-dependence applies to all relevant constructions in all languages, as far as is known. There is a simple explanation, the only one known: linear order is not available to the internal computational system that yields syntactic structures and their semantic interpretation. If so, then linear order is a peripheral part of language, presumably introduced in externalization to satisfy conditions imposed by the sensorimotor modality that is employed (and, in fact, sign, with different sensorimotor options, uses somewhat different arrangements than speech). These sensorimotor properties may be largely or completely independent of language, thus telling us little or nothing about language.

Proceeding, the next question is why language should lack linear order, except peripherally as a reflex of the sensorimotor interface. There is a simple and plausible assumption that yields this consequence: language design is optimal; its operations follow principles of *minimal computation (MC)*.

Specifically, the computational system of language is based on the simplest computational operation O for a recursive system: given objects X and Y already constructed, form Z = O(X, Y) without modifying X and Y or imposing any new structure in Z. In short, O is simply set-formation. In recent literature, O is called "Merge." The expressions constructed by Merge therefore lack order, and order will not be available for operations on Merge-created structures.

Expression (3) can be constructed in two different ways by iterated Merge, in one case merging *fly* and *instinctively* and in the other case merging *swim* and *instinctively* before these constructed elements are merged into the larger expression. Hence the ambiguity. The same is true of (4). In the case of (5) and (6), however, the construal rule that associates the initial adverb with the verb, again adhering to MC, will seek the closest verb, where distance is structural, linear order being unavailable. That yields the unambiguous interpretations of (5) and (6). The rules of externalization happen to place the verb in the more remote position, for reasons that apply quite independently.

Notice that the argument is the same for the standard cases in the literature on structure-dependence: auxiliary-raising, as in (7) but not (8), where *t* (*trace* in earlier literature) marks the position where the auxiliary is understood:

(7) Will birds that fly *t* swim

(8) *Will birds that *t* fly swim

Under MC, (7) is the only possibility if linear order is unavailable, while (8) would be selected if both linear order and hierarchical structure were available. The thought that (8) would express if linear order were available requires a paraphrase in language. Essentially the same argument holds for all cases of structure-dependence, in a wide variety of constructions in all languages.

The same optimal assumptions about the architecture of language yield a variety of other conclusions, some quite straightforward, some more interesting. One straightforward conclusion is that assignment of semantic roles should be order-independent; for example, the verb-object relation should receive the same interpretation in a head-initial language SVO or a head-final language SOV. That too appears to be the case over a broad range.

More interesting conclusions follow if we pursue the same reasoning further. Consider the sentences (9)–(10):

(9) [The boys expect the girls to like each other]

(10) which girls do [the boys expect to like each other]

In (9), the anaphor *each other* selects the local antecedent *the girl*, as expected under MC. In (10), however, it does not select the local antecedent *the boys*, within the bracket that is analogous to (9), but rather the remote antecedent *which girls*.[12] If we continue to assume optimal design under MC, hence that grammatical operations observe locality (minimal distance), then it follows that *which girls* is in fact the local antecedent for the anaphor. Accordingly, though what reaches the sensorimotor system is (9), the syntactic object that reaches the mind is something like (11):

(11) Which girls do [the boys expect which girls to like each other]

Here the bracketed element is identical with (9) except that *which girls* replaces *the girls.*

The question is why language is designed in this way.

Once again, the answer is provided by the assumption that the computational rules are optimal, based on Merge. By simple logic, there are two possible cases of Merge, which we can describe as follows. Assume a workspace containing objects already constructed (including the minimal "atoms" of the lexicon). Select X from the workspace, then select Y to Merge to X, where Y has already been constructed. Y can either be in the workspace, external to X, or it can be a part of X (technically, a *term* of X)—external Merge (EM) and internal Merge (IM), respectively.

Sentence (9) is formed by repeated EM, yielding the appropriate hierarchical structure. To form (11), first apply repeated EM to form (9′) = (9) with *the girls* replaced by *which girls.* Next apply IM merging *which girls* with (9′), yielding (11) with the appropriate hierarchical structures and with the two copies of *which girls* that yield the correct semantic interpretation.[13]

Note that there are no such notions as *Re-merge* or *Copy*; just Merge in the simplest form.

Another principle of MC yields (10) for externalization: pronounce as little as possible. At least one copy must be pronounced or there is no indication that the operations took place. Looking further, we find that either the structurally highest or lowest is chosen, depending on the construction and the language, but not other copies, for reasons that have a simple explanation.[14]

The property of displacement with deletion (Move) is ubiquitous in language, and was long considered to be a curious imperfection. That was an error (mine in particular). On the contrary, we can now see that it would be an imperfection of language if IM were not available. An approach to the phenomenon that bars IM has a double burden of justification: it must justify the stipulation barring IM and must also justify whatever new mechanisms are designed to yield what comes free under IM, assuming MC. The "copy theory of movement" illustrated in (9)–(11) yields quite intricate semantic interpretations (called "reconstruction" in earlier work).

Throughout, the results follow from the assumption that the design of language keeps to the overriding conditions MC. For these cases at least, UG reduces to providing a combinatorial operation to permit recursive generation of structures that provide semantic-pragmatic interpretations (and secondarily, can be externalized).

The construal of the anaphor, as in (9)–(11), keeps to minimal structural rather than minimal linear distance, as illustrated in (12), again suggesting that linear order is not available for the internal computational system:

(12) Women with children like each other

Further inquiry into anaphoric relations yields many intricacies, discussed in a rich and expanding literature, but elementary properties such as these appear to hold quite generally, in one or another form.

I mentioned that the two types of modules—input, central—suggest two different ways of approaching the nature of language: as a parsing system (with production and the linkage to input left to the side) or as an internal cognitive system accessed by various uses of language. The considerations just reviewed bear directly on this question.

Let's continue to keep to the assumption that whatever the computational system of language is, it keeps to the overriding principle MC as far as possible. That makes good sense on general grounds of scientific method and also with regard to origin of language, a guiding concern since the early days of generative grammar, contrary to much misunderstanding.[15]

Suppose that language is fundamentally an internal generative cognitive system accessed by various uses of language. UG determines that language incorporates a combinatorial operation, and by MC, it is the simplest one possible (Merge). We then have an explanation for the properties of language illustrated earlier: (a) the ubiquitous property of displacement, along with important steps towards semantic interpretation of constructions with displacement; (b) apparent violation of locality with anaphora; (c) the universal property of structure-dependence of rules. Further recourse to the overriding principle of MC determines that what reaches the ear has gaps that have to be filled by the parser—in the case of (10), the missing phrase (*which girls*) that receives the same semantic role as the overt phrase *the girls* in (9) as subject of *like each other*, and by the same mechanism, serves as the local antecedent for *each other*. In this case the parsing problem is fairly simple, but locating the gap and filling it ("filler-gap problems") can be quite complex because of the deletion of the copies mandated by MC.

Suppose, in contrast, that language is fundamentally a parsing system. Then all of these properties remain a mystery. In the many varied cases of structure-dependence, for example, we would expect that parsing would make use of the simple computational procedure of minimal linear distance rather than the complex procedure of minimal structural distance, contrary to fact in all relevant cases in all languages. Similar observations hold for the other cases discussed.

Note again that language design seems to pose numerous problems for parsing, in particular, the familiar filler-gap problems illustrated in a simple form in (10). The same conclusion is supported by numerous other familiar cases: structural ambiguity, garden path sentences, many island properties. These seem to arise by allowing rules to run freely, posing problems for parsing—and hence also for communication, which, for many reasons including these, does not appear to have the central role assigned to it in much modern doctrine. In fact, in all cases that I know of where communicative and computational efficiency conflict, the latter is selected, as in the examples illustrated earlier.

The evidence, then, strongly suggests that language is fundamentally an internal generative module providing the means for construction and expression of thought, with ancillary operations of externalization reflecting properties of the sensorimotor system, pretty much along traditional lines.

Fodor cites Hilary Putnam's 1961 suggestion (p. 50) that "there are grammatical transformations because communicative efficiency is served by the deletion of redundant portions of messages, etc." At the time, there were, understandably, many such suggestions about why language should have the odd property of displacement (hence grammatical transformations, or some other mechanism to deal with the "imperfection"). The situation has been different for some years, ever since it has been understood that displacement and its analysis in terms of IM is to be expected on the simplest assumptions, and that problems would arise if languages lacked this property. We can rephrase Putnam's suggestion in current terms as the thesis that deletion rules apply in externalization to enhance computational efficiency. Insofar as that it is plausible, they enhance the efficiency of *production*, but at the same time cause difficulties for parsing by posing filler-gap problems, as in the case of the obligatory deletion illustrated in (10). The suggestion again suggests that communication is a peripheral aspect of language.

It is important to recognize that there is compelling evidence from neuroscience and psycholinguistics supporting the conclusion that linear order is not available for the computational system.[16] It should be clear that if these conclusions about the general architecture of I-language are generally accurate, then a good deal of the technical work in linguistics must be reconsidered,[17] along with much general thinking about language and its functions and evolution.

These conclusions, if correct, imply nothing about the significance of the modular approach to parsing, and input operations generally, that Fodor develops. Rather, they place the study of parsing within the general domain of perception, with application to language a special case that may not be particularly informative about the nature of language.

Parsing is a form of behavior, and accordingly involves many different factors, of which the role of the language is only one. Hence the study of parsing seeks to identify the contribution of the language of the person carrying out this activity and to extricate it from the complex. As Fodor puts the point (135), "something like a representation of a grammar for L must be contained" within the parser, even for assigning tokens to types, surely beyond.[18] And the same L must be contained within the production system. That raises the question of what L is, if it is not an I-language in the sense of the acquisition-based approach to modularity. The latter approach focuses directly on the person's I-language, and is free to use all sorts of evidence to determine what this system is, without limit, as in the sciences generally. But if L is not a central module of the kind discussed here, questions arise about what it is, how we discover its properties, and how it fits into the general cognitive architecture.

The inquiry into parsing requires that we distinguish performance from competence; we distinguish actual behavior from generation by the linguistic system "contained" within the parser, in Fodor's terms. This distinction is often regarded as contentious, though it should not. Whatever organic system we are investigating, we want to determine its intrinsic nature and how this enters into its various uses—in this case, to determine how a person's I-language enters into parsing and other uses of language.

The distinction, which is implicit in traditional grammar, came to the fore as soon as the earliest efforts were undertaken to construct generative grammars. A familiar example is embedding, in the interesting case, with nested dependencies. As observed 50 years ago, without external aids (time, paper, pencil, etc.), sentences can be recognized as clearly grammatical with about six nested dependencies, while disruption of one of the dependencies (say, by replacing an occurrence of "if" by "either") renders it ungrammatical. With external aids there is of course no relevant bound on nesting.[19] Linguistic competence is not bounded by memory, though performance, such as parsing, of course must be. There is a simple explanation for the fact that parsing decays with increased nesting, and reaches a limit (without external aids) at about 7: Miller's famous "magic number" (Miller 1956). Actual speech naturally tends towards parataxis, so embedding rarely goes beyond 2. Hence the I-language property of unbounded nesting (like core properties of language generally) is not acquired by some kind of data-processing but rather derives from inherent properties of the language faculty, from UG, a part of Hume's hidden hand of Nature that enters into all forms of learning and growth.[20]

The situation is similar to arithmetical competence, which for some reason is considered less contentious. No one is confused about the fact that we can only add small numbers "in our heads," but can go on indefinitely with external aids. In brief, both language and arithmetic are based on the Turing architecture that Fodor describes, part of their essential nature, possibly with common roots (see Chomsky 2010).

It remains to consider the apparent contradiction between the postulation of central modules and Fodor's thesis that central systems lack any modular structure, but rather are "Quinean and isotropic." The contradiction is only apparent. Fodor is concerned with the central processes of fixation of belief, which indeed have the properties he describes. But knowledge of language (linguistic competence, having an I-language) is not some kind of structure of beliefs.

My one real disagreement with Fodor's account is his opening section (3ff.) on what he calls "neocartesianism," "what [Chomsky] means": namely, that the I-language a person acquires is "a body of innate *propositional attitudes*" (129). But I have never meant anything of the sort, and agree with Fodor that the idea makes little sense.[21] A person whose I-language has the properties discussed may have all kinds of beliefs about the expressions used as illustration here, or about his or her language. Some might be true, some false, but they are not what constitutes the language that the person has mastered and uses, any more than in the case of the visual system or insect navigation.

The confusion pretty clearly arises from Fodor's interpretation of the phrase "knowing a language," the normal locution in English (not other languages) for what in more technical terms we might call having internalized an I-language. As Fodor remarks, "knowledge is—or so many philosophers tell us—inter alia a normative notion having much to do with standards of justification." It is true that "so many philosophers tell us" in discussion of propositional knowledge, but the comment clearly does not hold of normal English usage, including the case in question. When one says, for example, "I know many of my cousins, I know their flaws and foibles, I know some of the reasons for them, I partially know their languages but I don't know the rules of verbal morphology though of course I know the rule of structure-dependence," and so on, there is no reason to seek a tortured, irrelevant, and hopeless account in terms of knowing-that or knowing-how, of propositional content, networks of belief, and so on. That's not what the phrase "knowing X" means. And invoking subdoxastic beliefs (whatever their merit in other contexts) does nothing here but deepen the confusion.

Fodor observes that "Chomsky himself is quite prepared to give up the claim that the universal linguistic principles (say, structure-dependence) are innately *known* in favor of the explicitly neologistic (hence sanitized) claim that they are innately 'cognized,'" but he misconstrues the reasons. It is simply an effort to avoid pointless debates with philosophers who insist on taking refuge in Wittgenstein's fly-bottle instead of using the terms of ordinary language with their own meanings (as in "knowing X"), or, as is commonly done even in the hard sciences, using these terms intelligibly if sometimes laxly in informal discourse. My point was much the same as Turing's in his famous paper introducing the imitation game, where he warned that the question whether machines can think "is too meaningless to deserve discussion"—along with such questions as whether kites fly, submarines swim, Chinese rooms translate, and so on. These are questions of ordinary usage in one or another language, or, sometimes, of what metaphorical extensions we choose to make. They are not substantive questions.

When we put these misinterpretations aside, there is no contradiction between the postulation of acquisition-based central modules and Fodor's rejection of central modules for fixation of belief.

In his discussion of the "Quinean" and "isotropic" character of internal systems—meaning that any evidence is in principle relevant—Fodor states that "some linguists" deny this property for language, claiming that "no data except certain kinds of facts about the intuitions of native speakers could, in principle, be relevant to the (dis)confirmation of grammatical theories." If so, they are severely mistaken. The only advocacy of this restriction that I know of is by Quine, who repeatedly insists that "there is nothing in linguistic meaning" (which he construes to extend to properties of syntax and semantics generally) "beyond what is to be gleaned from overt behavior in observable circumstances," proceeding to restrict the latter to "querying sentences for assent and dissent."[22]

Quine's restrictive stipulations contrast sharply with the practice and principles of generative grammar from its modern origins, which always insisted that evidence of any kind is in principle relevant to "the (dis)confirmation of grammatical theory," including evidence from other languages, available once we recognize the role of the species property UG.

Quine's restriction of relevant evidence is part of a much broader thesis, which might merit a few words in the light of its great influence and what it tells us about the tenor of the times. Quine's guiding principle in this domain is summarized clearly in his *Pursuit of Truth*: "in psychology one may or not be a behaviorist, but in linguistics one has no choice. Each of us learns his language by observing other people's verbal behavior and having his own faltering verbal behavior observed and reinforced or corrected by others. We depend strictly on overt behavior in observable situations" (Quine 1990, 37). An analogous argument would be that the study of the visual system must restrict itself to the visual stimuli that determine the specific form that the visual system assumes. Of course, that argument would be dismissed at once: though indeed the mature visual system is a function of input stimuli (and as well known, it can vary substantially depending on stimulation in early infancy), the outcome depends on many factors, including genetic endowment, and the scientist studying the visual system is free to consider these and indeed whatever evidence might be relevant to how the organism grows. But these options are barred in principle to the linguist, on the tacit assumption that the language faculty cannot in principle have any basis in human biology—that there can be nothing like UG. The linguist cannot in principle then learn anything about English from the study of Chinese, or from psycholinguistics, or neuroscience, or any other source. The central system of language (if that's what it is—what else could it be?) violates the Quinean and isotropic properties of central systems.

Note that (dis)confirmation of a *theory of language* (*or of particular I-languages*), relying on any evidence in principle, is not to be confused with the operations of language acquisition. In this case, to quote Fodor again, "the neural mechanisms . . . develop according to specific, endogenously determined patterns under the impact of environmental releasers"; and as in the case of growth and development of other subsystems of the organism, only certain "environmental releasers" trigger and shape the process.

To summarize briefly, I think Fodor is right to recognize two mental processes that are "basically a reflex": the input modules that are his topic and acquisition of language (with the qualification mentioned earlier), the latter providing a central module that falls together with others, but is not a system of propositional attitudes acquired by fixation of belief. This central module is accessed for production (occasionally externalized) and parsing. The latter, like all of performance, is a mixed system guided in some manner by the internal language but involving many other factors. The central module itself is a biological object, whose nature we seek to discover, using any evidence available, with no such restrictions as those that Quine imposes. The two approaches suggest two ways of seeking the fundamental nature of language. There is substantial evidence,

I think, favoring the latter, which has something of a traditional flavor. If the approach outlined here is on the right track, then considerable rethinking of the nature and use of language is in order, both within technical linguistics and in reflections on its nature and use.

NOTES

1. Statistically, by far the majority of the normal use of language. There is reason to suspect that most of it is inaccessible to consciousness. For some comments, see Chomsky (2013b,c).
2. The term "I-language" was introduced in Chomsky (1986), after Fodor's book appeared. The purpose of the terminological change was to overcome the ambiguity, which had often been misleading, and to clarify what was meant by "grammar" in the relevant sense.
3. Bizzi and Ajemian (2015).
4. Albright (2015).
5. See Curtiss (2013) for review of a wide variety of evidence.
6. Gallistel (1999a,b).
7. For example, Churchland (2013). In fact, "field linguists"—that is, linguists who work with the wide variety of languages that have come under investigation since the early days of generative grammar—have repeatedly demonstrated the opposite: that languages that appear to vary widely on the surface are in fact cast to much the same mold when investigated in depth.
8. Tomasello (2009). His comments suggest that he may be misinterpreting UG in the manner discussed in note 9.
9. Sometimes these beliefs are based on confusion between UG and "language universals," that is, properties found quite generally in language, like Greenberg's famous universals. Such generalizations are, of course, expected to have exceptions, which, like the generalizations themselves, are a valuable stimulus to research. Another common claim is that UG consists of only "tendencies," which, if there were any reason to believe it, would leave us in an even more difficult position: what is the genetic basis for the "tendencies"? Fodor uses the term "linguistic universals" in the sense of UG, but he was writing before the current confusions infected the fields.
10. This is the one case that has been subjected to extensive efforts to account for the facts by general learning mechanisms. All efforts are irremediable failures (Berwick et al., 2011), though the failure is in fact much deeper than discussed there: the wrong question is being addressed. The right question is *why* the principle holds for all constructions in all languages. The methods proposed would work just as well for a linguistic system in which the simpler linear computation held. The studies keep to the case of auxiliary inversion, a limitation that suggests (erroneously) that adequate data might be available to the child. The illusion is quickly dispelled by construal examples such as (3)–(6). One common fallacy is that the results follow from the fact that hierarchy is available—as is linear order, in fact far more saliently in presented data.
11. Quite commonly, linear and structural distance coincide. That would follow for "head-first" languages like English if the process of linearization is determined

by Richard Kayne's Linear Correspondence Axiom, which linearizes in terms of hierarchy. Kayne explores the matter far beyond, but we can keep to this case here.

12. This is one of the many kinds of examples that refute the proposal of Chater and Christiansen (2010) that anaphoric relations are simply "an instance of a general cognitive tendency to resolve ambiguities rapidly in linguistic and perceptual input," hence do not involve language-specific properties derived from UG. This is another of the very few attempts to deal with some non-trivial property of language in such terms. It should be noted that there is valuable work integrating UG and general learning mechanisms. For example, Yang (2002).
13. I ignore here the insertion of *do*.
14. For discussion, see Chomsky (2013a).
15. Cf. Lenneberg, op. cit. For some discussion, see Chomsky (2014), Berwick and Chomsky (2016).
16. Musso et al. (2003), following the paradigm of Smith and Tsimpli (1995). For replications, see Moro (2013), Smith (2004), Costa and Lobo (2015).
17. And, correspondingly, pursuit of these conclusions must deal with a great deal of linguistic work that appears to be inconsistent with them.
18. See Chomsky (1965), I.2.
19. Miller and Chomsky (1963), Chomsky (1965). Self-embedding has much narrower restrictions. See these sources and Chomsky (1964) for some early proposals on a parsing principle which, it seemed, might also account for what was later called "the wh-island constraint."
20. There is a great deal of rather surprising confusion about these matters in current technical literature. See the introduction to the 2015 reprinting of Chomsky (1965) for some discussion. See Chomsky (2015a) on fallacious arguments in the technical literature seeking to refute the trivially obvious observation of Chomsky (1956) that unbounded nested dependencies cannot be accommodated by finite automata, the standard models of the time.
21. Separately, I think Fodor's interpretation of "Cartesianism" in similar terms is dubious. The Cartesian concept of innate ideas, discussed briefly earlier, does not seem to be properly interpreted in terms of propositional attitudes. I think we can also question Fodor's interpretation of Hume's "epistemic boundedness" as relying on his "Empiricist theory of *meaning*" (124). Hume's conclusion that "mysteries of nature" lie "in that obscurity, in which they ever did and ever will remain" has quite distinct sources. See Chomsky (2009, 2013c).
22. Quine (1975, 1992, p. 46).

REFERENCES

Albright, T. (2015). Perceiving. *Daedalus*, Winter *144*(1), 112–122.

Berwick, R., & Chomsky, N. (2016). *Essays on the evolution of language*. Cambridge, MA: MIT Press.

Bizzi, E., & Ajemian, R. (2015). A hard scientific quest: Understanding voluntary movements. *Daedalus*, Winter *144*(1), 123–132.

Chater, N., & Christiansen, M. (2010). Language acquisition meets language evolution. *Cognitive Science*, *34*, 1131–1157.

Chomsky, N. (1956). Three models for the description of language. *I.R.E. Transactions on Information Theory IT-2*, 113–124.

Chomsky, N. (1964). *Current issues in linguistic theory.* Berlin: Germany: Mouton de Gruyter.

Chomsky, N. (1965/2015b). *Aspects of the theory of syntax.* Cambridge, MA: MIT Press. Reprinted with new introduction, 2015.

Chomsky, N. (1975). *Reflections on language.* New York: Pantheon Books.

Chomsky, N. (1986). *Knowledge of language.* New York, NY: Praeger.

Chomsky, N. (2009). The mysteries of nature: How deeply hidden? *Journal of Philosophy, 106*(4), 167–200. Reprinted in Chomsky (2016a).

Chomsky, N. (2010). Some simple evo devo theses: How true might they be for language? In R. Larson, V. Deprez, & H. Yamakido (Eds.), *The evolution of human language: Biolinguistic perspectives* (pp. 45–62). Cambridge, England: Cambridge University Press.

Chomsky, N. (2013a). Problems of projection. *Lingua* 130, 33–49.

Chomsky, N. (2013b). What is language? *Journal of Philosophy, 110*(12), 645–662. Reprinted in Chomsky (2016a).

Chomsky, N. (2013c). What can we understand? *Journal of Philosophy, 110*(12), 662–700. Reprinted in Chomsky (2016a).

Chomsky, N. (2015a). A discussion with Naoki Fukui and Mihoko Zushi. In *Sophia Linguistica, 64.* Tokyo, Japan: The Sophia Linguistic Institute for International Communication (SOLIFIC), Sophia University.

Chomsky, N. (2016a). *What kind of creatures are we?* New York, NY: Columbia University Press.

Chomsky, N. (2016b). Language architecture and its import for evolution. In R. Berwick & N. Chomsky (Eds.), *Essays on the evolution of language.* Cambridge, MA: MIT Press.

Churchland, P. (2013). Introduction (to new edition). In W. V. O. Quine (1960/2013), *Word and object.* Cambridge, MA: MIT Press.

Costa, J., & M. Lobo (2015). Testing relativized minimality in intervention effects: The comprehension of relative clauses with complex DPs in European Portuguese. Ms., FCSH/Universidade Nova de Lisboa [presented at Romance Turn, University of Islas Baleares].

Curtiss, S. (2013). Revisiting modularity: Using language as a window to the mind. In M. Piattelli-Palmerini & R. Berwick (Eds.), *Rich languages from poor inputs.* New York, NY: Oxford University Press.

Fodor, J. A. (1983). *The modularity of mind.* Cambridge, MA: MIT Press.

Gallistel, C. R. (1999a). Neurons and memory. In M. S. Gazzaniga (Ed.), *Conversations in the cognitive neurosciences*, 2nd ed. Cambridge, MA: MIT Press.

Gallistel, C. R. (1999b). The replacement of general-purpose learning models with adaptively specialized learning modules. In M. S. Gazzaniga (Ed.), *The cognitive neurosciences*, 2nd ed. Cambridge, MA: MIT Press.

Hoffman, D. (1998). *Visual intelligence.* New York:, NY: W. W. Norton.

Lenneberg, E. (1967). *Biological foundations of language.* Hoboken, NJ: John Wiley & Sons.

Miller, G. A. (1956). The magical number seven, plus or minus two: Some limits on our capacity for processing information. *Psychological Review, 63*, 81–97.

Miller, G. A., & N. Chomsky (1963). In R. Duncan Luce, R. W. Bush, & E. Galanter, (Eds.), *Handbook of Mathematical Psychology,* Vol. 2 (pp. 269–322). Hoboken, NJ: John Wiley & Sons.

Moro, A. (2013). *The equilibrium of human syntax: Symmetries in the brain*. Leading Linguists Series. Abingdon-on-Thames, England: Routledge.

Musso, M., Moro, A., Gluache, V., Rijntjes, M., Bücheli, J. C., & Weiller, C. (2003). *Nature Neuroscience 6*, 774–781.

Quine, W. V. O. (1975). Mind and verbal dispositions. In S. Guttenplan (Ed.), *Mind and language*. New York, NY: Oxford University Press.

Quine, W. V. O. (1992). *Pursuit of truth*, revised ed. Cambridge, MA: Harvard University Press. (Originally published 1990.)

Smith, N. (2004). *Chomsky: Ideas and ideals*. Cambridge, England: Cambridge University Press.

Smith, N., & Tsimpli, I. (1995). *The mind of a savant: Language learning and modularity*. New York, NY: Oxford University Press.

Tomasello, M. (2009). Universal grammar is dead. *Behavioral and Brain Sciences*, *32*, 5.

2

Exploring the Limits of Modularity

MERRILL F. GARRETT

Context is the name of a problem, not a solution.

Plus ça change, plus c'est la même chose.

This is about modularity . . . again. It's been a long debate, as Jerry Fodor so adroitly framed it in the centuries old garb of phrenology (Fodor, 1983). He did it with his usual insouciant grasp of what might both irritate and inform. Well done.

Standard psycholinguistic studies of word and sentence recognition have played out around a debate between proponents of "modular" and "interactive" perspectives on real-time language use. The theoretical and empirical tension is between studies that indicate limited penetration of non-linguistic background information on basic sentence processing and studies that indicate an early influence of such information. A related theoretical divide exists for experimental work in pragmatics. Here, the question has been to determine when (or whether) the "literal meaning" (the form driven interpretation) of a sentence is computed when supporting or canceling contextual information is present. I'll comment briefly on experimental studies pertinent to both areas with an eye to describing briefly how the interplay between syntactic processes and non-linguistic knowledge might be viewed.

My remarks on these matters will have three threads. One thread is the conflict in experimental findings in the psycholinguistics literature. Syntactic processing effects persist despite available situational and contextual constraints that are in principle sufficient to resolve temporary ambiguity delays and/or garden path errors. And on the other hand, we have multiple reports of interactive effects between basic sentence processing and non-linguistic background information. A second thread is the somewhat similar circumstance to be found in the experimental literature on pragmatics: conflicting reports of the effect of contextual influence on utterance interpretation are well represented. The final thread suggests a rationalization of conflicting findings in both standard psycholinguistic and experimental pragmatic research. It relies on an appeal to interaction between the two major language processing machines: language comprehension on the one hand, and language production on the other. The core idea is that production processes are intrinsically developed to derive language specific realizations of discourse and environmental contexts, and these may be harnessed to filter the products of comprehension mechanisms. A key feature of the argument for complementary roles of the systems is a degree of modular processing design to be found in both systems. This proposal is one described in Garrett (2000); my remarks here summarize and add to the earlier arguments.

WORKING THE PARSER: OLD NEWS AND SOME NEW NEWS

Speakers speak and listeners listen. Writers write and readers read. Most of the time, things turn out well. But sometimes not. Language in the wild does not lack for failed analyses—for misunderstandings flailing about in a scramble to establish communicative intent. But we do survive and communicate. Sometimes the consequence of structural processing does not fit the ultimate interpretive aim of the source. Some hoary and mildly amusing examples:

> (Imagine rapid incremental presentation, phrase by phrase, word by word, etc.)
>
> "Time flies like an arrow. . . for breakfast."
>
> "Big George . . . weighed . . . 300 lbs . . . . of grapes" (David Hakes, ca 1970)
>
> [Picture of a hippo here] "Hip . . . po . . . pot . . . comes in very large bales."

If information needed to resolve a temporary structural ambiguity is not available in the time frame "required" for a parsing operation, the mechanisms of parsing based on grammatical categories and their sequential deployment (whatever their detail) apply, and move on, come what may—or, we can put the brakes on and bail out with appeal to higher powers.

Here are some reminders: Long-standing observations from language disorder arising from injury and/or disease demonstrate that significant syntactic capacity can survive profound compromise of general cognitive functions required for coherent thought (e.g., see Kempler et al., 1987). Moreover, aspects of effective parsing can be readily demonstrated in normal speakers when potentially helpful semantic and contextual constraint is denied—as in parsing effects for jabberwocky, nonsense strings, and dire cases of implausibility. These sorts of effects have been tested and attested in a variety of ways, including standard behavioral accuracy and response time measures (see, e.g., Forster, 1979; Forster, 1987), electrophysiology (e.g., Hagoort, Brown and Osterhout, 1999; Neville et al, 1991) and brain imaging (e.g., Friederici and Kotz, 2003).

This much seems clear. The bare bones parsing system is up to the task. There is a syntactic machine that structures the speech/print stream effectively without much in the way of higher order information. How much structure? That raises levels of representation and process questions. A variety of two stage parsing proposals have been offered over the years, ranging, for example, from the noteworthy "Sausage machine" of Frazier and Fodor (1978) to a more recent hybrid by Bever, Sanz, and Townsend (1998), among several others. The ideas that I will ultimately pursue have something of this flavor. For the moment, I set that aside. The elementary phrasal organization of utterance suffices for immediate discussion.

The capacity to work without support of meaning variables is a *partial necessity* for a modularity argument. It's not decisive because, for example, empty (semantic/contextually driven) slots in a procedural design could work to default values. From a radical interactive perspective, one could argue that the basic machine is designed to use meaning values (however they may be derived) in its parsing actions, but the system can function in the absence of such information.

The time course of processing, and the conditions under which given types of contextual information can become available for application are, thus, essential ingredients in this mix. If an informational source that could constrain a specific processing decision can be shown as intrinsically slower to be retrieved than the presumptive modular information set (i.e., a syntactically driven procedure), then its failure to exert influence on processing may be explained by that fact alone. Presumptive availability of whatever information is proposed as combining directly with the syntactically defined information set is essential. Related to this are obvious and important differences between the application of constraints that inhere in the meaning of individual lexical items and those that depend on contingent facts merely associated with the interpretation of a word, or that cannot be determined from an individual lexical entry at all but instead arise via the interpretation of phrase, sentence, and discourse. The import of a processing effect that relies, for example, on the fact that dogs are animate physical objects differs from one that relies on our knowledge that they may tend to have fleas or do not read newspapers. The likely time course for access to such different informational classes will vary. And much to the point: our interpretation of its significance for modularity claims will vary.

Most compelling is the demonstration that a putatively relevant constraint is present but *not* effective. That argues against the incorporation of such a meaning parameter in the foundational parsing procedures. A remark by Forster encapsulates this. It is as telling today it was at its inception:

> "... if syntactic processing is not autonomous, but guided by the assessment of plausible semantic relations [...] between key lexical items, then there should be no task that is simultaneously sensitive to syntactic effects and insensitive to plausibility." (Forster, 1979, p. 44)

Here, he referred to results of his same-different matching task results. In that task, rapidly presented pairs of word strings must be judged for whether their constituent elements are the same or differ in one or more words. Lexical, syntactic and meaning variables in the test strings all influence response times. Forster reported conditions for which clear effects of syntax were observed, but for which plausibility variation did not affect the presence or magnitude of those effects. There are many other, and more recent, examples of similar import, but this earlier work is quite sufficient, and time has not dimmed its excellence of execution or relevance. My objective is just to pose the challenge of a comprehensive response to two classes of experimental evidence, with modular and non-modular import, not to execute, here or later, any comprehensive contemporary review.

So let's look at the other side of the experimental coin. We find plenty of results supporting the claim that basic parsing procedures incorporate contextually driven interpretative bias. I will offer a wider range of these examples since they are the focus of my alternative account. Early work by Tyler and Marslen-Wilson (1977) gets right at the core type of contextual constraint. They manipulated the interpretation of an ambiguous phrase like "landing planes" as a plural noun phrase (NP) ("landing planes are very noisy near an airport") or a singular gerundive nominal ("landing planes is dangerous in a storm"). An immediately prior context ("If you are standing near the runway . . ." vs. "If you are not a trained pilot . . .") affected decision time for a choice between the variants of the linking verb "is/are," which was visually presented following the ambiguous phrase. Decisions were faster for targets compatible with the prior context phrase. The influence of the context arose within the few hundred milliseconds required to choose one of the linking verbs, and by assumption, the syntactic analyses for the ambiguous phrase. Contextual constraint must be projected from plausible world scenarios, and this is a key feature of a strong interaction claim. (See Marslen Wilson & Tyler, 1987, for review of other relevant reports of the time.)

Work by Crain and Steedman (1985), and a substantial range of work by others thereafter, implicates a quite specific discourse feature in parsing. That influential research showed interpretation of a verb phrase as a relative clause modifier rather than a main verb was affected by the discourse setting: contexts with dual protagonist settings versus those with single protagonist settings (for NP interpretation) promoted relative clause analyses—that is, fitting with continuation (a) rather than (b) in the example.

> the teachers taught by the Berlitz method . . .
> a) . . . were very successful / b) . . . but couldn't get jobs anyway

Readiness to analyze material following the initial NP (e.g., "the teachers") as a relative modifier increased when a relative could function to distinguish among referents in the discourse. The result projects a contingent fact from the context to a parsing decision.

The Crain and Steedman study is one of many focused on Tom Bever's justly celebrated "horse raced past the barn" example (Bever, 1970). Studies of argument structure effects are also prominent—a number of studies have varied subcategorization and thematic structure (see, e.g., Carlson & Tannenhaus, 1988). There are experimental reports for effects that ignore subcategorization (e.g., postulation of a trace following an intransitive verb; Mitchell, 1994, among others), and others that report early application of such information. But interactions of lexical biases with plausibility are of most interest. Trueswell and Tannenhaus (1994), and others, combined subcategorization bias with thematic role assignment and reported plausibility effects: suitability of NP's to their thematic roles affected processing time. Analysis of the detailed character and possible representations of such constraints is essential, of course, but for my purposes, these sorts of results are good grist for the problem solving mill that I favor.

A substantial number of experiments examine prepositional phrase attachment. Various investigators (see, e.g., Taraban and McClelland, 1988) have emphasized effects of plausibility relations among the lexical constituents in biasing preference for interpreting a prepositional phrase as an instrumental adverbial ("the spy saw the policeman with the binoculars"), or as a reduced relative clause modifier ("the spy saw the man with the revolver"). More telling are studies that record the time course of eye movements. These have assumed a prominent role in this research area (the "visual world paradigm": Tannenhaus et al., 1995), and the procedure has been exercised to study influences on prepositional phrase attachment. A typical experiment measures gaze patterns recorded for listeners listening to and executing spoken instructions pertinent to elements of a visual test array. Contingencies in the visual array (e.g., numbers of elements of a given type, relative locations, and qualitative properties of objects, etc.) affect the analysis of spoken test sentences. So, in particular, observation of the layout of a visual array can affect attachment preferences for prepositional phrases. Apprehension of the features of a visual scene can translate to an interpretive bias, and that to sentence analysis. It's clear that some mechanism brings the conceptual force of the visual information to bear on language with great rapidity. Is it a parsing effect in the conventional sense? We will return to this issue.

These remarks highlight a mixed experimental landscape with a few, but well-known and productive, examples. Multiple structural types have been the focus of activity. Some seem more susceptible to interpretive manipulation than others—for example, prepositional phrase attachment. By contrast, the effects of semantic constraint on direct object/sentence complement ambiguities (as in "the child knew the answer to the problem was in the book" vs "the child knew the answer to the difficult question by heart") are murkier. Experiments that focus on implicit structural elements (e.g., empty categories or structural traces of varying sort) present still another variable scene that I have not commented on. But across the full spectrum of such research, we need consistent accounts of why biasing effects vary across experiments and structural types. What property of language processing systems accounts for significant materials and task specificity in the research outcomes? Are the determinants of the variability principled, or adventitious—accidents of strength and timing. Some answers are no doubt mundanely methodological. But more is going on, I think. Before essaying my suggestion for a production system perspective, I want to add one more stone to the soup.

EXPERIMENTAL PRAGMATICS AND INTIMATIONS OF MODULARITY

Why pragmatics? We know that people supply their own "contextual assumptions" when we don't stipulate. And our contextual reach when we do so is impressive—the effects on interpretation that go beyond the semantic force inherent in sentence form can rely on deep background information. Some pragmatic processing theories are, nevertheless, committed to the claim that there is contextual penetration of processing at the earliest sentence analysis level. Such theories, for example, reject the idea that literal sentence meaning is computed in contexts that do not require it (see Harnish, 1994, for some comment). Pragmatics covers several subareas, and the extent to which they have been experimentally attacked is variable. I briefly note features of some research in three areas: metaphor, indirection, and scalar inference.

Metaphor has been a parade case for study of non-literal language. It's evident that metaphoric extension is fast, flexible, and in some sense "automatic" (e.g., Gildea & Glucksberg, 1983). But it is not free. Good experimental evidence indicates that metaphoric interpretation takes more processing resource than non-metaphoric language (e.g., Coulson &Van Petten, 2002). And metaphors do not block literal interpretation (e.g., Pynte et al., 1996; Tzuyin Lai, Curran, & Menn, 2009). See Janus and Bever (1985) for an instructive report on experimental issues in the study of metaphoric language. Not everyone agrees (not even all those I have cited, of course), and I don't suggest there is no room for scientific debate. My point is only that there is a good case to be made for processing models that incorporate literal interpretation as necessary, and distinct from pragmatic extension.

Indirection—as in "Could you pass the salt" taken as a request to "pass the salt"—has been the subject of several experimental studies. Clark and Lucy (1975) reported their experimental findings in terms of a model that first recovers literal meaning as a context free operation based on sentence form, with the result checked for appositeness to contextual criteria. Depending on the contextual fit, further processes would be initiated and the extended meaning achieved. Clark (1979) modified that treatment by weighting the meaning components, and with literal meaning computed but not necessarily temporally prior. Gibbs (1979) made sharply contrasting claims that do not include a literal meaning priority. For my purposes, however, a study by Foldi (1987) is nicely apposite to my tone (it's old, and the results were clear). She studied comprehension of indirect requests in right hemisphere damaged patients (RHD) compared with left hemisphere damaged (LHD) in a picture description/decision task. Some RHD patients have been clinically remarked as "literal minded." Research prior to Foldi's suggested limitations on the performance of RHD patients for metaphor and idiom interpretation—in both instances, they showed a preference for literal interpretation (e.g., Gardner, et al., 1975; Winner & Gardner, 1977). And some RHD patients were alleged to be "humor impaired"—they don't get the point of jokes (Bihrle, Brownell, & Gardner,1986). Bear in mind, although the dominant pattern of localization for language functions shows a left hemisphere focus, that description oversimplifies matters. Language is, we all know, not a single system—left dominant features of language are manifest most strongly for morphological and syntactic structure. The potential for some aspects of language use to be significantly influenced by the RH is quite real. Foldi's experiment, indeed, showed that RHD patients predominantly made literal interpretations in her task as compared to pragmatically appropriate indirect request interpretations; the latter were the strongly preferred responses by LHD and normal controls. The RHD patients did not spontaneously make the step from literal interpretation to the pragmatic; pragmatic extension mechanisms were not available or not automatically invoked. Whatever the precise account of the RHD failures, the dissociation between direct and indirect speech acts was clear and does not readily fit accounts that treat the integration of context with sentence analysis in ways that assign no special status to literal and direct meaning.

The third example concerns scalar inference—the inference from use of the weaker term (e.g., "some") in a quantified scale to reject the stronger ("all"). It's a powerful intuition for normal adult speakers. If I were to say to you (as some experimental studies have posed it) "Some elephants have trunks," you'd likely be inclined to suppose I was joking, or up to no good, in uttering a remark ignoring the palpable truth that all elephants have trunks (barring the most unpleasant of misadventures). This, general area, and certainly the some/all contrast, has been the focus of several detailed experimental investigations to assess whether the extended meaning based on the scalar inference is immediate at the level of primary parsing and interpretation, or represents an added layer of processing. Precise measures of the time course of scalar interpretation are required (see Katsos & Cummins, 2010, for a useful review). My chosen example is a study by Huang and Snedeker (2009) that cleanly illustrates the message useful to the current discussion. Their study is instructive in two ways. They contrasted performance in children and adults. Young children have been reported to accept the "some and possibly all" interpretation (e.g., Papafragou & Musolino, 2003). Huang and Snedeker's examination of this contrast gives detailed evidence for the time course associated with making the scalar inference. Timing and control of gaze direction for the referents of stimulus sentences in picture displays was measured (with a variant of the visual world paradigm referred to in the first section). The test sentences were locally ambiguous (vis-à-vis the reference space). So listeners would hear one of the variants as indicated in this example:

"Point to the girl with [two, three, all, some] of the socks."

Display pictures (that followed the test expression) might include a boy with two socks, a girl with two socks and a girl with three soccer balls. So for quantifiers "two," "three" and "all," decision could be fixed at the quantifier. If semantic interpretation of quantifiers were immediate, there would be no need to await the phonetic disambiguation re socks and soccer balls. For the "some" condition, matters are contingent: if the inference from "some x" to "not all x" were made immediately upon encounter with the quantifier, the picture of the girl with two socks would also be uniquely determinate (she had "some but not all" of the socks). The soccer ball picture would be ruled out because that girl had all the soccer balls. But if the scalar inference were not made in the "some" condition (either underspecified or semantically interpreted as "some X and possibly all"), disambiguation would come only at the phonetic change between "socks" and "soccer." The outcome showed clear differences among quantifier conditions. For adults, gaze moved to the appropriate targets by 400 milliseconds after quantifier onset for "two," "three," and "all" conditions; the "some" condition was delayed (about 800 milliseconds after onset), but still well before the phonetic disambiguation point. That pattern showed the scalar inference was made, but later than the semantic interpretation of the quantifiers. The pragmatic process was rapid and contextually appropriate but followed basic sentence interpretation. For the children, the evidence differed: they were similar to adults except that they did *not* make the scalar inference, waiting until phonetic disambiguation for the "some" quantifier. There is some resonance in the child/adult differences just described and the RHD/LHD patients difference in response to indirection noted earlier: the pragmatic extension imposes additional cognitive demand, whether informational or procedural. It's fast, it's flexible, it's pervasive; but it's not free.

The evidence in this research area indicates some separation of contextual constraint from the analysis of sentence form. The recruitment of background information—relatively speaking—takes "extra time" over and above that required for securing the sentence form that underlies an utterance. That description is apt for a number of experimental studies in pragmatics. But there is variation in the nature of the effects, and the impact of these influences is very rapid. We want to be able to deal with both those effects. A virtue of the production based application of context constraints that I outline next is that has the potential to accommodate multiple types of processes in a natural way.

LANGUAGE PRODUCTION STUDY

Language production models are intended to describe the real-time integration of a spoken form that expresses the meaning a speaker wishes to convey on a given occasion of utterance. Production is driven at the outset by an interpretation of conceptual content: the communicative intent of a speaker. On the other hand, comprehension is driven initially by an acoustic or orthographic input that is the fodder for interpretation. A natural assumption is that the primary processes of sentence analysis for comprehension and sentence construction for production respectively should reflect their controlling inputs. And so they seem to—up to a point. Where things depart from that assumption is where the lines of the modularity discussions begin to be drawn in both modeling domains.

The study of language production began to emerge in the 1970s based on normal speech error data and to some extent on language pathology (Fromkin, 1971). That base was rapidly expanded by more observational and experimental work. Levelt (1989) provided a comprehensive interpretation and elaboration of language production theory that remains relevant today. Research over the ensuing decades enriched the observational base with new experimental procedures, and tested general architectural claims for production systems. Figure 2.1 gives a summary picture of a production system organization based on a wide range of data types.

The multi-stage process represented in Figure 2.1 captures several robust features of speech error distributions and results of a variety of experimental studies. Abstract lexical representations (i.e., phonologically uninterpreted objects, referred to as "lemmas" in Figure 2.1) are retrieved and integrated into a syntactic representation, followed by processing that determines the phonological form of utterance elements and associated prosodic structures. Semantic control applies at the initial stage of lexical and phrasal selection, but is not evidenced in the mechanisms of phrasal integration or phonological interpretation. Error data supporting this claim include patterns of word substitution errors, as well as several types of movement errors (e.g., sound and word exchanges, anticipations or shifts in location). These indicate the computationally effective representations of sentence elements at different points in the sentence formulation process. Many detailed constraints of syntax, morphology, phonology and prosody strongly affect speech error patterns, but semantic constraints on error interactions of sentence elements at syntactic and phonological levels are not evident (Bierwisch, 1982). Powerful semantic similarities affect word substitution errors, but these are best understood as errors in the initial selection of lexical content; that is, errors that occur prior to the operations that incorporate semantically selected elements into syntactic and prosodic structures (see, e.g., Garrett, 1980, 1993a, for review of these arguments).

Levelt and several collaborating colleagues experimentally attacked many core questions in this area (see Levelt, 1989 for much of the foundational theory and research tool development). Issues of lexical retrieval for syntactic integration were a key target for investigation (see Levelt, Roelofs, & Meyer, 1999, for a summary report of that influential work). Experimental investigations with picture/word interference tasks examined the staging of processes at conceptual/semantic levels, lemma levels, and word form levels (see Figure 2.1). Constraints on the time course for activation of the different classes of structure associated with lexical targets, vis a vis semantic and phonological information, converged with patterns seen in studies of speech error data. Work with similar implications for a two stage lexical retrieval system emerged from the study of tip-of-the-tongue states (see, e.g., Vigliocco, et al., 1997). To this, one may add work by van Turrenout, Hagoort, and Brown (1998) with electrophysiological measures. In picture naming tasks, they found syntactically controlled lexical responses

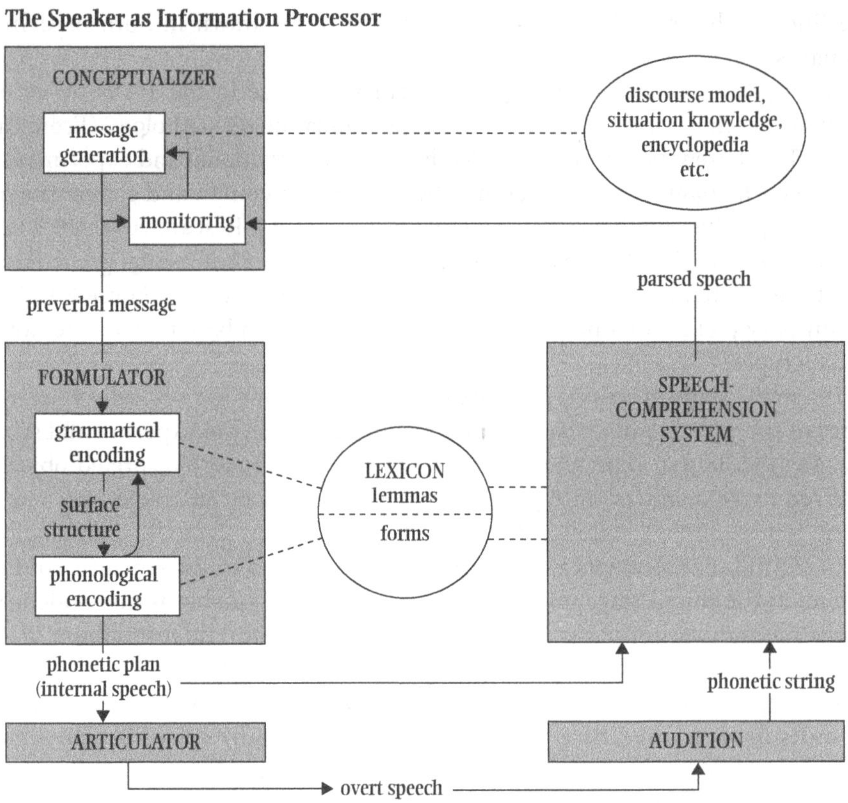

Figure 2.1 Speech production model (after Levelt, 1989).

preceded phonologically controlled responses as evidenced by differences in timing of motor readiness potentials.

NB: though the ordering of lexical recovery as just sketched is well supported, there is also good evidence of overlap in the time course of processes as outlined. Investigation of the degree and role of feedback relations between systems responsible for providing lexically specific semantic, syntactic and phonological information are relevant. Gary Dell and colleagues have closely studied this area (see, e.g., Dell, Schwartz, Martin, Saffran, & Gagnon, 1997) and have combined modeling and experimental work showing conditions under which lemma and word form representations may interact in normal and language disordered populations. Cutting and Ferreira (1999) also provided experimental evidence for conditions that promote a feedback link from form to meaning based representations and these observations are important for our understanding of the time course of lexical retrieval during production. The feedback links, however, do not compromise the staged retrieval design or arguments for a separation of syntactic and semantic control in the integration of phrasal structure.

Turning more directly to syntactic integration, experimental research by Kathryn Bock and colleagues creatively examined ways in which interpretive and lexical constraints interact with sentence construction (see, e.g., Bock & Levelt, 1994; Bock, 2004, for reviews). It is a wide ranging program of study, with work on memory (e.g., Bock & Warren, 1985), agreement processes (e.g., Bock & Eberhard, 1993) and work on "syntactic priming" (Bock, 1986).

The syntactic priming work is most immediately germane. These experiments studied changes in the rate at which a given syntactic type is produced in a picture description episode by manipulating the syntactic form of a (*semantically unrelated*) sentence generated on a *just preceding trial*. So, for instance, a number of studies find that passives occur in the picture description task more often when preceded by a passive prime than an active prime. Similar patterns of effect arise for double object constructions vs. *to*-datives. The carry-over across trials in such circumstances is prima facie based on syntactic configuration and not meaning representation. Further to this point, Bock and Loebell (1990) provided more sharply focused evidence for a processing stage/representation sensitive to phrasal configuration but not to the lexical content, and most particularly, not to the semantic relations among content elements. They reported syntactic priming mechanisms treat as equivalent objects the *by*-phrases in sentences like:

"the plane landing *by the control tower*," versus "the plane landed *by the pilot*" Thematic roles differ but syntactic configuration is preserved, and similar priming effects ensue.

A more recent study by Konopka and Bock (2009) reinforces these implications. They compared non-idiomatic and idiomatic phrasal verbs for their efficacy in a syntactic priming paradigm. Prime and target sentences used phrasal verbs with particles adjacent to the verb (e.g., "pull off a sweatshirt") or shifted to the slot following the direct object (e.g., "pull a sweatshirt off"). Idiomatic primes, (e.g., pull off a robbery) were semantically opaque vis-à-vis the literal interpretation of the verb particle construction. The correspondence of the two types of primes is only configurational. And here, too, the outcome was driven by configural overlap: idiomatic and non-idiomatic primes produced significant structural effects, and to comparable degree.

Note that these findings arise via the use of lexical and structural priming techniques that are sensitive to influence by interpretive constraints. So, for example, animacy and thematic role have been examined in similar tasks (Bock, Loebell, & Morey, 1992). Priming can influence the relative likelihood that a given NP will occupy a particular argument slot. But such effects do not interact with the priming of phrasal configurations. This outcome comports precisely with the fact that speech error distributions indicate a computational separation between the semantically driven selection of lexical elements that are to be embedded in a sentence structure and the integration of the structure itself. Note that this claim is fully compatible with lexically driven encoding schemes, but not with those that eliminate a distinguishable syntactic representation. The priming patterns and the speech error data provide prima facie evidence for abstract structure that encodes syntactic configuration but not lexical content.

The details of timing and local interaction among components in the production systems remain active research areas. But the case for a significant degree of modular structure in the language generation system with global outline as in Figure 2.1 is well supported. Against this background, we consider the issues launched in the first three sections: rationalizing contrasting experimental claims for interaction and modularity in comprehension.

LANGUAGE PRODUCTION AS A COMPREHENSION FILTER

The core conundrum as I've sketched it is that for comprehension studies, there is good reason to claim both modular and non-modular processing profiles across the experimental landscape. This is not soccer. We can't declare a tie. How do we rationalize this? Is a wild and wooly methodological scramble the only path? If, for example, one wished to accept the claims for a constraint based processing system that is relevantly non-modular, then some plausible account should be on offer for the several experimental circumstances (with different methodologies) in which semantic, conceptual, and situational constraints have no apparent impact on parsing. The constraint system can readily accommodate interactions; a persuasive account of the several circumstances in which such are not forthcoming is challenging.

Quite apart from the behavioral measures on the psycholinguistic experimental scene, the electrophysiological and brain imaging profiles of language processing that have emerged over the past two or more decades make an all-out embrace of full interaction for syntactic processing questionable. The evidence for distinct brain responses to different types of information processing demands is strong and multi-faceted. And finally, we can fold the variable profile for pragmatics into the picture. The issue there is the incredible openness of the information that must be identified for relevance and recruited in the time frame of normal speech and reading rates. Recovery of the relevant information in that time frame is, to say the least, implausible.

The proposal on offer here appeals to the intrinsic design demands of the language production system as the means for application of the several sorts of required background information. The normal functions of production require such capacity, and in comparable time frames to that of comprehension. The proposal relies on the integration of language comprehension and language production systems in ways that preserve their individual identity, but incorporates production processes in the routine function of comprehension. Using the comprehension system to monitor for error in production outputs is a well-established feature of language processing modeling (see Levelt, 1989, for a review). The complementary idea (occasionally suggested: e.g., Forster, 1979; Garrett, 1993b) is that production systems might play a similar role with respect to comprehension, namely to monitor the adequacy of interpretations delivered by the recognition system. The current proposal is more aggressive, assuming that the production system can "filter" the generation of alternative analyses in the parsing system. There is good logical and experimental reason supporting the view that recognition systems respond to local structural ambiguity by temporarily maintaining multiple analysis paths that are rapidly pruned based on posterior context, or on recruitment of higher order interpretive constraints. The suggestion is that the production system may provide a means for resolution at such choice points and a source for predictive devices to induce structural preference in sentence comprehension. The viability of this idea depends on a capacity of the language production system to combine the initial lexical elements and elementary phrasal structures identified by the recognition system with existing discourse constraints and thereby generate candidate sentence structures within the required time frame. A brief evaluation of some aspects of this position follows (see Garrett, 2000 for more detailed discussion of some of the following points.)

Lexically based production routines match many features of contemporary parsing study (e.g., subcategorization and thematic structure; computational efforts at merging parsing and generation systems). Lexically driven encoding for production is compatible with "lexically driven" approaches to parsing. "Lexical preferences" as determinants of choice among parsing options—as contrasted with general principles of configural economy or structurally driven choice (e.g., minimal attachment, late closure, etc.) do not require online access to the underlying forces driving those preferences.

The similarity of the compositional operations required of a parser and those required for a lexically driven sentence encoding model shows up in the computational literature. There are a number of efforts to develop systems that support both analysis and generation. See, for example, Kempen (1999) for a review of such a research program; see also Stone and Doran (1997) for relevant work with tree adjoining grammars.

Though the operations of phrasal composition may be very similar, controlling inputs to the two types of processors are distinct. In human language production, lexical nomination is by message level constraint. In first stage comprehension, it is by outputs of the lexical recognition system. A production architecture that separates semantic control process from direct involvement in the phrase building operations may enable phrase building by the *producer* to engage lexical inputs from either conceptual/semantic or form driven systems.

This way of talking might suggest that it is literally the very *same* machinery that computes the phrasal structures for both production and comprehension. Kempen, et al (2012) has argued for this strongest position in a recent paper. It's an interesting potential implementation, but one with some significant logistical issues to be solved. For example, accepting this position would seem to compromise the effectiveness of error checking systems. If, in fact, production and comprehension systems perform mutual error monitoring functions, then independent sources for the compared signals is a necessity. Important aspects of the systems operations might be very similar—but there must be two of them with different drivers.

The speed of lexical recognition and structure projection is compatible with rapid engagement of context effects via production. The "production as comprehension filter" proposal requires early and accurate identification of lexical targets and the use of their structural information in time frames that could match time constants for comprehension performance. Some very clever work by Marslen-Wilson (1973) uncovered and exploited a phenomenon that might be taken as a kind approximate "existence proof" for this. These are "close shadowing" performances—which refers to an ability to sustain repetition latencies for normal connected prose of 250–300 milliseconds with quite good accuracy—the loop from ear to mouth is closed under that time frame. He discovered that some (~1 in 5) persons could do this. What's in the loop? There was clear engagement of aspects of the syntactic and semantic force of the shadowed materials, though precisely what types and detail of representation may be recovered could not be sharply fixed. To this, we emphasize that lexical recognition speed is fast—and hence the necessary information for launching a production based filtering operation is available early. How fast? Very. Work by Zwitserlood (1989) is instructive. Using a cross-modal priming paradigm, she provided evidence for the multiple activation of words compatible with the initial phonetic segments of test words and an effective linking of lexical targets to sentence environments at a moment contemporaneous with the occurrence of the terminal phonetic segments of the test words. Substantial later work with other methodologies reinforces this. There is, of course, much we don't know about this general set of issues, but the evidence on boundary conditions is not discouraging.

Production profiles predict "comprehension" performance. Experimental comprehension investigations that look for effects of frequency of structural configuration or other collocational factors often design test material using sentence completion procedures. The values used are production values. Apart from that, statistical measures of relative frequency of occurrence rely on corpora of spoken or written language output. The conflation of production and comprehension performance is clear. Other things being equal, the postulation of a production based parsing filter suits existing accounts of structural preference in comprehension. This is by no means to say that there could be no differences in the statistical regularities relevant to comprehension and production. Different ways to arrive at such estimates have attracted the interest of computational and experimental investigators (e.g., Mitchell, 1997; Gibson, 2006; Kim, Srinivas, & Trueswell, 2002). However, I should note that, from somewhat different perspective, McDonald (2013) has argued in favor of a central role for production processing in the development of grammar and of comprehension strategies. Her framework appeals to exigencies of production processing as the prime determinant for usage patterns across variant language forms. The parser's development inherits preferences dictated by the production driven landscape. This does put production contingencies at the top of the structural preference food chain from the outset of our experience with language.

Turning from statistical matters, here is an interesting example of detailed convergence of performance profiles in the two domains. It arises from work on agreement error patterns in speaking. Bock and Miller (1991) and several follow-up studies demonstrated stable patterns in the breakdown of mechanisms for number agreement in English. The significant facts are these: in a sentence completion task, they found number mismatch between a head noun and a local noun—for instance, as italicized in the example—enhances the likelihood of number error

The *baby* on the *blankets* . . . (is/*are playing with a puppy.)

compared to a control with no mismatch. The detail is that error rate is substantially greater when the mismatch is between a plural local "distractor" and a singular head—the reverse mismatch is significantly weaker as an error inducer. The kicker is that the same effect can be observed in a "recognition" task. Nicol, Forster, and Veres (1997) evaluated the same contrasts in a reading task. A key aspect of their approach was that their test sentences did *not* have errors in them—they were fully grammatical. The NP's in their test sentences were located in the ways that matched the positions of interfering NPs in the production task. They reported elevated reading times for the sentences with mismatched interfering NP's (e.g., "baby"/ "blankets") compared to those with NP's matching in number (e.g. baby/'blanket'). But these effects were limited to the singular head NP/plural interfering NP condition. This matched the detail of the production performance. Processing difficulties in the coordination of number marking occur in production and in comprehension and do so in very similar ways.

Context, discourse, and plausibility effects. The kind of information available and the time available to extract it are crucial issues. Here we consider higher order sorts of information with clear impact on comprehension. For the production system role being examined here, some sort of scenario would be the natural source to work from. Isolated sentences with no context are not good candidates for the operation of production based constraint mechanisms. But it is, of course, a feature of experimentation aimed at a test of interpretive and discourse factors on comprehension to use carefully designed contextual environments for target sentences. There's no other way to do it. So, in various of the Marslen-Wilson and Tyler and Crain and Steedman studies cited earlier (and many similar not cited), a mini-discourse context is provided for the test sentences. Those conditions are ideal for the application of production based constraints on the analysis of target structure. From this, one may expect that well established discourse based influences on parsing may be among the fastest acting sources of constraint—faster perhaps than some inferential constraints based on the lexical content of sentences. The gaze tracking studies noted earlier (e.g., Tannenhaus et al., 1995) also readily fit a production perspective. Those data seem very plausibly driven to a significant degree from the primary production system, where interpretation draws on a structured environment and responds to commentary of structurally defined range. The task demands a match of auditory inputs to a small range of potential descriptions. Language production mechanisms fall naturally into such experimental environments.

Finally, I note Forster's (1979) characterization of the plausibility effects he systematically investigated. He attributed the effects to the ease of generation of a possible construal of the relations implied by the sentence. Plausibility studies of the sort he initiated are of interest as vehicles for production system study. The intractability of plausibility effects under repeated presentation (Forster, 1987) is intriguing, given that repetition might reasonably be expected to greatly diminish perceptual and comprehension influences.

Language disorders. The selectivity of language impairment is a recurrent theme in discussions of modularity. What more specific issues might arise in the context of the production as comprehension filter idea? Agrammatism is an obvious candidate for discussion. It is, as seen (typically) in Broca's aphasia an expressive disorder often associated with clinical evaluations of good comprehension. If production is a significant ingredient in normal comprehension, what should one make of the agrammatic dissociation of the two? On balance, it's not easy to say because the nature of the impairment is quite variable, and its different forms can be mapped onto production/comprehension interactions in multiple ways. The beginning of an odyssey of agrammatic dissection was Caramazza and Zurif's (1976) demonstration that at least some such patients' comprehension success did not rely on syntax, and when appropriately tested experimentally, showed syntactic limitations (see Caplan, 1995, for review). A preserved lexical semantic capacity, and associated inferential processes, give leverage on accounts of comprehension success despite syntactic loss. In any event, underspecified impairments in both production and comprehension do not helpfully constrain ways in which syntactic processes might be linked across the two systems.

Two other reported features of agrammatic disorders deserve mention. One comes from work initiated by Linebarger, Schwartz, and Saffran (1983): That work showed some agrammatic producer/comprehenders could succeed in well-formedness judgment tasks for test items they misinterpreted in comprehension tests. They had "paradoxical syntactic capacity." So although they failed to understand, for example passives, they correctly distinguished syntactically well-formed from ill-formed instances of that structure. This implies what, in fact, what I would wish to affirm, namely, that the input parser may deliver a well formed product that lacks a reliable semantic interpretation. The linkage between sentence form and interpretation is disturbed in such patients. And this could take different forms given the framework being considered here. If integration of a recognition based representation with one that is production based were impaired, an erroneous semantically driven production target could be accepted even if the input representation is accurate. That might occur given a failure in semantically driven production machinery that engages phrasal construction. That mechanism, once engaged, may operate without immediate semantic control (viz., as the production evidence reviewed in previous section suggests). The possible breakdowns would be in links between syntax and interpretation, not in mechanisms of phrasal construction per se. And that could occur in comprehension system, production system, or both.

A complementary profile to the one just discussed is patients with agrammatic speech but no comprehension limitation—that is, apparently genuine instances of "paradoxical comprehenders"—unlike the patients tested by Caramazza and Zurif. In such cases, the production as comprehension filter hypothesis entails that, if basic production capacity is compromised, it should reduce the speed and/or accuracy with which interpretive constraints constrain syntactic analysis. Thus reports of agrammatic patients who display no comprehension deficit in experimentally controlled evaluations of their sensitivity to syntactic detail (see, e.g., Miceli et al., 1983; Kolk, Van Grunsven, & Keyser, 1985) are a challenge. Here is where the core aspects of the production proposal must be kept in mind. Production driven syntax is assumed to provide rapid access to contextually derived interpretive constraints. But detailed syntactic representations built by a "data driven" parser are also available and interpretable. So it is necessary to know to what extent comprehension performance in such patients is responsive to the normal range and time course of contextual constraints.

"Agrammatic" is a description of performance by patients with significant variation in underlying deficit and physical impairment. Until we know how a given agrammatic output symptom relates to underlying production processes, its significance is uncertain. There is potentially useful leverage on the proposal for a production filter on comprehension in examination of this and other production disorders.

Acquisition profiles. A truism of language acquisition is that kids understand syntactically complex speech of others before they themselves talk much, and certainly long before their own spontaneous speech is comparably elaborated re syntax (see, e.g., McKee, McDaniel, & Snedeker, 1998 and references therein). Such an apparent dissociation, raises questions similar to those from the performance profile of agrammatic aphasics described earlier: production output is limited while comprehension can draw on more sophisticated syntax. The proposal for an integrated production and comprehension system that I am suggesting calls for some rationalization of this disconnect. For openers, one may question the extent to which children's early production systems are actually lacking in capacity. There is good evidence that children's utterances are more complex than superficial appearance suggests (see, e.g., McKee & Emiliani, 1992). Further, elicited production tasks provide evidence that children's production systems are, even at early stages, adult-like in many respects (see McDaniel, McKee, & Garrett, 2010). Young children's production profile may be reduced in range not because they cannot bring conceptual content to bear on syntactically complex linguistic representation, but rather because of late stage limitations on the organization and control of phonologically detailed outputs. The literature has suggestive reports of trade-offs between length and complexity (see McDaniel, Mckee, & Garrett, 2017, for comment). On these various grounds, a disconnect of production and comprehension in child language may arise elsewhere than in the components needed to support a production filter on parsing of the sort I wish to advance. Close examination of the emergence of links between production and comprehension during language development actually looks like a good place to dig for ways to test that idea.

SUMMING UP

The general question: "is human information processing modular or interactive?" is a poser that generates hang-ups. It's silly to say either is the sole answer. It's clearly both. Indeed, the power of human cognition springs from the artful combination of the outputs of diverse dedicated processors. Data fusion across cognitive domains is a trick that humans are very good at.

The question that bedevils the research enterprise in language is, of course, where the boundaries are drawn for specific specialized systems. From the perspective of the proposal outlined here, a radical interaction position on comprehension finds itself in the somewhat odd position of claiming an early and pervasive intermingling of syntactic, semantic, and background world knowledge variables in that piece of the language system focused on the perceptual construal of language inputs. And this is in the face of strong evidence for a significant degree of modular organization for phrasal construction in production systems—systems with primary access to the full conceptual and discourse background underlying the generation of utterance form. What would impel "ignoring" such riches in the organization of the phrasal generation machinery? But I don't think that is what is going on. The organization of the phrase building system may be an architectural finesse in production that enhances its interaction with comprehension processes. On the comprehension side, getting the interpretation right is what is hard, not working out the potential forms that could be extracted from the perceptual data. This is where the application of resources incorporated into the language production machinery can play a role. As a filter on the perceptual products of the comprehension machine, it is a natural optimizing of the resources needed for bringing the things we know about the world to bear on what we say and what we hear.

But I can almost hear the muttering in the background: *Everybody's* account must include a means of coupling conceptual structure and background information to language—fast, versatile, pretty much automatic. You bet. Do I know how to do that? No. And I don't have a relevance ray-gun to point at the hard problem either. So some might say my story is just a kind of hijacking of the key capacity for my own ends without delivering on the hard problem of how any system does the crucial work. Fair enough. In the end, it may turn out to be changing which walnut shell the pea is under. But I think there is something to be gained by trying out different ways of looking at the problem. And taking syntax out of the direct mix by running it through a production loop has appeal. It calls for a different approach to combining interpretive demands with specific language structure and puts some rough boundaries on linkages among the components that get it done.

REFERENCES

Bever, T. G. (1970). The cognitive basis of linguistic structures. In J. R. Hayes (Ed.), *Cognition and the development of language*. New York, NY: John Wiley & Sons.

Bever, T. G., Sanz, M., & Townsend, D. (1998). The emperor's psycholinguistics. *Journal of Psycholinguistic Research, 27*(2), 261–284.

Bierwisch, M. (1982). Linguistics and language error. In A. Cutler (Ed.), *Slips of the tongue*, 29–72. Amsterdam, Netherlands: Mouton.

Bihrle, A., Brownell, H., & Gardner, H. (1986). Comprehension of humorous and non-humorous materials by left- and right-brain damaged patients. *Brain and Cognition, 5*, 399–411.

Bock, J. K. (1986). Syntactic persistence in language production. *Cognitive Psychology, 18*, 355–387.

Bock, J. K. (2004). Psycholinguistically speaking: Some matters of meaning, marking, and morphing. In B. H. Ross (Ed.), *The psychology of learning and motivation* (Vol. 44, pp. 109–144). San Diego, CA: Elsevier.

Bock, J. K., & Eberhard, K. (1993). Meaning, sound, and syntax in English number agreement. *Language and Cognitive Processes, 8*, 57–99.

Bock, J. K., & Levelt, W. J. M. (1994) Language production: Grammatical encoding. In M. Gernsbacher (Ed.), *Handbook of psycholinguistics* (pp. 945–984). San Diego, CA: Academic Press.

Bock, J. K., and Loebell, H. (1990). Framing sentences. *Cognition, 35*, 1–39.

Bock, J. K., Loebell, H., & Morey, R. (1992). From conceptual roles to structural relations: Bridging the syntactic cleft. *Psychological Review, 99*, 150–171.

Bock, J. K., & Miller, C. A. (1991). Broken agreement. *Cognitive Psychology, 23*, 45–93.

Bock, J. K., & Warren, R. (1985). Conceptual accessibility and syntactic structure in sentence formulations. *Cognition, 21*, 47–67.

Caplan, D. (1995). Issues arising in contemporary studies of disorders of syntactic processing in sentence comprehension in agrammatic patients. *Brain and Language, 50*, 325–338.

Caramazza, A., & Zurif, E. (1976). Dissociation of algorithmic and associative processes in language comprehension: Evidence from Aphasia. *Brain and Language, 3*, 572–582.

Carlson, G., & Tannenhaus, M. (1988). Thematic roles and language comprehension. In W. Wilkens (Ed.), *Thematic relations, syntax, and semantics*, Vol. 21 (pp. 263–300). New York, NY: Academic Press.

Clark, H. H. (1979). Responding to indirect speech acts. *Cognitive Psychology, 11*, 430–477.

Clark, H. H., & Lucy, P. (1975). Understanding what is meant from what is said: A study in conversationally conveyed requests. *Journal of Verbal Learning and Verbal Behavior, 14*, 56–72.

Coulson, S.,VanPetten, C., 2002. Conceptual integration and metaphor: An event-related potential study. Memory and Cognition, *30*, 958–968.

Crain, S., & Steedman, M. (1985). On not being led up the garden path: The use of context by the psychological processor. In D. Dowty, L., & A. Zwicky (Eds.), *Natural language parsing* (pp. 320–358). Cambridge, England: Cambridge University Press.

Cutting, J. C., & Ferreira, V. S. (1999). Overlapping phonological and semantic activation in spoken word production. *Journal of Experimental Psychology:Learning, Memory, and Cognition, 25*, 318–344.

Dell, G., Schwartz, M., Martin, N., Saffran, E., & Gagnon, D. (1997) Lexical access in aphasic and non-aphasic speakers. *Psychological Review, 104*, 801–838.

Fodor, J. A. (1983). *The modularity of mind*. Cambridge, MA: MIT Press.

Foldi, N. (1987). Appreciation of pragmatic interpretations of indirect commands: Comparisons of right and left hemisphere brain-damaged patients. *Brain & Language, 38*, 88–108.

Forster, K. I. (1979). Levels of processing and the structure of the language processor. In W. Cooper & E.C.T. Walker (Eds.), *Sentence processing* (pp. 27–85). Englewood, N.J.: Erlbaum.

Forster, K. I. (1987). Binding, plausibility, and modularity. In J. Garfield (Ed.), *Modularity in knowledge representation and natural-language understanding* (pp. 63–82). Cambridge, MA: MIT Press.

Frazier, L., & Fodor, J. D. (1978). The sausage machine: A new two-stage parsing model. *Cognition, 6*, 291–325.

Friederici, A. D., & Kotz, S. A. (2003). The brain basis of syntactic processes: Functional imaging and lesion studies. *NeuroImag, 20*, 8–20.

Fromkin, V. (1971). The non-anomalous nature of anomalous utterances. *Language, 47*, 27–52.

Gardner, H., Ling, P., Flamm, L., & Silverman, J. (1975). Comprehension and appreciation of humour in brain-damaged patients. *Brain, 98*, 399–412.

Garrett, M. (1980). Levels of processing in sentence production. In B. Butterworth (Ed.), *Language production: Vol 1. Speech and Talk* (pp. 177–220). London, England: Academic Press.

Garrett, M. (1993a). Errors and their relevance for theories of language production. In G. Blanken, J. Dittmann, H. Grimm, J. Marshall, & C. Wallesch (Eds.), *Linguistic disorders and pathologies: An international handbook* (pp. 72–92). Berlin, Germany: Walter de Gruyter.

Garrett, M. (1993b). The structure of language processing: Neuropsychological evidence. In M. Gazaniga (Ed.), *Cognitive neuroscience* (pp. 881–899). Cambridge, MA: MIT Press.

Garrett, M. (2000). Remarks on the architecture of language processing systems. In Y. Grodzinsky, L., Shapiro, & D. Swinney (Eds.), *Language and the brain* (pp. 31–68). San Diego, CA: Academic Press.

Gibbs, R. (1979). Contextual effects in understanding indirect requests. *Discourse Processes, 2*, 1–10.

Gibson, E. (2006). The interaction of top-down and bottom-up statistics in the resolution of syntactic category ambiguity. *Journal of Memory and Language, 54*, 363–388.

Gildea, P., & Glucksberg, S. (1983). On understanding metaphor: The role of context. *Journal of Verbal Learning and Verbal Behavior, 22*, 577–590.

Hagoort, P., Brown, C., & Osterhout, L. (1999) The neurocognition of syntactic processing. In C. Brown & P. Hagoort (Eds.), *The neurocognition of language* (pp. 273–316). Oxford, England: Oxford University Press.

Harnish, R. M. (1994). Mood, meaning and speech acts. In S. L. Tsohatzidis (Ed.), *Foundations of speech act theory* (pp. 407–459). London/New York: Routledge.

Huang, Y., & Snedeker, J. (2009). Semantic meaning and pragmatic interpretation in 5-year-olds: Evidence from real-time spoken language comprehension. *Developmental Psychology, 45*, 1723–1739.

Janus, R. A., & Bever, T. G. (1985). Processing of metaphoric language: An investigation of the three-stage model of metaphor comprehension. *Journal of Psycholinguistic Research, 14*, 473–487.

Katsos, N., & Cummins, C. (2010). Pragmatics: From theory to experiment and back again. *Language and Linguistics Compass, 4*, 282–295.

Kempen, G. (1999). *Human grammatical coding*. Cambridge, England: Cambridge University Press.

Kempen, G., Olsthoorn, N., & Sprenger, S. (2012). Grammatical workspace sharing during language production and language comprehension: Evidence from grammatical multitasking. *Language and Cognitive Processes, 27*, 345–380.

Kempler, D., Curtiss, S., & Jackson, C. (1987) Syntactic preservation in alzheimer's disease. *Journal of Speech and Hearing Research, 30*, 343–350.

Kim, A., Srinivas, B., & Trueswell, J. C. (2002). The convergence of lexicalist perspectives in psycholinguistics and computational linguistics. In P. Merlo & S. Stevenson (Eds.), *Sentence processing and the lexicon: Formal, computational and experimental perspectives* (pp. 109–135). Philadelphia, PA: John Benjamins.

Kolk, H., Van Grunsven, J., & Keyser, A. (1985). On parallelism between production and comprehension in agrammatism. In M. L. Kean (Ed.), *Agrammatism* (pp. 165–206). New York: Academic Press.

Konopka, A. E., & Bock K. (2009) Lexical or syntactic control of sentence formulation? Structural generalizations from idiom production. *Cognitive Psychology, 58*, 68–101.

Levelt, W. J. M. (1989). *Speaking: from intention to articulation*. Cambridge, MA: MIT Press.

Levelt, W. J. M., Roelofs, A., & Meyer, A. S. (1999). A theory of lexical access in speech production. *Behavioral and Brain Sciences, 22*, 1–75.

Linebarger, M., Schwartz, M., & Saffran, E. (1983). Sensitivity to grammatical structure in so-called agrammatic aphasics. *Cognition, 13*, 361–392.

Marslen-Wilson, W. (1973) Linguistic structure and speech shadowing at very short latencies. *Nature, 244*, 522–523.

Marslen-Wilson, W., & Tyler, L. (1987). Against modularity. In J. Garfield (Ed.), *Modularity in knowledge representation and natural-language understanding*. Cambridge, MA: MIT Press.

McDaniel, D., McKee, C., & Garrett, M. (2010). Children's sentence planning: Syntactic correlates of fluency variations. *Journal of Child Language. 37*, 59–94.

McDaniel, D., McKee, C., & Garrett, M. (2017). Children's performance abilities: Language production. In E. Fernandez & H. Cairns (Eds.),: *Handbook of Psycholinguistics* (pp. 479–503). Hoboken, NJ: Wiley-Blackwell.

McDonald, M. C. (2013). How language production shapes language form and comprehension. *Frontiers in Psychology, 4*, 226.

McKee, C., McDaniel, D., & Snedeker, J. (1998). Relatives children say. *Journal of Psycholinguistic Research, 27*(5), 573–596.

McKee, D., & Emiliani, M. (1992). Il Clitico: Cé ma non si vede. *Natural Language and Linguistic Theory, 10*, 415–437.

Miceli, G., Mazzucchi, A., Menn, L., & Goodglass, H. (1983). Contrasting cases of Italian agrammatic aphasia without comprehension disorder. *Brain and Language*, *19*, 65–97.

Mitchell, D. (1994). Sentence parsing. In M. Gernsbacher (Ed.), *Handbook of psycholinguistics* (pp. 375–409). San Diego: Academic Press.

Neville, H., Nicol, J. Barss, A., Forster, K., & Garrett, M. (1991). Syntactically based sentence processing classes: Evidence from event related potentials. *Journal of Cognitive Neuroscience*, *3*, 151–165.

Papafragou, A., & Musolino, J. (2003). Scalar implicatures: Experiments at the semantics–pragmatics interface. *Cognition*, *86*, 253–282.

Nicol, J., Forster, K., & Veres, C. (1997). Subject-verb agreement processes in comprehension. *Journal of Memory and Language*, *36*, 569–587.

Pynte, J., Besson, M., Robichon, F., &Poli, J. (1996). The time-course of metaphor comprehension: An event-related potential study. *Brain and Language*, *55*, 293–316.

Stone, M., & Doran, C. (1997). Sentence planning as description using tree-adjoining grammar. Proceedings of the 35th Annual Meeting of the Association for Computational Linguistics and Eighth Conference of the European Chapter of the Association for Computational Linguistics, ACL 98, 198–205.

Tannenhaus, M., Spivey-Knowlton, M., Eberhard, K., & Sedivy, J. (1995). Interpretation of visual and linguistic information in spoken language comprehension. *Science*, *268*, 1632–1634.

Taraban. R., & McClelland, J. L. (1988). Constituent attachment and thematic role assignment in sentence processing: Influences of content-based expectations. *Journal of Memory and Language*, *27*, 597–632.

Trueswell, J., & Tannenhaus, M. (1994). Toward a lexicalist framework for constraint based syntactic ambiguity resolution. In C. Clifton, K. Rayner, & L. Frazier (Eds.), *Perspectives on sentence processing* (pp. 155–179). Hillsdale, NJ: Erlbaum.

Tzuyin Lai, V., Curran, T., & Menn, L. (2009). Comprehending conventional and novel metaphors: An ERP study. *Brain Research*, *1284*, 145–155.

Tyler, L., & Marslen-Wilson, W. (1977). The on-line effects of semantic context on syntactic processing. *Journal of Verbal Learning and Verbal Behavior*, *16*, 683–692.

Van Turennout, M., Hagoort, P., & Brown, C. (1998). Brain activity during speaking: From syntax to phonology in 40 msec. *Science*, *280*, 572–574.

Vigliocco, G., Antonini, T., & Garrett, M. (1997). Grammatical gender is on the tip of Italian tongues. *Psychological Science*, *8*, 314–317.

Winner, E., & Gardner, H. (1977). The comprehension of metaphor in brain-damaged patients. *Brain*, *100*, 717–729.

Zwitserlood, P. (1989). The locus of effects of sentential-semantic context in spoken word processing. *Cognition*, *32*, 25–64.

3

The Modularity of Sentence Processing Reconsidered

FERNANDA FERREIRA AND JAMES NYE

The idea that the sentence processing system is modular has fallen out of fashion. The proposal got off to a promising start with the publication of results in the early to mid-1980s suggesting that word meanings are activated without regard to their global contexts, and that sentence structures are assigned to words at least initially without consideration of whether the structure would map on to a sentence interpretation that made sense given prior knowledge or given the contents of the immediate linguistic or visual context. Eventually, the modular view of sentence processing became strongly associated with what was termed the "two-stage" model of comprehension, a model which assumed that an initial syntactic analysis or parse was created by implementing a couple of simple parsing operations, and that this initial parse was then revised if it either did not lead to a globally connected syntactic structure, or if it led to a meaning that did not fit the comprehender's expectations and goals. By the late 1980s, connectionist approaches to cognition were becoming increasingly popular, and although one of their most salient properties is their flexibility, connectionism became strongly associated with interactive architectures, and those were assumed to be nonmodular. About a quarter century of research has since been directed at trying to show that sentence processing is not modular, and that instead the interpretation assigned to a sentence is influenced by all kinds of knowledge and sources of information ranging from visual context to beliefs about the intentions and even social background of the speaker. The nonmodular view is so widely accepted at this point that it is now almost mandatory to end scholarly papers and presentations with the observation that the findings support a highly interactive system in which knowledge sources freely communicate. It has been a very long time since anyone in the field came forward with any sort of argument in support of the modularity hypothesis.

In this chapter, we will review the evidence that is meant to support this overall consensus in the field that sentence processing is nonmodular. We will begin by summarizing the original (1983) modularity proposal. We will briefly examine the important features of a module as described in the (1983) book (*The Modularity of Mind*—henceforth, *TMOM*), focusing specifically on how those properties were interpreted by researchers working on sentence processing. Then, we will summarize a large literature that emerged in response to the idea that sentence processing might be modular. The organization will be thematic: We will consider first the debate concerning the use of what might be described as intra-linguistic information, including prosody and lexical information. From there, we will consider the debates focused around the use of context, including both visual and discourse context. We will argue that although some of the simplest and most obvious versions of modularity might be implausible, it is a distortion to assert that the data undermine modularity in sentence processing entirely. Indeed, seen in a fresh light, the results of the bulk of studies conducted over the last 25 years can be taken as evidence for a more refined and detailed view of sentence comprehension, which retains many of the features of a modular system. The point is to use the findings from the studies to inform how we understand what the sources of information are and how they are organized, activated, and combined. We will also suggest that, in many cases, the claims for nonmodularity have simply been exaggerated—particularly those based on experiments using the so-called *visual world paradigm* (VWP).

An interesting new development in the field of sentence processing is the advent of new approaches emphasizing the shallowness of sentence comprehension. These approaches go under a few different names, including *good-enough language processing* (Ferreira et al., 2002), *shallow processing* (Sanford & Sturt, 2002), *late assignment of syntax theory* (Bever, Sanz, & Townsend, 1998), *analysis-by-synthesis* (Fodor, Bever, & Garrett, 1969; Garrett, 2000), and *noisy channel/rational communication* (Levy, 2008; Gibson, Bergen, & Piantadosi, 2013) models of processing. The common assumption is that comprehenders simplify, misinterpret, or alter the input to end up with an interpretation that is more compatible with semantic expectations. These models have been difficult to categorize with respect to the modularity hypothesis. On the one hand, the idea that comprehenders use simple tricks or heuristics to obtain at least an initial interpretation seems compatible with modularity, particularly the features relating to shallowness. In addition, the models are consistent with other approaches to cognition that emphasize the limited use of information for speed and sometimes, even, for more accurate performance (Gigerenzer, 2004; Kahneman, 2011). On the other hand, because these models suggest that the system is biased toward plausibility or, in more current terminology, because they emphasize the role of "priors" in the Bayesian sense, they seem to emphasize nonmodular aspects of the system; they seem to highlight the idea that the sentence processing system is driven by semantic considerations and above all wants to create interpretations that are semantically or pragmatically compelling. One of our goals will be to try to sort through these possibilities and make the case that these approaches are consistent with a modular approach to sentence processing, if we emphasize shallowness rather than encapsulation.

THE MODULARITY OF MIND

As is now well known, the modularity thesis assumes that some cognitive systems have the following features. First, there are what we might call the more biological properties: Modular systems are associated with *neural specialization*; for example, specific areas of the brain seem to respond selectively to linguistic input (Fedorenko, Duncan, & Kanwisher, 2012). In addition, modular systems *emerge during development* with little in the way of individual variation. Although recent research on child language has tended to emphasize major differences in vocabulary and some other aspects of language competence in children from different social and economic backgrounds (Hoff, 2006), it remains clear that core language capacities emerge in almost all children at about the same time and in roughly the same sequence (Gleitman & Wanner, 1982). Modules also tend to become *selectively impaired* when an individual suffers from a biologically based disorder such as dyslexia or when a person experiences brain damage (e.g., aphasia; Joanisse, Manis, Keating, & Seidenberg, 2000; Dick, Bates, Wulfeck, Utman, Dronkers, & Gernsbacher, 2001; Sitnikova, Salisbury, Kuperberg, & Holcomb, 2002; Caramazza & Zurif, 1976).

The second set of module properties have to do with what we'll describe as superficiality: Modules deliver *shallow outputs*, which in the case of language can be taken to mean that what the sentence processing system delivers to systems that operate further along the information processing stream is merely the conditions for an interpretation; for example, the system that must determine what action to be performed based on a spoken utterance does not have information about the presence of gaps or traces in the syntactic representation from which the interpretation was derived. Similarly, people have "*limited central access*" to the internal operations of the sentence processing system; they might obtain an interpretation for a sentence, but they can't reason about the sources of that interpretation or the intermediate representations that were constructed to obtain it. This set of properties concerning superficiality have received less attention than the others, but we will argue that they are at least as significant, and that they relate closely to the newer models of sentence processing that were mentioned earlier—models which assume that the sentence processing system often engages in shallow processing.

The final set of properties of a module are the ones that have been the target of the great empirical scrutiny, particularly in the area of sentence processing. These are the features that relate most closely to issues of information flow in a cognitive system, and map on to the older distinction between so-called "top-down" and "bottom-up" streams of information flow (Zekveld, Heslenfeld, Festen, & Schoonhoven, 2006; Field, 2004). Most important of these is that a modular system must exhibit *information encapsulation*: a module can access its inputs and its own databases and processes, but it cannot access anything outside the module. Its operations are also therefore *domain-specific*: the module consults a narrow range of information and that database is stated in a proprietary vocabulary related to the domain of processing. And because of this domain specificity and information encapsulation, the system can operate *automatically* (mandatory operation) and *quickly*.

Fodor in (2000) reinforces the importance of information encapsulation by describing it as being at the "heart of modularity" (p. 63). For a system to be a module, it must consult only a limited computational database when it analyzes input. It is also perhaps for this reason that most empirical investigations of whether a system is modular, and in particular whether the sentence processing system is modular, have tended to focus on demonstrating that a piece of information assumed to be outside the module does or does not affect processing in that domain. But what the notion of information encapsulation should also highlight is the importance of determining the information sources that are assumed to be used by a particular module. In other words, delineating the representational domain of a putative module is critical to determining whether its operations conform to modularity. In the area of language comprehension, this point was never properly confronted before the claims for anti-modularity started to be made. For example, some of the earliest studies were focused on demonstrating that the sentence processing system takes into account information about prosody when it makes syntactic decisions. The idea was that because prosodic information was stated in a different vocabulary from syntax, it should not be able to affect the computation of a parse tree. The problem with this argument, however, is twofold: First, and more obviously, if a prosodic analysis is *input* to the module that performs syntactic analyses, then prosodic effects on parsing are to be expected and in no way violate the modularity thesis. Second, and perhaps a bit more controversially, if a representational format is proposed which blends syntactic and prosodic information, then again, prosodic influences on syntax are compatible with modularity, as are syntactic influences on prosody. This point will be discussed in more detail.

Finally, it is important to recognize that, in *TMOM*, Fodor also argued that modularity should be construed is a matter of degree: "One would thus expect—what anyhow seems to be desirable—that the notion of modularity ought to admit of degrees. The notion of modularity that I have in mind certainly does" (p. 37). A system is modular "to some interesting extent" if it exhibits some of the properties summarized earlier; not all of them need to be present. At the same time, as we have also seen, the one property that seems necessary for a system to be described as modular is information encapsulation, at least for Fodor.

THE "TWO-STAGE MODEL" OF SENTENCE PROCESSING

For a variety of historical reasons, almost from the beginning, the idea that the sentence processing system might be modular became almost entirely conflated with testing a particular model of parsing—the so-called "two-stage model" first developed by Lyn Frazier (Frazier & Fodor, 1978) and then elaborated by her colleagues, including the first author (Ferreira & Clifton, 1986; Rayner, Carlson, & Frazier, 1983; Frazier, Pacht, & Rayner, 1999). Thus, in the interests of full disclosure, we acknowledge that the first author is strongly associated with this model, and both authors believe it is a compelling and empirically valid approach to explaining sentence comprehension. Nonetheless, it is important to recognize the historical coincidence that at the same time that *TMOM* was published, the two-stage model was also dominant. That model made several critical architectural assumptions from the perspective of evaluating the modularity hypothesis in this cognitive domain: First, the model assumed that a single parse is constructed for any sentence based on the operation of two simple principles: *Minimal attachment*, which constrains the parser to construct no potentially unnecessary syntactic nodes, and *late closure*, which causes the parser to attach new linguistic input to the current constituent during a parse, rather than going back to a constituent created earlier or postulating the existence of a new constituent. In addition, the two-stage model in its 1980s form assumed that the only information that the parser had access to when building a syntactic structure was its database of phrase structure rules. It therefore could not consult the syntactic information associated with lexical items. For example, in the sequence *Mary knew Bill* the noun phrase (NP) *Bill* would be assigned the role of direct object because that analysis is simpler than the alternative subject-of-complement-clause analysis, and the information that *know* takes sentence complements more frequently than direct objects could not be used to inform the initial parse.

Similarly, decisions concerning the creation of the initial parse could not be influenced by prosodic information either. For example, given something like *Because Mary left Bill*, the NP *Bill* would be syntactically integrated as a direct object, even in the presence of a major intonational phrase boundary after *left*. Of course, during this period when the two-stage model and modularity were both relatively new, the question how prosody might affect parsing had to be put largely on hold because there were few good techniques available for studying the earliest stages of spoken sentence comprehension. And, as was argued in *TMOM*, the modularity of a system cannot be assessed with offline measures or techniques that provide information about the final stages of processing; to assess modularity, it is necessary to tap into early online processing. Yet another historical coincidence is that, in the 1980s, eye movement monitoring systems started to become affordable and easier to use, and so more and more psycholinguistic laboratories acquired some type of eyetracking device. But, at this point, eyetracking was applied almost exclusively to investigations of visual language processing (reading), and reading was assumed not to involve prosody in any serious way. (This assumption would change, of course, with the "implicit prosody" hypothesis of reading, but that is a topic for a different volume.) Eventually, researchers did venture into the field of spoken language processing and studies examining prosody in parsing were conducted. We will discuss those studies shortly.

In summary, the modularity thesis was tested against a specific model of sentence processing—a model which assumed that the parser proposes analyses serially and consults only phrase structure rules to make syntactic decisions. Eventually, evidence against the two-stage model would be construed as evidence against modularity as well, even though obviously other architectures for sentence processing are conceivable and even plausible. Moreover, findings that challenged assumptions such as the lack of access to subcategory information were not used to inform and update the assumptions about how any hypothetical sentence processing module might be organized or might operate; instead, they were taken as evidence against modularity itself. Having set the stage for the tests of modularity in this way, we now turn to experimental work designed to evaluate the modularity of sentence processing, keeping in mind that they were also, simultaneously, tests of the so-called two-stage model of parsing.

EVALUATING THE USE OF LANGUAGE-INTERNAL SOURCES OF INFORMATION

We begin with the question whether lexical information, and in particular, information linking elements such as verbs with the kinds of constituents with which they may occur, affects initial parsing. On the surface, it would appear to be rather odd to think this information would not be used, because in many theories of grammar, verb subcategorization information is stated in a syntactic vocabulary (Chomsky, 1965; Gahl & Garnsey, 2006; Hare, McRae, & Elman, 2003). For example, the information that the verb *put* must occur with both a noun phrase and a prepositional phrase can be represented as something like *put[__NP PP]*. As *TMOM* emphasizes, to establish whether a system is modular, it is critical to understand what its proprietary databases are. If we assume that a parser builds syntactic structures using syntactic information, then it would not seem unreasonable to assume that verb subcategorization information would be integral to the parser's operations. And, indeed, the earliest studies examining this question suggested that it is. Following from linguistic arguments based mainly on intuition data (Ford, Bresnan, & Kaplan, 1982), Mitchell and Holmes (1985) investigated this question by looking at the processing of sentences such as *The historian suspected the manuscript of his book had been lost*. They found that participants took less time to read the phrase *had been lost* when it co-occurred with *suspected* rather than with a verb such as *read*, which was presumed to occur because *suspected* takes sentential complements more frequently (see also Ford, Bresnan, & Kaplan, 1982). This result could be interpreted as evidence that the parser consults two sources of syntactic information during construction of its initial parse: phrase structure rules and verb subcategorization frames. It is not obvious that it stands as evidence against the modularity hypothesis.

Soon afterward, however, Ferreira and Henderson (1990) conducted a follow-on study designed to address a limitation of the Mitchell and Holmes (1985) experiments: Because Mitchell and Holmes employed a phrase-by-phrase reading task, it was possible that the reading times conflated initial and reanalysis processes. Self-paced reading requires participants to make a decision on each displayed chunk concerning whether to push a button to receive the next chunk or stay put in order to get more processing time. Ferreira and Henderson therefore designed a similar experiment but used the eye movement monitoring technique, which has exceptional temporal resolution (a sample of the eye position is taken approximately every millisecond) and spatial resolution. They found that verb bias had no effect on early eyetracking measures (e.g., first fixation and gaze durations) but did influence global measures such as total reading time. They concluded that the parser does not consult verb-specific syntactic information, but that such information is used in later stages to revise a misanalysis. They also viewed the results as confirmation of the two-stage model of parsing, which assumed this basic architecture.

Following publication of Ferreira and Henderson (1990), a large number of studies were conducted designed to challenge these conclusions (Wilson & Garnsey, 2009; Trueswell & Kim, 1998). Although some findings consistent with theirs were also reported (Pickering & Traxler, 1998), the field eventually coalesced around the idea that verb information indeed informs initial parsing. Moreover, this idea was also taken as evidence against the original two-stage model, which is appropriate. However, in addition, the finding that verb information influences early parsing processes was also taken as evidence against modularity. But as our arguments thus far should make clear, we believe this conclusion is far too broad. One can easily imagine a modular theory of sentence processing in which the sources of information consulted to derive an initial parse include all the syntactic rules or principles relevant to projecting phrase structure, including verb subcategory information. In short, evidence for lexical guidance of early parsing decisions is not evidence against modularity, because the lexical information is plausibly internal to the syntactic module.

Next, let us consider the question how prosodic information might influence sentence processing. The starting point for most studies published in the topic is that syntactic and prosodic structures are related, and in particular, major syntactic boundaries such as those separating clauses are usually marked by phrase-final lengthening and changes in pitch (Ferreira, 1993). Some clause-internal phrasal boundaries are also marked, although much less reliably (Allbritton, McKoon, & Ratcliff, 1996)—for example, in the sentence *John hit the thief with the baseball bat*, the higher attachment of *with the baseball bat*, which supports the instrument interpretation, is sometimes (but not always) associated with lengthening of *thief*. The logic of the research enterprise was as follows: If certain prosodic "cues" signal syntactic structure, then the parser might be able to use this information to avoid "going down the garden-path"—that is, it might be able to avoid misanalyzing the sentence structure. Of course, it is not obvious that the use of this information would constitute a violation of modularity, but that was the motivation for some of this research.

One of the earliest studies to consider this question was conducted by Beach (1991), and it claimed to show that prosodic information affects parsing. What the experiments actually demonstrated is that metalinguistic judgments about sentence structure were influenced by the availability of durational and pitch information linked to the final structures of the sentences. The obstacle to drawing any strong inferences concerning modularity at this stage in the history of the field was the unavailability of tasks for measuring online spoken language processing. The phoneme monitoring task had been abandoned in the 1980s (prematurely, as argued by Ferreira & Anes, 1994). The field still awaited the widespread use of electrophysiology to measure online processing of visual and auditory stimuli, and eyetracking had not yet been adapted to the investigation of spoken language. A couple of decades later, these techniques have yielded a wealth of information about the comprehension of utterances, and one of the ideas on which there is now a general consensus in the field is that prosody indeed influences the earliest stages of parsing. To take just one recent example, Nakamura, Arai, and Mazuka (2012) conducted an auditory study using temporarily ambiguous Japanese sentences and the visual world paradigm to investigate how contrastive intonation affected parsing decisions. Their results suggest that prosody can affect early stages of spoken sentence processing, leading comprehenders even to anticipate upcoming structure. Numerous other studies led researchers to similar conclusions (Price, Ostendorf, Shattuck-Hufnagel, & Fong, 1991; Kjelgaard & Speer, 1999; Millotte, Wales, & Christophe, 2007).

Now, how shall we evaluate these results and interpretations in light of the modularity hypothesis? If we conflate the two-stage model of parsing and the modularity hypothesis, then we must conclude that sentence processing is nonmodular. But we could instead update a model offered more than 25 years ago in light of this sort of evidence relating to prosody, as indeed the proponents of the two-stage model have (Carlson, Frazier, & Clifton, 2009; Frazier, Carlson, & Clifton, 2006). However, even if evidence is presented to refute specific models of modularity, this should not be taken as evidence against modularity as a whole, but only one potential form of modularity. Our argument is that, when considering modularity, it is important to establish not only what information sources are *internal* to the module, but also what information is *input* to that module. In the case of sentence processing, it seems reasonable to assume that prosodic cues or prosodic representations might be input to the sentence analyzer—that is, in terms of the more traditional bottom-up/top-down processing distinction, it seems plausible that prosodic analysis would take place *before* syntactic parsing. This idea makes some sense, as the flow of information during comprehension seems to be from sensory to conceptual, and prosodic features such as loudness, duration, and pitch are more sensory/perceptual than information about syntactic categories. Thus, prosody may indeed influence the earliest stages of parsing, but this does not undermine modularity.

THE USE OF CONTEXT AND PLAUSIBILITY INFORMATION DURING SENTENCE PROCESSING

Although investigations of verb subcategorization information and prosody are important for understanding the nature of sentence processing, it is not clear that they're useful for evaluating the modularity hypothesis, as we have argued. What is clearly relevant and indeed critical is information that certainly appears to be nonsyntactic. One of the earliest analyses came from Crain and Steedman (1985). They observed that many of the sentence forms treated as syntactically dispreferred by the two-stage model are also presuppositionally more complex. For example, consider the sentence *The evidence examined by the lawyers turned out to be unreliable.* According to the two-stage model, minimal attachment leads the parser to initially treat *examined* as a main verb, which causes the parser to be garden-pathed when the *by*-phrase is encountered. The parser must then reanalyze the structure as a reduced relative (see Fodor & Ferreira, 1998, for proposals concerning syntactic reanalysis). Similarly, the prepositional phrase attachment ambiguity in a sentence such as *John hit the thief with the stick* allows for two interpretations: initially, the *with*-phrase is interpreted as an instrument, but the *with*-phrase may instead serve as a modifier. As in the case of the reduced relative ambiguity, in this case too, the more complex syntactic analysis involves modification while the simpler analysis does not.

Crain and Steedman (1985) pointed out that these modification interpretations are not just syntactically more complex; they're presuppositionally more complex as well. Felicitous use of a complex phrase such as *the evidence examined by the lawyer* requires that there be more than one type of evidence in the discourse so that the modifier can be used to pick out the correct referent. This analysis appeals to the Gricean Maxim of Quantity (Grice, 1975), which states that speakers should not include unnecessary information in their utterances (but see Engelhardt, Bailey, & Ferreira, 2006). They argued further that null contexts favor the minimal attachment interpretation because, without a context specifying a set of objects denoted by the head noun, the listener will assume the presuppositionally simpler interpretation. Crain and Steedman presented intuitive evidence that sentences with reduced relative clauses were easy to process in proper contexts, contrary to what the two-stage model would predict.

The problem with the Crain and Steedman (1985) argument, of course, is that offline judgments are not adequate for assessing modularity, because they measure only the output of any putative module. Certainly a sentence such as *The evidence examined turned out to be unreliable* sounds better in context than by itself (as does almost any sentence), but that observation gives us no insight into the processes that support the intuition. For that reason, Ferreira and Clifton (1986) conducted an eyetracking study to assess whether the effect of context was mainly to influence offline interpretations, or if it indeed intervened in the initial syntactic decisions of the parser. Their data were consistent with the idea that context did not affect initial parsing decisions. Supportive contexts led to shorter global reading times and more accurate question-answering behavior, but early measures of processing revealed that processing times for reduced relative and prepositional modification structures were longer than for their structurally simpler counterparts.

To the best of our knowledge, the findings from this 1986 study still hold. The only serious challenge came from Altmann and Steedman (1988), who elaborated on the Crain and Steedman (1985) proposal and also reported a set of self-paced reading experiments that purported to provide contrary results. This in turn led to a debate between Altmann and Steedman, on the one hand, and Clifton and Ferreira, on the other (1988). However, as Clifton and Ferreira argued, it is unclear that self-paced reading data can trump eyetracking results because the self-paced reading measure has far poorer temporal and spatial resolution, and therefore is biased against detecting early effects of syntactic manipulations.

More interesting than this debate about techniques, however, are the actual details of the Altmann and Steedman (1988) theoretical proposal. We believe the importance of the position they took in that paper has not been adequately appreciated in the 25 years since the paper's publication. Altmann and Steedman argued for a sentence comprehension system with two important properties. The first is that their parser consulted a syntactic database very different from the one assumed in the two-stage model. The important difference is that the representational format for structural information was Steedman's Combinatory Categorial Grammar, which combines syntactic and semantic information (and even some aspects of prosody and intonation; see Steedman, 2000; Steedman & Baldridge, 2011). Thus, if the parser consults a database of structural information contained in that sort of vocabulary, then effects of certain semantic manipulations on initial parsing are not inconsistent with modularity. This argument is the same as the one we made earlier regarding the use of verb subcategorization information: If the information is part of the module's proprietary database, then use of that information cannot constitute a violation of modularity.

But the second property is even more important: Altmann and Steedman (1988) argued for what they termed a weakly interactive architecture. What this architecture amounts to is a system in which "syntax proposes" and "semantics disposes." Crucially, on this model, alternative structural analyses are activated in parallel, and context retains the interpretation that is most contextually appropriate. This sort of mechanism is the same as the one that had been suggested in earlier work to explain the processing of lexical ambiguity (e.g., *bank*), and was specifically discussed in *TMOM* as an example of how a modular system might work. The idea is that, bottom-up, all alternatives are retrieved and made available to subsequent modules that then choose the one that is most suitable. In the case of lexical ambiguity, both meanings of *bank* are activated (and not necessarily equally strongly; modulation of activation according to frequency is also perfectly compatible with bottom-up processing), and the meaning that fits the context is retained while the other meaning either decays or is inhibited by executive cognitive systems. Similarly, all syntactic structures might be computed or retrieved, and the one that post-sentence processing systems like are retained while the others either decay or are inhibited. The important point, then, is that this type of interaction with context does not violate modularity, as Altmann and Steedman themselves emphasized with their description of their model as merely "weakly interactive."

A related debate has centered around another potential influence on initial parsing decisions—semantic plausibility. Ferreira and Clifton (1986) not only looked at the effects of discourse context on parsing; they also focused on plausibility information linked to animacy. The critical contrasting cases are *the evidence examined* versus *the defendant examined*. With the animate noun *defendant*, the verb *examined* is naturally interpreted as the thing doing the examining; but with the inanimate noun *evidence*, the same syntactic analysis leads to an anomalous interpretation. Ferreira and Clifton reported that the animacy information did not block the garden-path, which led them to argue for a strongly modular architecture. This conclusion has been the target of numerous challenges (Altmann & Steedman, 1988; McClelland, 1987; MacDonald, Pearlmutter, & Seidenberg, 1994; MacDonald, 1993), and at this point, the consensus seems to be that animacy does indeed influence initial parsing (but see Clifton, Traxler, Mohamed, Williams, Morris, & Rayner, 2003). And, in turn, this view is taken to be evidence against modularity. Again, however, animacy is a very basic type of semantic information which some languages treat as a grammatical feature (Dahl & Fraurud, 1996). If the lexical entries for nouns include a simple +/– animacy feature, then it is not implausible to think that a modular parser might be able to access that information in a lexical entry and match it to a lexico-syntactic rule stating that the subject of an agentive verb such as *examine* must be animate. In addition, our arguments concerning the propose/dispose architecture also hold: If syntactic alternatives are constructed in parallel and then selected on the basis of plausibility, then what we have is what Altmann and Steedman (1988) called weak interaction, which is compatible with the modularity thesis. Once again we see that a result incompatible with the two-stage model of parsing (which assumes serial analysis plus reanalysis rather than a propose/dispose architecture) was taken as evidence against modularity itself.

MODULARITY AND THE VISUAL WORLD PARADIGM

The early 1990s saw the creation of a new paradigm for studying sentence processing—the VWP. The idea behind the paradigm is simple: From reading studies, it was known that what the eyes fixate on and how much time is spent during a fixation are closely tied to attention and processing (Rayner, 1977). The VWP extends this logic to spoken language processing by pairing spoken utterances with simple visual displays containing mentioned and unmentioned objects. The "linking hypothesis" (Tanenhaus, Magnuson, Dahan, & Chambers, 2000) is that as a word is heard, its representation in memory becomes activated, and this in turn automatically triggers eye movements toward the named object as well as objects semantically and even phonologically associated with it (Huettig & McQueen, 2007). The acceptance and widespread adoption of the task occurred because it lined up with several trends in cognitive science: First, there was an emerging emphasis on cognition and action—that is, on trying to capture how cognitive processes might be used to guide intelligent action and behavior. Second, the idea of multimodal processing was also catching on, with many cognitive scientists wanting to understand the way different cognitive systems might work together—in this case, the auditory language processing system and the visuo-attention system associated with object recognition (Henderson & Ferreira, 2004; Jackendoff, 1996). Third, there was growing interest in auditory language processing generally, and in the investigation of how prosodic information might be used during comprehension (Bear & Price, 1990). And, most relevant to one of the themes of this volume, there was dissatisfaction with the lack of experimental paradigms for empirically evaluating the modularity hypothesis. Reading techniques were of course useful and often quite powerful, but not all questions regarding language processing can be studied with reading (e.g., the use of overt prosody), and some researchers were bothered by the idea that reading is not as fundamental or primary a mode of language as is spoken language. Thus, the VWP was enthusiastically adopted. By now, hundreds of studies have been reported making use of it in one way or another (for summaries, see Huettig, Rommers, & Meyer, 2011; Huettig, Olivers, & Hartsuiker, 2011; Ferreira, Foucart, & Engelhardt, 2013).

The report that triggered the widespread use of the VWP and that is also viewed as having fatally undermined the idea of a modular sentence processing system is Tanenhaus et al. (1995), reported in more detail in Spivey, Tanenhaus, Eberhard, & Sedivy (2002). This study adapted the Altmann and Steedman (1988) ideas concerning presuppositional support to the domain of visual contexts and spoken sentences that could be evaluated against them. To illustrate the study, consider the imperative sentence *Put the apple on the towel in the box.* At the point at which the listener hears *on the towel*, two interpretations are possible: Either *on the towel* is the location to which the apple should be moved, or it is a modifier of *apple*. The phrase *into the box* forces the latter interpretation because it is unambiguously a location. Referential Theory specifies that speakers should provide modifiers only when modification is necessary to establish reference (e.g., we do not generally refer to a big car if only one car is discourse-relevant). From referential theory, it follows that if two apples are present in the visual world and one of them is supposed to be moved, then right from the earliest stages of processing, the phrase *on the towel* will be taken to be a modifier, because the modifier allows a unique apple to be picked out. The listener faced with this visual world containing two referents should therefore immediately interpret the phrase as a modifier and avoid being garden-pathed, and this is indeed what the data seem to show (Farmer, Cargill, Hindy, Dal, & Spivey, 2007; Novick, Thompson-Schill, & Trueswell, 2008; Spivey, Tanenhaus, Eberhard, & Sedivy, 2002; Tanenhaus et al., 1995; Trueswell, Sekerina, Hill, & Logrip, 1999).

However, in recent work we have argued that the VWP is in many ways highly unsuited to the task of assessing modularity (Ferreira, Foucart, & Engelhardt, 2013). Of course, there are numerous other significant questions concerning sentence processing for researchers to ask, and for those questions, the VWP is quite useful (Huettig, Rommers, & Meyer, 2011). But recall once again the argument in *TMOM* that evaluating modularity requires an experimental approach that allows the measurement of online processing, and it should not encourage subjects to adopt atypical strategies for dealing with the experimental situation that might have little to do with normal sentence processing. Now consider how the original Tanenhaus et al. (1995) study was set up. Subjects were allowed to watch as an experimenter laid out a 2 × 2 arrangement of real objects to be manipulated in response to auditory instructions. Two quadrants contained the target and the distractor object and the other two quadrants contained two potential goal locations. Listeners then heard either a syntactically ambiguous or unambiguous instruction containing a prepositional phrase modifier. With this setup, the amount of time available to preview the visual context could be several seconds, and this time interval was not controlled. It seems likely that, during the preview period, listeners might start to generate fairly specific expectations about the form and content of the upcoming utterance, especially since all the utterances consisted of a transitive verb followed by a noun phrase and at least one prepositional phrase. After experience with some trials, the participant may form a template or underspecified form of the upcoming utterance. Thus, both the visual display and the sentences conform to predictable patterns, which participants can learn after a small number of trials (Fine & Jaeger, 2013).

To address these concerns about the suitability of the VWP for evaluating modularity in language processing, we conducted three experiments examining the effects of depriving subjects of a preview of the visual world, and we conducted a production experiment to determine how accurately naïve participants could guess the sentence likely to occur with a particular visual display (Ferreira et al., 2013). We found that participants were not garden-pathed in any condition when they were denied preview of the visual world prior to hearing the sentences, and we also reported that participants were surprisingly good at anticipating which object they would be asked to move and which objects would serve as potential locations. From these results we concluded that listeners engage in a fairly atypical mode of processing in VWP experiments with visual world previews and utterances that are highly similar to each other over all experimental trials: rather than processing utterances incrementally, they instead form an underspecified representation of what they are likely to hear next based on the content of the visual world. They then evaluate that prediction against the utterance itself. Now, it is certainly possible that humans sometimes process language in this way, but most people would agree that typical processing situations are quite a bit more open-ended.

For these reasons, then, we are not convinced that the VWP can provide strong evidence against modularity. Again, the technique is superb for getting at many important questions about how language is processed, but it is not clear that it is suited for determining to what extent sentence processing is characterized by information encapsulation or domain-specificity.

MODULARITY AND SHALLOW PROCESSING

In the last fifteen years or so, a new framework for thinking about sentence comprehension has emerged. There are many variants with important distinctions among them, but what they share is the idea that comprehenders sometimes end up with an interpretation that differs from the actual input received—the interpretation is either simpler (construal), somewhat distorted (late assignment of syntax theory; good-enough processing), or outright inconsistent (noisy channel approaches) with the sentence's true content. These models have been difficult to pigeon-hole with respect to the modularity thesis. To try to sort out this issue, we feel it is important to shift the emphasis away from the features of modularity having to do with information encapsulation and toward the features that emphasize shallow outputs and limited central access to the internal operations of a module. Typically, psycholinguists have assumed that the output of any parsing or sentence processing module is a syntactic representation, which is turned over to "central" systems that relate to knowledge and belief. But we could assume instead that the output of the module is an interpretation, with structure-building operations being used to create it. If we adopt these assumptions, then we might not be surprised to discover that people can end up with interpretations that are simpler than the input would seem to mandate, and that might even be nonveridical.

To see how this argument works, let's begin with the mildest form of these models—the ones that assume representations that reduce the input in some way. One implementation is to allow representations to be underspecified (Sanford & Sturt, 2002). Consider construal (Frazier & Clifton Jr, 1997): A major assumption of the construal model is that syntactic structures are not always fully connected—adjunct phrases in particular (e.g., relative clauses) may instead simply get associated with a certain processing domain, "floating" until disambiguating information arrives. The parser thus remains uncommitted (Pickering, McElree, Frisson, Chen, & Traxler, 2006; Traxler, Pickering, & Clifton, 1998) concerning the attachment of the relative clause and the interpretation of the noun phrase and sentence that would follow from any particular attachment (see Frisson & Pickering, 2001; Sanford & Graesser, 2006; Sturt, Sanford, Stewart, & Dawydiak, 2004; Frisson S., 2009 for evidence favoring underspecified representations). A more radical possibility is that the attachment decision is strategically postponed, which is what the good enough language processing (henceforth, GE) theory predicts. Swets, Desmet, Clifton, Ferreira (2008) tested this idea by presenting participants with either fully ambiguous sentences (*the maid of the princess who scratched herself was embarrassed*) or disambiguated controls (*the son of the princess who scratched himself/herself was embarrassed*). The twist they introduced was to manipulate whether participants were required to answer easy or difficult comprehension questions following each sentence.

The rationale was that, with easy questions, readers would not be motivated to resolve the ambiguity; with no interpretive consequences, they would be happy to leave the relative clause unattached. In contrast, with challenging questions, subjects would know they were being "called out" on their understanding of the sentences, and therefore attachment decisions were incentivized. The findings supported these predictions: they found a reading time advantage for sentences with ambiguous relative clauses relative to disambiguated controls when they were followed by easy questions, suggesting that they were easier to process due to the omission of the attachment operation. In contrast, when readers expected to receive questions probing their interpretation of the relative clause, critical regions of the sentences were read more carefully, and the ambiguity advantage was reduced. Other studies support the idea of underspecified representations for global syntactic structures (Tyler & Warren, 1987), semantic information (Frazier & Rayner, 1990), and coercion structures (Pickering, McElree, Frisson, Chen, & Traxler, 2006).

Another line of work explores psycholinguistic analogues of the so-called Moses illusion. The now-famous Moses illusion involves asking people a question such as *How many animals of each sort did Moses take on the ark*. Amusingly, most people answer "two" instead of pointing out that the presupposition behind the question is incorrect (Erickson & Mattson, 1981). The illusion is presumed to occur because Moses and Noah share a large number of semantic features, and semantic processing is often too shallow to allow the distinguishing features to be activated and integrated (see also Barton & Sanford, 1993). Sanford and Sturt (2002) suggest that shallow processing is linked to the focus-presupposition structure of a sentence: elements that are in semantic focus are processed deeply, but those that are assumed or backgrounded are processed more shallowly, leading to these kinds of semantic illusions. This proposal is reminiscent of one offered by Cutler and Fodor (1979), who found in phoneme monitoring studies that phonemes in words which are part of the focus of a sentence are detected more quickly than those that are in words located in the presupposed portion.

More radical variants of shallow processing models are those that allow the comprehension system to generate an interpretation that is even more discrepant from the input. Researchers in the field of text processing and cross-sentence integration have shown that readers are sometimes remarkably insensitive to contradictions in text (Otero & Kintsch, 1992), and also often fail to update their interpretations when later information undermines a fact stated earlier—for example, a character described initially as guilty of a crime but described later as exonerated remains tainted by the original charge in people's memory representations for the story (Albrecht & O'Brien, 1993). These ideas from text processing were exported to the sentence processing literature in a series of experiments showing that people did not seem to fully recover from garden-paths (Christianson, Hollingworth, Halliwell, & Ferreira, 2001). Participants were asked to read sentences such as *While the woman bathed the baby played in the crib* and then they answered a question such as *Did the woman bathe the baby?*. The surprising finding was that most people answered "yes," even though the meaning of the reflexive verb *bathe* requires that the object be interpreted as coreferential with the subject in an intransitive structure (see also Slattery et al.; Ferreira, 2013). It appears that comprehenders are not entirely up to the task of syntactic reanalysis, and sometimes fail to revise either all pieces of the syntactic structure or all elements of the semantic consequences of the initial, incorrect parse. In addition, the more semantically compelling the original, garden-path interpretation, the more likely people are to want to retain it rather than revise it to the one consistent with the global grammatical form.

Townsend and Bever (2001) offered up a model of sentence comprehension very different from either the traditional two-stage model or the connectionist models of sentence processing that had become popular in the 1990s. The Townsend and Bever model implements an architecture similar to what has been suggested for decision-making (Gigerenzer, 2004; Kahneman, 2003), which distinguishes between a so-called System 1 and System 2 (or Type 1 and Type 2) for reasoning. System 1 is fast, automatic, and operates via the application of simple heuristics—"quick and dirty" rules that usually deliver a reasonably good result. System 2, on the other hand, is slow, attention-demanding, and that is able to consult a wide range of beliefs—essentially anything the organism knows and has stored in memory. Notice how closely this architecture echoes the one suggested in *TMOM*, where System 1 would map on to modular systems and System 2 would map on to the central reasoning system. Of course, one important difference is that Fodorian modules are assumed to be computational—for example, the modular parser consults a detailed, complex syntactic database when building an interpretation, rather than relying on a small set of simple heuristics. Nonetheless, the points of overlap are intriguing.

In Townsend and Bever's (2001) model, which they refer to as LAST (late assignment of syntax theory), sentences are essentially processed twice: first, heuristics are accessed which yield a quick and dirty meaning, and then syntactic computations are performed on the same word string to yield a fully connected, syntactic analysis. The second process ensures that the meaning that is obtained for a sentence is consistent with its actual form. Townsend and Bever also assume that the first stage is nonmodular and the second modular; this is to account for the use of semantics in the first stage, and the use of essentially only syntactic constraints in the second. However, this type of two-stage model can be construed in such a way that the first stage is modular, as long as the heuristics are essentially "reflexes"—as long as they are simple syntactic tricks that are blindly applied to the input without the benefit of consultation with other sources of knowledge. Two models similar in spirit to LAST but which assume a modular architecture for the first stage are the one offered by Ferreira (2003) and Garrett (2000). The Ferreira model assumes that the first stage consults just a couple of heuristics—a version of the "NVN" strategy, in which people assume an agent-patient mapping of semantic roles to syntactic positions, and an animacy heuristic, in which animate entities are biased toward subjecthood. The 2003 Ferreira model captures the results of a series of experiments in which participants appeared to frequently misinterpret passive sentences, particularly when they expressed an implausible event with reversible semantic roles (e.g., the dog was bitten by the man = the dog bit the man). The application of heuristics in the first stage yields the dog-bit-man interpretation; a proper syntactic parse will deliver the opposite, correct interpretation, but the 2003 model assumes that it is fragile and susceptible to interference from the more frequent interpretation. Garrett (2000) offers a more explicitly analysis-by-synthesis model which incorporates the production system to yield what are widely believed to be top-down effects. A first pass, bottom-up process uses basic syntactic information to yield a simple parse which in turn allows for a rudimentary interpretation; then the language production system takes over and uses that representation to generate the detailed syntactic structure that would support the initial parse and interpretation.

Finally, a family of models has been proposed that assume people engage in rational behavior over what they understand to be a noisy communication channel. The channel is noisy both because listeners sometimes mishear or misread due to processing error or environmental contamination, and because speakers sometimes make mistakes when they talk. Thus, a rational comprehender whose goal is to recover the intention behind the utterance will normalize the input according to Bayesian priors. A body of evidence from research using event-related potentials (ERPs) helped to motivate these ideas (Van Herten, Kolk, & Chwilla, 2005; Kim & Osterhout, 2005). In these experiments, it is reported that subjects who encounter a sentence such as *The fox that hunted the poachers stalked through the woods* experience a P600 rather than an N400 upon encountering the semantically anomalous word, even though an N400 would be expected given that it is presumed to reflect problems with semantic integration. There is still not a great deal of consensus on what triggers P600s, but an idea that has been gaining traction is that it reflects a need to engage in some type of structural reanalysis or revision. The idea, then, is that when a person encounters a sentence that seems to say that the fox hunted the poachers, they "fix" it so it makes sense, resulting in a P600. Other models have taken this idea and developed it further (Gibson, Bergen, & Piantadosi, 2013; Levy, 2011; Levy, Bicknell, Slattery, & Rayner, 2009). These models seem less compatible with modularity than the other "shallow processing" approaches discussed earlier, because the information that is accessed to establish the priors can potentially be anything, ranging from biases related to structural forms all the way to beliefs concerning speaker characteristics (e.g., that a person with an upper-class speech style is unlikely to refer to his tattoo; Van Berkum, van den Brink, Tesink, Kos, & Hagoort, 2008). However, these noisy channel models have not yet been rigorously tested using a methodology that allows early processes to be distinguished from later ones. For example, it remains possible that comprehenders create a simple quick-and-dirty parse in a manner compatible with modularity and then consult information outside the module to revise that interpretation, right down to actually normalizing the input. Indeed, models designed to explain the comprehension of sentences containing self-repairs (*turn left uh right at the light*) assume mechanisms that allow input to be deleted so that the speaker's intended meaning can be recovered in the face of disfluency (Ferreira, Lau, & Bailey, 2004).

CONCLUSION

We began this chapter on the modularity of sentence processing with a summary of the main features of modules, because it is essential to appreciate that modularity is about more than information encapsulation—other key features include speed, automaticity, shallow outputs, and limited central access. If information encapsulation is treated not as simply one of a cluster of features but rather as "the heart of modularity," then the challenges to the notion that sentence processing is modular will continue to resonate in the cognitive science community, despite the arguments we've made here that many studies purporting to show interactivity can be reconciled with modularity. The key, we argued, is to appreciate two points. First, the so-called "two-stage" model associated with Frazier and colleagues (including the first author) is only one kind of modular model for sentence processing, so evidence against the two-stage model is not evidence against every instantiation of a modular model. And second, whether an influence of some piece of information constitutes a violation of information encapsulation depends critically on what information is contained in the "capsule." If we assume the sentence processing module can consult phrase structure rules only, then effects of even information such as verb subcategorization frames will be construed as disconfirming encapsulation. But if we accept that one of the aims of theory construction in the field of sentence processing is to develop an explanatory model of how the system works, then one key goal will be to determine what sources of information are in fact part of the sentence processing module. The goal would then be to determine what the proprietary databases are that the sentence processing module must consult. Certainly almost everyone would agree that information about what speakers from different social classes are likely to say probably does not belong in a parsing module, but information about verb subcategorization and even animacy are a different matter entirely. Moreover, the assumption of seriality relating to ambiguity resolution should be open to empirical scrutiny and revision as well; as we argued, a system with parallel consideration of alternative parses is compatible with modularity, and indeed mimics the architecture proposed as a bottom-up account of how lexical ambiguity is processed.

We would like to offer a further suggestion, and that is to emphasize the modularity features that cluster around shallowness rather than those that focus on encapsulation. We could assume that the output of the sentence processing module is not a parse in the sense of a detailed syntactic structure, but is rather the conditions for interpretation—a representation that includes information about thematic roles, focus-presupposition structure, and so on, but does not retain highly articulated syntactic forms or traces of movement operations. Complex, detailed syntax might get accessed and used by the module that creates an interpretation, but those detailed syntactic representations also are likely discarded once they serve their role of allowing a propositional interpretation to be built (Sachs, 1967). In addition, the module would be able to consult simple frequency based heuristics such as the NVN strategy (Townsend & Bever, 2001). And if the heuristics deliver a compelling interpretation faster than the syntactic algorithms do (as in some cases of garden-path reanalysis, which can be time-consuming and often require accessing infrequent forms; MacDonald, Pearlmutter, & Seidenberg, 1994), then the systems subsequent to the sentence processing module may decide to proceed with what they have rather than waiting for more detailed analyses to be performed. These tendencies would result in phenomena such as the Moses illusion, garden-path misinterpretations, and misinterpretations of implausible passives. Moreover, if that interpretation still seems unsatisfactory in a Bayesian sense, then post-sentence processing modules may engage in the sort of normalization and correction that would be expected on a rational view of communication.

We end by returning to our opening observation: Modularity might be out of fashion, but this is not because the evidence against it is particularly compelling. Instead, we suspect that many researchers simply grew weary of the limited set of questions that were being asked in the context of testing modularity against one specific model of sentence processing, and so they decided to shift their energies to broader questions such as dialogue, embodiment, and language-vision interactions. This shift in focus has been positive for the field because so much more is known now than even ten years ago. However, much of what we've learned is not relevant to evaluating modularity, and these new approaches and findings are quite possibly compatible with it.

REFERENCES

Albrecht, J. E., & O'Brien, E. J. (1993). Updating a mental model: Maintaining both local and global coherence. *Journal of Experimental Psychology: Learning, Memory, and Cognition, 19*(5), 1061–1070.

Allbritton, D. W., McKoon, G., & Ratcliff, R. (1996). Reliability of prosodic cues for resolving syntactic ambiguity. *Journal of Experimental Psychology: Learning, Memory, and Cognition, 22*(3), 714–735.

Altmann, G., & Steedman, M. (1988). Interaction with context during human sentence processing. *Cognition, 30*(3), 191–238.

Barton, S. B., & Sanford, A. J. (1993). A case study of anomaly detection: Shallow semantic processing and cohesion establishment. *Memory & Cognition, 21*(4), 477–487.

Beach, C. M. (1991). The interpretation of prosodic patterns at points of syntactic structure ambiguity: Evidence for cue trading relations. *Journal of Memory and Language, 30*(6), 644–663.

Bear, J., & Price, P. (1990). Prosody, syntax and parsing. *Proceedings of the 28th annual meeting on Association for Computational Linguistics* (pp. 17–22). Pittsburgh, PA: Association for Computational Linguistics.

Bever, T. G., Sanz, M., & Townsend, D. J. (1998). The emperor's psycholinguistics. *Journal of Psycholinguistic Research, 27*(2), 261–284.

Caramazza, A., & Zurif, E. B. (1976). Dissociation of algorithmic and heuristic processes in language comprehension: Evidence from aphasia. *Brain and Language, 3*(4), 572–582.

Carlson, K., Frazier, L., & Clifton, C. (2009). How prosody constrains comprehension: A limited effect of prosodic packaging. *Lingua, 119*(7), 1066–1082.

Chomsky, N. (1965). *Aspects of the theory of syntax* (Vol. 11). Cambridge, MA: The MIT Press.

Christianson, K., Hollingworth, A., Halliwell, J. F., & Ferreira, F. (2001). Thematic roles assigned along the garden path linger. *Cognitive Psychology, 42*(4), 368–407.

Clifton, C., Traxler, M. J., Mohamed, M. T., Williams, R. S., Morris, R. K., & Rayner, K. (2003). The use of thematic role information in parsing: Syntactic processing autonomy revisited. *Journal of Memory and Language, 49*(3), 317–334.

Crain, S., & Steedman, M. (1985). On not being led up the garden path: The use of context by the psychological parser. In D. Dowty, L. Karttunen, & A. Zwicky, *Natural language parsing: Psychological, computational, and theoretical perspectives* (pp. 320–358). Cambridge, UK: Cambridge University Press.

Cutler, A., & Fodor, J. A. (1979). Semantic focus and sentence comprehension. *Cognition, 7*(1), 49–59.

Dahl, Ö., & Fraurud, K. (1996). Animacy in grammar and discourse. In T. Fretheim & J. Gundel, *Reference and referent accessibility* (pp. 47–64). Amsterdam: John Benjamins.

Dick, F., Bates, E., Wulfeck, B., Utman, J. A., Dronkers, N., & Gernsbacher, M. A. (2001). Language deficits, localization, and grammar: Evidence for a distributive model of language breakdown in aphasic patients and neurologically intact individuals. *Psychological Review, 108*(4), 759–788.

Engelhardt, P. E., Bailey, K. G., & Ferreira, F. (2006). Do speakers and listeners observe the Gricean Maxim of Quantity? *Journal of Memory and Language, 54*(4), 554–573.

Erickson, T. D., & Mattson, M. E. (1981). From words to meaning: A semantic illusion. *Journal of Verbal Learning and Verbal Behavior, 20*(5), 540–551.

Farmer, T. A., Cargill, S. A., Hindy, N. C., Dale, R., & Spivey, M. J. (2007). Tracking the continuity of language comprehension: Computer mouse trajectories suggest parallel syntactic processing. *Cognitive Science, 31*(5), 889–909.

Fedorenko, E., Duncan, J., & Kanwisher, N. (2012). Language-selective and domain-general regions lie side by side within Broca's area. *Current Biology, 22*(21), 2059–2062.

Ferreira, F. (1993). Creation of prosody during sentence production. *Psychological Review, 100*(2), 233–253.

Ferreira, F. (2003). The misinterpretation of noncanonical sentences. *Cognitive psychology, 47*(2), 14–203.

Ferreira, F., & Anes, M. (1994). Why study spoken language? In M. A. Gernsbacher, *Handbook of Psycholinguistics* (pp. 33–56). San Diego, CA: Academic Press.

Ferreira, F., Bailey, K. G., & Ferraro, V. (2002). Good-enough representations in language comprehension. *Current Directions in Psychological Science, 11*(1), 11–15.

Ferreira, F., & Clifton, C. (1986). The independence of syntactic processing. *Journal of Memory and Language, 25*(3), 348–368.

Ferreira, F., Foucart, A., & Engelhardt, P. E. (2013). Language processing in the visual world: Effects of preview, visual complexity, and prediction. *Journal of Memory and Language, 69*(3), 165–182.

Ferreira, F., & Henderson, J. M. (1990). Use of verb information in syntactic parsing: Evidence from eye movements and word-by-word self-paced reading. *Journal of Experimental Psychology: Learning, Memory, and Cognition, 16*(4), 555–568.

Ferreira, F., Lau, E. F., & Bailey, K. G. (2004). Disfluencies, language comprehension, and tree adjoining grammars. *Cognitive Science, 28*(5), 721–749.

Field, J. (2004). An insight into listeners' problems: Too much bottom-up or too much top-down? *System, 32*(3), 363–377.

Fine, A. B., & Jaeger, T. F. (2013). Evidence for implicit learning in syntactic comprehension. *Cognitive Science, 37,* 578–591.

Fodor, J. (1983). *The modularity of mind: An essay on faculty psychology.* Cambridge, MA: The MIT Press.

Fodor, J. A. (2000). *The mind doesn't work that way: The scope and limits of computational psychology.* Cambridge, MA: MIT press.

Fodor, J. A., Bever, T. G., & Garrett, M. (1969). *The development of psychological models for speech recognition.* Report ESD-TR-67-633 of the Electronic Systems Division, US Air Force. Bedford, MA: Hanscom Field.

Fodor, J. D., & Ferreira, F. (1998). *Reanalysis in sentence processing.* Dordrecht, The Netherlands: Kluwer.

Ford, M., Bresnan, J., & Kaplan, R. (1982). A competence-based theory of syntactic closure. In J. Bresnan, *The mental representation of grammatical relations* (pp. 727–796). Cambridge, MA: MIT Press.

Frazier, L., & Clifton Jr., C. (1997). Construal: Overview, motivation, and some new evidence. *Journal of Psycholinguistic Research, 26*(3), 277–295.

Frazier, L., & Fodor, J. D. (1978). The sausage machine: A new two-stage parsing model. *Cognition, 6*(4), 291–325.

Frazier, L., Carlson, K., & Clifton, C. (2006). Prosodic phrasing is central to language comprehension. *Trends in Cognitive Sciences, 10*(6), 244–249.

Frazier, L., Pacht, J. M., & Rayner, K. (1999). Taking on semantic commitments, II: Collective versus distributive readings. *Cognition, 70*(1), 87–104.

Frazier, L., & Rayner, K. (1990). Taking on semantic commitments: Processing multiple meanings vs. multiple senses. *Journal of Memory and Language, 29*(2), 181–200.

Frisson, S. (2009). Semantic underspecification in language processing. *Language and Linguistics Compass, 3*(1), 111–127.

Frisson, S., & Pickering, M. J. (2001). Obtaining a figurative interpretation of a word: Support for underspecification. *Metaphor and Symbol, 16*(3-4), 149–171.

Gahl, S., & Garnsey, S. M. (2006). Knowledge of grammar includes knowledge of syntactic probabilities. *Language, 82*(2), 405–410.

Garrett, M. (2000). Remarks on the architecture of language processing systems. In Y. Grodzinsky, & L. Shapiro, *Language and the Brain: Representation and Processing* (pp. 31–69). San Diego, CA: Academic Press.

Gibson, E., Bergen, L., & Piantadosi, S. T. (2013). Rational integration of noisy evidence and prior semantic expectations in sentence interpretation. *Proceedings of the National Academy of Sciences, 110*(20), 8051–8056.

Gigerenzer, G. (2004). Fast and frugal heuristics: The tools of bounded rationality. *Blackwell Handbook of Judgment and Decision Making* (62–88). Oxford, England: Blackwell.

Gleitman, L. R., & Wanner, E. (1982). *Language acquisition: The state of the art.* Cambridge, England: Cambridge University Press.

Grice, P. (1975). Logic and conversation. In P. Cole & J. Morgan, *Syntax and semantic: Speech acts* (Vol. 3, pp. 41–58). New York: Seminar Press.

Hare, M., McRae, K., & Elman, J. L. (2003). Sense and structure: Meaning as a determinant of verb subcategorization preferences. *Journal of Memory and Language, 48*(2), 281–303.

Henderson, J. M., & Ferreira, F. (2004). Scene perception for psycholinguists. In J. M. Henderson & F. Ferreira, *The interface of language, vision, and action: Eye movements and the visual world* (pp. 1–58). New York, NY: Psychology Press.

Hoff, E. (2006). How social contexts support and shape language development. *Developmental Review, 26*(1), 55–88.

Huettig, F., & McQueen, J. M. (2007). The tug of war between phonological, semantic and shape information in language-mediated visual search. *Journal of Memory and Language, 57*(4), 460–482.

Huettig, F., Olivers, C. N., & Hartsuiker, R. J. (2011). Looking, language, and memory: Bridging research from the visual world and visual search paradigms. *Acta Psychologica, 137*(2), 138–150.

Huettig, F., Rommers, J., & Meyer, A. S. (2011). Using the visual world paradigm to study language processing: A review and critical evaluation. *Acta Psychologica, 137*(2), 151–171.

Jackendoff, R. (1996). The architecture of the linguistic-spatial interface. In P. Bloom, M. A. Peterson, L. Nadel, & M. F. Garrett (Eds.), *Language and space. Language, speech, and communication* (pp. 1–30). Cambridge, MA: MIT Press.

Joanisse, M. F., Manis, F. R., Keating, P., & Seidenberg, M. S. (2000). Language deficits in dyslexic children: Speech perception, phonology, and morphology. *Journal of Experimental Child Psychology, 77*(1), 30–60.

Kahneman, D. (2003). Maps of bounded rationality: Psychology for behavioral economics. *The American Economic Review, 93*(5), 1449–1475.

Kahneman, D. (2011). *Thinking, fast and slow.* New York: Farrar, Straus & Giroux.

Kim, A., & Osterhout, L. (2005). The independence of combinatory semantic processing: Evidence from event-related potentials. *Journal of Memory and Language, 52*(2), 205–225.

Kjelgaard, M. M., & Speer, S. R. (1999). Prosodic facilitation and interference in the resolution of temporary syntactic closure ambiguity. *Journal of Memory and Language, 40*(2), 153–194.

Levy, R. (2008). A noisy-channel model of rational human sentence comprehension under uncertain input. In *Proceedings of the conference on empirical methods in natural language processing* (pp. 234–243). Pittsburgh, PA: Association for Computational Linguistics.

Levy, R. (2011). Probabilistic Linguistic Expectations, Uncertain Input, and Implications. *Studies of Psychology and Behavior, 9*(1), 52–63.

Levy, R., Bicknell, K., Slattery, T., & Rayner, K. (2009). Eye movement evidence that readers maintain and act on uncertainty about past linguistic input. *Proceedings of the National Academy of Sciences, 106* (50), 21086–21090.

MacDonald, M. C. (1993). The interaction of lexical and syntactic ambiguity. *Journal of Memory and Language, 32*(5), 692–715.

MacDonald, M. C., Pearlmutter, N. J., & Seidenberg, M. S. (1994). The lexical nature of syntactic ambiguity resolution. *Psychological Review, 101*(4), 676–703.

McClelland, J. L. (1987). The case for interactionism in language processing. In M. Coltheart, *Attention and performance XII: The psychology of reading* (pp. 1-36). London, England: Erlbaum.

Millotte, S., Wales, R., & Christophe, A. (2007). Phrasal prosody disambiguates syntax. *Language and Cognitive Processes, 22*(6), 898–909.

Mitchell, D. C., & Holmes, V. M. (1985). The role of specific information about the verb in parsing sentences with local structural ambiguity. *Journal of Memory and Language, 24*(5), 542–559.

Nakamura, C., Arai, M., & Mazuka, R. (2012). Immediate use of prosody and context in predicting a syntactic structure. *Cognition, 125*(2), 317–325.

Novick, J. M., Thompson-Schill, S. L., & Trueswell, J. C. (2008). Putting lexical constraints in context into the visual-world paradigm. *Cognition, 107*(3), 850–903.

Otero, J., & Kintsch, W. (1992). Failures to detect contradictions in a text: What readers believe versus what they read. *Psychological Science, 3*(4), 229–235.

Pickering, M. J., McElree, B., Frisson, S., Chen, L., & Traxler, M. J. (2006). Underspecification and aspectual coercion. *Discourse Processes, 42*(2), 131–155.

Pickering, M. J., & Traxler, M. J. (1998). Plausibility and recovery from garden paths: An eye-tracking study. *Journal of Experimental Psychology: Learning, Memory, and Cognition, 24*(4), 940–961.

Price, P. J., Ostendorf, M., Shattuck-Hufnagel, S., & Fong, C. (1991). The use of prosody in syntactic disambiguation. *The Journal of the Acoustical Society of America, 90*, 2956–2970.

Rayner, K. (1977). Visual attention in reading: Eye movements reflect cognitive processes. *Memory & Cognition, 5*(4), 443–448.

Rayner, K., Carlson, M., & Frazier, L. (1983). The interaction of syntax and semantics during sentence processing: Eye movements in the analysis of semantically biased sentences. *Journal of Verbal Learning and Verbal Behavior, 22*(3), 358–374.

Sachs, J. S. (1967). Recopition memory for syntactic and semantic aspects of connected discourse. *Perception & Psychophysics, 2*(9), 437–442.

Sanford, A. J., & Graesser, A. C. (2006). Shallow processing and underspecification. *Discourse Processes, 42*(2), 99–108.

Sanford, A. J., & Sturt, P. (2002). Depth of processing in language comprehension: Not noticing the evidence. *Trends in Cognitive Sciences, 6*(9), 382–386.

Sitnikova, T., Salisbury, D. F., Kuperberg, G., & Holcomb, P. J. (2002). Electrophysiological insights into language processing in schizophrenia. *Psychophysiology, 39*(6), 851–860.

Slattery, T. J., Sturt, P., Christianson, K., Yoshida, M., & Ferreira, F. (2013). Lingering misinterpretations of garden path sentences arise from competing syntactic representations. *Journal of Memory and Language, 69*(2), 104–120.

Spivey, M. J., Tanenhaus, M. K., Eberhard, K. M., & Sedivy, J. C. (2002). Eye movements and spoken language comprehension: Effects of visual context on syntactic ambiguity resolution. *Cognitive Psychology, 45*(4), 447–481.

Steedman, M. (2000). Information structure and the syntax-phonology interface. *Linguistic Inquiry, 31*(4), 649–689.

Steedman, M., & Baldridge, J. (2011). Combinatory categorial grammar. In R. Borsely & K. Borjars, *Non-Transformational Syntax* (pp. 181–224). Oxford, England: Blackwell.

Sturt, P., Sanford, A. J., Stewart, A., & Dawydiak, E. (2004). Linguistic focus and good-enough representations: An application of the change-detection paradigm. *Psychonomic Bulletin & Review, 11*(5), 882–888.

Swets, B., Desmet, T., Clifton, C., & Ferreira, F. (2008). Underspecification of syntactic ambiguities: Evidence from self-paced reading. *Memory & Cognition, 36*(1), 201–216.

Tanenhaus, M. K., Magnuson, J. S., Dahan, D., & Chambers, C. (2000). Eye movements and lexical access in spoken-language comprehension: Evaluating a linking hypothesis between fixations and linguistic processing. *Journal of Psycholinguistic Research, 29*(6), 557–580.

Tanenhaus, M. K., Spivey-Knowlton, M. J., Eberhard, K. M., & Sedivy, J. C. (1995). Integration of visual and linguistic information in spoken language comprehension. *Science, 268*(5217), 1632–1634.

Townsend, D. J., & Bever, T. G. (2001). *Sentence comprehension: The integration of habits and rules* (Vol. 1950). Cambridge, MA: MIT Press.

Traxler, M. J., Pickering, M. J., & Clifton, C. (1998). Adjunct attachment is not a form of lexical ambiguity resolution. *Journal of Memory and Language, 39*(4), 558–592.

Trueswell, J. C., & Kim, A. E. (1998). How to prune a garden path by nipping it in the bud: Fast priming of verb argument structure. *Journal of Memory and Language, 39*(1), 102–123.

Trueswell, J. C., Sekerina, I., Hill, N. M., & Logrip, M. L. (1999). The kindergarten-path effect: Studying on-line sentence processing in young children. *Cognition, 73*(2), 89–134.

Tyler, L. K., & Warren, P. (1987). Local and global structure in spoken language comprehension. *Journal of Memory and Language, 26*(6), 638–657.

Van Berkum, J., van den Brink, D., Tesink, C., Kos, M., & Hagoort, P. (2008). The neural integration of speaker and message. *Journal of Cognitive Neuroscience, 20*(4), 580–591.

Van Herten, M., Kolk, H. H., & Chwilla, D. J. (2005). An ERP study of P600 effects elicited by semantic anomalies. *Cognitive Brain Research, 22*(2), 241–255.

Wilson, M. P., & Garnsey, S. M. (2009). Making simple sentences hard: Verb bias effects in simple direct object sentences. *Journal of Memory and Language, 60*(3), 368–392.

Zekveld, A. A., Heslenfeld, D. J., Festen, J. M., & Schoonhoven, R. (2006). Top–down and bottom–up processes in speech comprehension. *Neuroimage, 32*(4), 1826–1836.

4

The Unity of Consciousness and the Consciousness of Unity

THOMAS G. BEVER

> *Truth is stranger than fiction. . . . because fiction is obliged to stick to the possibilities.*
>
> *Truth isn't*
>
> —Mark Twain, "*Following the Equator*"

A SENTENCE IS LIKE A (MINIATURE) OPERA

Music is often analyzed in relation to language to give perspective on the structural and formal aspects of language. But even the simplest sentence surpasses what music can tell us about it. A sentence in everyday use combines a stream of sound, with rhythm and pitch variations, with memorized units of meaning, an organizing structure that recombines those meaning units into a transcendental unified meaning that includes informational representations, general connotations, and specific pragmatic implications unique to the conversational context. In other words, each sentence is a miniature opera of nature.

Children grow up surrounded by one opera after another, and miraculously learn to create their own. This is achieved in the context of experiencing only a small number of fully grammatical sentences, many ungrammatical ones, and very little specific feedback on their mistakes. This situation is generally referred to as "the poverty of the stimulus," which is the basis for the argument that much of linguistic structure must be innately prefigured (Chomsky, 1959,1965,1975,1980). Fodor (1981) broadened the implications of this argument beyond language into cognition in general: "The [argument from the poverty of the stimulus] is *the* existence proof for the possibility of a cognitive science"

(p. 258). Nonetheless, the flagship case of the argument continues to be the speed with which children learn language in erratic environments with variable feedback.

In this chapter, I begin with one of the major components of what the child has to discover in learning to understand and then produce language—the perception and comprehension of natural units of composition in the serial string. Interestingly, this problem exists on virtually any grammatical theory, from taxonomic phrase structure all the way up to today's Minimalism. Every view of what language is, going back centuries has some notion of serial hierarchical phrasing as a fundamental component. In phrase structure grammars, describing the phrase is the direct goal and what the child must discover; in generative theories that utilize "structure dependence" the child must discover the phrase in order to have access to the structure. In the next sections, I trace some research on how major language units are perceived, over the past decades, and then turn to the implications of recent studies of the acoustics of normal conversation, which show how deep and puzzling the problem of the poverty of the stimulus really is. The processing of normal conversation reveals a disconnect between the listener's representation of the sound and meaning of utterances. In critical cases it is possible to show that compressed or absent words are unintelligible until the listener hears *later* acoustic information Yet the listener perceives the acoustic presentation of the words as simultaneous with the comprehension of it. This is an instance of *creating* a conscious representation retrospectively.

I draw a number of morals from such facts in language processing: notably, the "poverty of the stimulus problem" is far graver than usually supposed—although the words in some child-directed speech are carefully pronounced, many are not. And children are also surrounded by the same kind of garbled and cue-poor instances from adult speech; this means that structure dependence must guide ongoing comprehension processes of externalized serial input, not used only to decide about the abstract structure of one's language during learning; every level of language experience involves some encoding: this supports the notion that ongoing processing occurs in a set of simultaneous parallel processes in a "computational fractal," that is, each level involves the interaction of associative-serial and structure dependent processes; thus, our conscious experience of language is in part reconstructive in temporarily time-free "psychological moments"—so language comprehension processes move forward *and backward*, even though the phenomenal experience is that it moves only forward.; this reconstructive analysis of our conscious experience of language may be typical of other modalities of our experience.

This leads us to distinguish the computational problem of language acquisition from the acoustic input problem. The computational problem concerns how children generalize in the right way from scant examples of complete, well-formed sentences with clearly presented words, how they alight on the right kind of structure dependent hypotheses. The acoustic input problem is that children (and adults) are often not presented with clear word-by-word inputs to learn and understand from. Rather children must have already solved a large part of the computational problem in order to resolve the acoustic input problem. This magnifies what we must assume is available to the child at a very young age, and geometrically complicates any attempts to model acquisition with statistical models unadorned by massive and keen prior structures and expectations.

WHERE IS THE UNIT OF LANGUAGE PROCESSING?

Psychology as a field often depends on resurgent methodology and continually mysterious phenomena: One of the most enduring methods and mysteries is the systematic mislocation of "clicks" presented during auditory presentation of sentences toward "phrasal" boundaries of some kind. The use of click-mislocation was pioneered by Ladefoged and Broadbent (1960), as a way of showing on-line segmentation of syllables. Its utility for exploring on-line complexity and the effect of "phrase" boundaries was initially explored by Garrett (1964). Fodor and Bever (1965) demonstrated the general role of relative depth of surface "phrase" breaks in determining the likelihood of click mislocation to or toward them; Garrett, Bever, and Fodor (1966) showed that the mislocation was not due to local intonational cues, but to the "phrasal" structure that listeners impose on what they are hearing (also demonstrated by Abrams and Bever, 1969, with a different technique). (For a contemporary demonstration of brain spectral activity corresponding to phrase construction without benefit of intonational or statistical cues, see Ding et al., 2016 and further discussion in this chapter).

A revealing aspect relevant for today's discussions is the fact that the citation of the two original click location essays has experienced a "U shaped function" with almost as many citations in the last five years as in the first five, and less than a third of that rate in the intervening years. This reflects the rediscovery of questions about what the "real" unit of ongoing language processing is.

Later studies attempted further to define what perceptual and comprehension units are revealed by click mislocations, "deep" structure units (Bever, Lackner, & Kirk, 1969) or "major" surface phrases (Chapin, Smith, & Abrahamson, 1972). Many click location studies required subjects to write out the sentence and indicate the click location—this invited the interpretation that the click mislocation effect was not perceptual at all, but some form of response strategy related to recapitulating the sentence. Bever (1973) explored this by having listeners mark the click location within a window in the text written out and presented right after hearing the stimulus. In critical cases there was no auditory click at all: to make it plausible that there was a click, the loudness of the actual clicks was varied. When a click was present, the usual effect of a major phrase boundary occurred: when there was no click, subjects' guesses were *not* systematically placed in the major phrase boundary. Using a different method, Dalrymple-Alford (1976) confirmed that click mislocation is not due to a response bias.

Two approaches to the question of the processing unit have continually surfaced and resurfaced over many years: each rests on one of the two ideas dominant in centuries of cognitive science: (a) the currency of mental life is statistically determined associations; (b) mental life is organized into categorical representations. The argument started with a closer examination of the 'psychological reality of linguistic segments namely the "phrase." During the 1960s much attention was being given to the idea that "phrases" could be defined in terms of serial predictability (Johnson,1970; Osgood,1968). On this view, "phrases" are behaviorally bounded by relatively low points of serial predictability: indeed it is generally the case that phrase-final (content) words are more predictable locally than phrase initial words. So behaviors that seem to reflect categorical phrasing might actually be reflecting variation in serial predictability. However, when syntactic structures are held constant while local predictability is varied, the high serial predictable points actually attract clicks perceptually (Bever et al., 1969). So probability governed segmentation does not account for the online perceptual formation of phrases.

Yet the conflict between some version of association and categorical structural assignment always finds new life. The connectionist postulation of hidden units, back propagation and other statistical devices, along with the rehabilitation of Bayesian statistics, resuscitated notions of mediated associations with complex descriptive power, enabling simulation of categorical structures as their limit (e.g., Rumelhart & McClelland, 1986). In this vein, great attention is given to "feed forward" models of perception in general and sentence processing in particular: the perceptual system is constantly making predictions of what is about to occur, so that much of language interpretation is actually rolling confirmation of specific kinds of rolling perceptual expectations. In the case of language, this can occur simultaneously at various levels from the acoustic to the semantic. The expectations are arguably a blend of probabilistic and categorical features in many domains; phonological, semantic, and syntactic. Canonical demonstrations of this are effects of left → right constraints during processing: something that occurs at point *a* affects the perception of something later at point *b*.

What I will explore in the next few pages is more recent evidence that parsing is not only "forward" it is also "downward," the construction of meaning units within short epochs. The crucial demonstration of this is evidence for backward constraints: something at point *b* in a sentence determines the percept at an earlier point *a*. Most critical to this argument is that the conscious awareness is of a constant forward moving perception, not a period of blank content suddenly filled in by something that comes later. That is, we perceive sentences in "psychological moments" in which the underlying computational processing can move back and forth, or more to the point, forth and back, before "reporting out" to conscious awareness.

THE UNITY OF PROCESSING UNITS AND THE CONSCIOUS EXPERIENCE OF LANGUAGE

Linguistic and psycholinguistic research on sentence structure and processing has implicitly assumed that the constituent words are given: that is, the syntactician's (and child's) problem is to determine the regularities that govern how the words and other syntactic units are arranged (and inflected when relevant); the psycholinguist's problem is to determine the processes that underlie how the words and units are composed together in production of sentences, mapped onto representations in comprehension of sentences, and learned in relation to their role in possible syntactic constructions. But outside of syntax classes and psycholinguistic experiments, the words in natural language are rarely clearly or fully presented—the acoustics of one word blends into another, and in many cases, large portions of a word or word sequence are actually not present at all: to borrow a term from phonology, the words are encoded together.

Some well-known facts about serial encoding at the phonological level may help us understand the situation at the syntactic level. First, it is well documented that unvoiced stop consonants in English, may actually not be given any acoustic power of their own. Thus, the final consonant in the words /top/, /tot/, /toc/ may be silent or all converge on glottal stop—yet we hear them quite clearly as distinct—it is the way that the preceding vowel changes as it quickly approaches the articulated position of the consonant. If we could hear the preceding vowels drawn out in time, they would be more like /TOuP/, /TOiT/, /TOaC/: the last bit of the vowel gives the clue as to where the tongue is heading before the vowel goes silent. Yet our conscious percept is that the consonant was actually uttered. This is an example of a "feed forward" activity, in which the material preceding the final silence or glottal stop makes a strong enough prediction about what will be "heard" so that it is actually perceived even when not in the signal itself.

But the influence of one part of a phonological sequence on another is not always "forward," it can be "backward" as well. It is well known that it is the timing of the onset of a post-consonantal vowel that communicates whether the preceding consonant is to be heard as voiced or unvoiced. Even more striking is that in some variants, the initial voiced consonant can also not be explicitly uttered: the difference between /bill, dill, gill/ can be only in the vowel transition following the initial occlusion of the vocal tract, just long enough to indicate voicing—it is the vowel transition away from the silent initial consonant (except for the voicing itself) that indicates what the preceding consonant *was*.

The moral is that at the phonological level, even when a word is uttered in isolated "citation" form, we automatically use early phonetic information to guide the conscious representation of what follows, *and conversely*.

It can be argued that at the level of individual words, this only shows that the unit of word recognition is larger than individual phonemes, for example, that listeners have prepackaged representations of entire syllables, or that different kinds of acoustic features work together in a "cohort" (see e.g., Marlsen-Wilson & Zwitserlood, 1989). This kind of argument may be possible in principle for words and phonology, since there is a finite number of syllables used in any particular language. But as is classically argued, such proposals of memorized units become much harder to rely on at phrasal and sentential levels, since the number of different phrases and sentences is enormous, arguably infinite in the latter case. So we might not expect both forward and backward processing interactions at these higher levels of language. But in fact, recent evidence suggests that this is the case in normal uses of language outside of the syntax classroom and laboratory.

UNCONSCIOUS COMPREHENSION PROCESSES WITH BACKWARD INFERENCES

The rapid and unconscious resolution of local ambiguity suggests that corresponding prospective and retrospective processes occur at the syntactic level. For this discussion, the most significant effect is the immediate role of retrospective processing that we are unaware of. If you hear a sentence like the following, in (1a, b), there can be evidence that the ambiguity of the lexically ambiguous phonetic sequence "pair/pear" creates momentary computational complexity reflected for example in decreased accuracy of a click immediately after the word (Garrett, 1964). But you are not aware of it, and have the strong impression that you assigned it the correct interpretation as you heard it. Swinney (1979) showed that both meanings of an ambiguous word facilitate an immediately following lexical decision task, even when there is a preceding disambiguating context, for example, as in (1c, d); but a few words later, only the contextually supported meaning facilitates the task.

(1) a. The pair of doves landed on our porch.
 b. The pear and apple landed on our porch.
 c. The doves in a pair landed on our porch.
 d. The apple and pear landed on our porch.

A series of investigations by Fernanda Ferreira and colleagues (e.g., Christianson, Hollingworth, Halliwell, & Ferreira, 2001; Christianson, Williams, Zacks, & Ferreira 2006).) complements Garrett's (1964) finding at the phrasal level. Even after a garden path in segmentation of a written sentence is corrected by later material in the sentence, listeners retain a semantic representation of the initial (incorrect) segmentation. So for example, in critical trials, they follow the sentence (2a) below with a question, to which the subjects have to report "yes" or "no" to the question in (2b)

(2) a. While Bill hunted the deer ran into the woods.
 b. Did Bill hunt the deer?
 c. Did the deer run into the woods?

Surprisingly, Christianson et al. (2001) found that about a quarter of the responses were "yes" to (2b) following (2a). At the same time, they found that the subjects almost always answered the question in (2c) correctly: so they argued that "the reanalysis processes got as far as identifying a subject for the main clause verb, but didn't finish up by revising the interpretation on which that same NP was once the object of the verb in the subordinate clause." What is important for my current focus is that when subjects answered (2b) correctly or not, they were quite confident in their answers. "subjects were quite poor at arriving at an interpretation licensed by the input string, yet surprisingly confident that they had correctly understood the sentences." (p. 380). Christianson et al. take this to be evidence that comprehenders construct representations that are "good enough" to contribute to ongoing comprehension, especially in normal discourse contexts (Ferreira & Henderson, 1991). Since most sentences do not have strong garden paths (especially in auditory mode), "good enough" representations are usually good enough. That is, people arrive at conceptually appropriate interpretations based on incomplete or incorrect analyses of which they are totally unaware. More recent studies support the view that subjects do in fact analyze the correct segmentation in the garden path structures on-line, even though their answers to probe questions indicate that they consciously retain the influence of the incorrect parse (Ferreira et al., 2002; Ferreira & Patson, 2007; Ferreira et al., 2009; Slattery et al., 2013).

A classic line of research on backward influences on processing started with the studies by Connine and colleagues (Connine et al., 1991). They showed that a word with an initial phonetically ambiguous consonant midway between being heard as a voiced or voiceless consonant would be perceptually disambiguated by later context. For example, a sequence phonetically midway between "tent" and "dent," is reported as "tent" when followed by ". . . . in the forest," and as "dent" when followed by ". . . . in the fender." Bicknell et al. (2016) report that the backward influence can extend over more than just the immediately following phrase (e.g., when the following context is either ". . . . was noticed in the forest" vs. ". . . .was noticed in the fender"). It is not clear from the methodologies used whether subjects believe they heard the critical word as disambiguated, or reasoned after the fact as to what the word must have been (for a discussion of these phenomena and related issues, see Bicknell et al., 2016.).

The preceding cases involve the role of apparent "backward" processing in which information that comes later in a sentence is used to specify or revise a prior analysis. A current line of experimental research by Brown, Dilley, and Tanenhaus (2012) complements the study of conversational ellipses and the role of both forward and backward processing. In their study subjects think they "heard" a word that was acoustically ambiguous, or even marginally present at all, based on later acoustic input. Farmer, Brown, and Tanenhaus (2013) apply Clark's (2013) model of hierarchically structured predictions to comprehension: the predictions guide the formation of representations of the world as new information becomes available.

> ". . . . Clark's framework predicts that expectations at higher levels of representation (e.g., syntactic expectations) should constrain interpretation at lower levels of representation (e.g., speech perception). According to this view, listeners develop fine-grained probabilistic expectations about how lexical alternatives are likely to be realized in context. . . . that propagate from top to bottom through the levels of a hierarchically organized system

> representing progressively more fine-grained perceptual information.... As the signal unfolds, then, the activation of a particular lexical candidate.... [is the one] most congruent with the acoustic signal. . . ." (Farmer, Brown, & Tanenhaus, 2013, p. 211)

This view of language comprehension emphasizes ongoing confirmation of hierarchically organized predictions, with error corrections when a given prediction is disconfirmed, shifting the interpretation of the prior material to an alternate hierarchical analysis. That is, material later in a sequence can revise the organization and interpretation of what came earlier, as a more subtle instance of the garden path phenomena explored by Ferreira et al. (2009). Brown et al. (2013) presented sentences with sequences like (3), and varied the length of the indefinite article, /a/ and the initial /s/ of the last word in the sequence. Using the "visual world" paradigm, they report that when the article /a/ is shortened and the /s/ is lengthened, subjects look at plural target pictures ("raccoons") even after the /s/, indicating that the interpretation of the ambiguous noun in the sequence /a raccoon s . . . . is/ determined on line by what follows it. That is, when the /s/ is lengthened, subjects first look at the picture with one raccoon; then as the lengthened /s/ is heard, they shift and look at the picture with several raccoons.

Ostensibly this reflects a reanalysis, in which the shortened /a/ is not treated as a separate word; it is attached as part of the final vowel of /saw/, or perhaps reanalyzed as a brief pause. This interpretation is strengthened by the complementary finding that when the /s/ is not lengthened, the shortened definite article is then perceived and interpreted as a word.

The focus of Brown et al. is on how their research shows that listeners are sensitive to variations in local speech rate, but for my purposes the phenomenon is an online demonstration of the use of late information in determining morphological analysis of earlier speech. (See also Farmer, Yan, Bicknell, & Tanenhaus 2015 for general discussion; and Brown et al., 2012 for an example that arguably involves truly "hallucinating a definite article that was not present at all, based on extending the /s/.) Importantly, Tanenhaus et al.'s view of how the comprehension of sentences proceeds is an example of a "top down" application of an interpretation, and perception in which an entire representation can be triggered by information at the end of the signal. This gives great weight to immediate access of contextual cues of a range of kinds, including actual syntactic hierarchical structure. (For more perspective on Tanenhaus's view on how representational levels interact during sentence comprehension, see Degen & Tanenhaus, 2015.)

(3) . . . . saw uh raccoon Swimming

But in normal conversation, many words aren't there at all. . . .

The preceding examples assume that all the words in the sentences are present to some degree. But in everyday speech, many acoustic details are slurred or even omitted. This can be demonstrated by showing that fragments several "words" long are impossible to recognize in isolation, but pop into complete clarity (for native speakers) when heard as part of an entire sentence (Pollack & Pickett, 1964; Greenberg et al., 1996; Greenberg, 1999; Arai, 1999; Arai & Warner, 1999; Johnson, 2004; Warner et al., 2009; Tucker & Warner, 2010).[1] Consider first an approximate transcription of an example from adults talking to each other in a normal conversation (this is an actual example provided by N. Warner, the reader can hear examples like it on her website: http://www.u.arizona.edu/~nwarner/reduction_examples.html).[2]

(4) [tjutʌ̃m]

(Hint: this corresponds to four words). It is completely incomprehensible by itself, but when a latter portion of the longer sequence is included it is comprehensible:

(5) [tju tʌ̃m ɾɨ tʰak̚ tɨ̃ mi]

Everyone immediately hears this as:

(6) Do you have time to talk to me?

The striking significance of this is that phenomologically listeners think they simultaneously hear the fragment and assign it its three word analysis. But we know this cannot be true since the fragment in isolation is incomprehensible. This suggests that backward processing at a local acoustic level is a normal part of comprehension and building representations of conscious experience of language.

But this example was the beginning of a sentence, so perhaps it is a special case, where there is no preceding context. However in an experimental paradigm Van de Ven (2011) found that the following context can contribute importantly to recognition of material in the middle of a sentence. In fact, the following example from a natural conversation supports the view that in some cases, the following context alone is sufficient to clarify a reduced word, while the preceding context alone is not sufficient.

(7) [tʃũɪ̃n:]

Try pronouncing this to yourself (hint: the production intent is 2 syllables). Now look at a longer sequence in which the example was embedded:

(8) [ɚ: ʌ: thɨzdɛ nʌit̚ (pause) ʌ̰mn wɪɹ tʃũɪ̃nɨ̃n:(ɨ) spa]

When listeners hear the surrounding material, the excerpt immediately pops into consciousness and what one "hears" is:

(9) . . . err Tuesday night, when we were chillin' in the spa.

Recently we tested this further: it turns out that even with all the material preceding [tʃũːn:] (as in "and err Tuesday night when we were. . . .") almost no one perceives it correctly. But if only the following material ("in the spa") is heard along with the sequence, then [tʃũːn:] is heard clearly as "chillin." First, such facts support the view that in everyday comprehension the minimal phonetic unit of comprehension is not the word, and that comprehension must be operating with parallel hypotheses at several interactive levels—syntactic and phonetic computations proceed in parallel with frequent cross checks at specific points. One can expect that where those cross checks occur will be the focus of ongoing research, now that we have tools that can chop running speech into a full range of possible units. An initial hypothesis is the *phase*, the unit of syntactic structure that has just enough content for semantic analysis (Chomsky, 1995, 2000, 2008). Phase theory is an active research area in linguistics, so the reader should be skeptical about details by the time this chapter is published, never mind a few years later. (See Boeckx, 2006 for a lucid explication of the technical issues and Citko, 2014 for a recent introduction.) So we can start with a particular hypothesis, as the latest idea on how different levels of a sentence are integrated in working units:

(10) *The unit over which local acoustic/phrasal/meaning integration occurs is the phase.*

However, we must note that "chillin" is involved in two prima facie phases: (a) the preceding material which includes a WH, subject, and auxiliary, which embeds the verb in a complex structure with at least several levels of hierarchical organization; (b) the following material, which embeds the verb in a more compact verbphrase only. The unique effectiveness of the following material leads to a hypothesis for further investigation, based on a single case, but one with some intuitive appeal:

(11) *The effectiveness of a phase in integrating distinct language levels is proportional to its structural simplicity.*

Further research will (I hope) verify or falsify these working hypotheses. A particular question is whether the role of the less complex phases is unique in the comprehension processes, or whether it reflects different degrees of reduction in the production processes. For example in (9) the failure of the preceding material to clarify the excerpt may be because as a NP-Verb phase it is actually less reduced in speech. So it is now an interesting research question whether phases are the "true" units of comprehension that the many "click" experiments attempted to define (Fodor & Bever, 1965; Garret et al., 1966; Bever et al., 1969), whether those effects depend on production processes, or whether the phase in fact is not the relevant factor that elicits segmentation effects. For example, there is new interest in how speakers maintain the predictability, (aka "information density") of their sentence output (e.g., Jaeger, 2006; Levy & Jaeger, 2007; Jaeger, 2010; Frank & Jaeger, 2008). This principle extends both to choice of phrases and words, and to use of contractions. For example, Frank and Jaeger show that local predictability can determine whether "you are" is contracted to "you're" in sentence production. Dell and Chang (2014) recently proposed a model that combines this approach with Macdonald's ideas that production patterns condition comprehension processes (Macdonald, 1999, 2013). Within a connectionist model of syntax production, they unify the processes of acquisition, production and comprehension based on serial predictability of words. The examples I have mentioned in this chapter suggest that for such a model to be adequate, the unit of predictability is not only serial word-by-word, but ranges within a larger unit. It stands to reason that more complex phases (e.g., NP-Verb) have more information and hence less internal predictability than simpler phases (e.g., VprepP). Thus, increased phonetic reduction in smaller phases (if true in general) could be due to structural or statistical factors in production. These alternatives open up the usual kind of research program in which a structural hypothesis (e.g., that the phase regulates speech production and phonetic reduction) competes with a statistical hypothesis (e.g., that units of mutual predictability regulate speech production and phonetic reduction). Specific experimental predictions are going to interact with each candidate theory of what phases are, so it is too rich an area to explore further here. But it does promise the possibility of an informative interaction between comprehension research, production research and using behavioral data to constrain theories of phases.

Implications for stages of comprehension and assigning syntax during processing: There is an intriguing interaction between the idea of analyzing serial sequences in whole chunks and Townsend's and my proposal about logical stages of alternating between associative and derivational processes during comprehension (Bever & Townsend, 2001; Townsend & Bever, 2001, chapters 5 and 8). We argued and reviewed evidence that comprehension processes necessarily integrate statistically valid patterns with computationally applied derivations, within an "analysis by synthesis" framework. On this model, pattern recognition templates can apply quickly to assign a likely meaning, to be complemented by derivational processes. This raised a question about when the derivational reconstruction of that input occurs: we answered this in the acronym for the model, LAST—late assignment of structure theory—making the point in contradistinction to other models, which either assume that structure must be assigned before meaning, or that derivational structures are actually not assigned at all. In that work, most attention was given to the analysis of sentence level comprehension and syntactic structure assignment. The discussion in this chapter gives some further organizational shape to the units within which pattern recognition and derivational processes can apply to assign meaning—our initial hypothesis for this is the phase, as described in (10). The demonstration of backward processes within such a unit supports the idea that comprehension proceeds in bursts that integrate learned patterns and composed structures.

The disconnect between unconscious processing and our conscious experience of normal conversational language calls into question the *immediacy assumption*—the theoretical preconception that a complete hierarchical layering of grammatical analyses is applied to language input *as we hear it* (Just and Carpenter, 1980; Marslen-Wilson, 1973, 1975). This assumption has been the bedrock of many distinct kinds of comprehension models (see Christiansen & Chater, 2016 for a review). The importance of backward processing of information I have reviewed shows that the assumption is false. I have focused on the importance of such processing for discriminating the speech signal. However, recent discussions have given a computational motivation for allowing indeterminate sequences to be held in immediate memory to be disambiguated or clarified by following input. On this model, there can be uncertainty immediately after each subsequence as to what it was: the subsequence is held in memory until the following material completes a larger pattern of analysis that embraces the critical subsequence (Levy et al., 2009; Kleinschmidt & Jaeger, 2015; for general discussions see Kuperberg & Jaeger, 2015; K. Bicknell et al., 2016). The critical point of consistency with the model in LAST is the notion that larger units immediately organize the local structure and ultimately the meaning of a lexical sequence. In the critical cases, an early indeterminacy is informed by its role in a larger unit of structure and meaning.

But this cannot be the whole story in the LAST framework. In our proposals, we noted that there must be a hierarchy of parallel levels during ongoing processing, each of which can involve integration of associative cues and structural assignments: this includes individual sounds, words, short phrases, phases, sentences and, arguably so-called "discourses" (see Townsend & Bever, 2001, chapters 5 and 8; Bever & Poeppel, 2010; Poeppel et al., 2007). Integrating Clark's notion of parallel hierarchical processes with analysis-by-synthesis, we can think of these parallel computations as organized into a "computational fractal" in which the same alternation and integration of the two major kinds of information occur within each local linguistic unit (e.g., syllable, word, phrase, phase. . . .): separate study of the processes at each level is a matter of "grain"—the size of each domain over which analysis by synthesis processing can occur.

This reinterpretation of our Analysis by Synthesis model moves toward a reconciliation between our view and the view that syntactic derivational structures are assigned serially from "left" to "right," as sentences are experienced. In this vein, Colin Philips has adduced arguments that such immediate structural assignment occurs, and also counter arguments to examples used by us to demonstrate the original analysis by synthesis proposals (for a review of his model and critique of ours, see e.g., Phillips & Lewis, 2013; Lewis & Tanenhaus, 2015). In discussing our proposal, Philips also notes that an important issue is one of "grain." Our proposal here is that such processes occur in units of layered levels starting with individual sounds, overlapping with those of increasing size—that is, the processing is simultaneously multigrained. As it stands, this proposal offers a resolution of the theoretical conflicts, in principle, though much remains to be spelled out. And of course, it is important to review how Philips' positive research findings that support his model might also fit within the modified, "computational fractal" framework I am presenting here: but that will have to await a new thorough analysis.

IMPLICATIONS FOR NOTIONS OF CONSCIOUS EXPERIENCE

A related phenomenon is our conscious, but apparently false perception in many cases, that we understand the speech as we hear it serially. This has been long noted in phonology, but most of the effects are extremely local, and hence subject to solution by simply enlarging the scope of the initial input to a bigger chunk, e.g., the syllable, or word, as I mentioned. However, even in this case there is a puzzle: listeners "think" consciously that they heard the individual sounds in words uttered in a citation form, in the order that they occurred. So even at the most basic level of speech perception, our conscious experience of a series of stimuli, actually involves some "backward" processing.

The significance of this sort of phenomenon is magnified in the case of phrasal and sentence level processing. For example, in the cases of "tyuv," and'chilln', where the critical (and incomprehensible) isolated sequence is *followed* by the crucial contextual material, we are not aware that we could not have analyzed the initial sequence until the later material was heard: rather we are convinced that we understood it as it was phonetically presented. This simple fact demonstrates that language comprehension may proceed in sequences of "psychological moments" in which actual processing moves both forward and backward, with some definition of phases specifying the domain of the interaction. This phenomenon has barely been touched in the language sciences, but is clearly fascinating and will have profound implications for consciousness theories, once it is better understood. Prima facie, it is an ultimate demonstration that even in language behavior (i.e., "externalization" of timeless linguistic structures) serial order may be less important than structure dependent organization.

There is a methodological dividend of the notion that there is a decoupling of the perceptual and comprehension processes and our consciousness of when they occurred. Throughout the literature on the post sentence location of clicks, when the reported location is not a phrase boundary, it systematically precedes the actual location. (This started with Fodor & Bever, 1965, and it has popped up several times; see also Townsend & Bever, 1991.) At first blush, this might be interpreted as a simple demonstration of the notion of "prior entry" (Titchener, 1908; Spence & Parise, 2009): an attended to stimulus is perceived earlier than others. It is possibly also related to demonstrations of "chronostasis" in which a more complex stimulus is slowed down relative to a simpler one. For example, Wundt reported a study in which a bell is perceived earlier than its actual location relative to a moving arrow across a series of numbers on a clock-face display. Wundt referred to the relative delay of the numbers as "positive time displacement" (Wundt, 1897, 1918). Correspondingly, in our studies, the subject's task in locating the clicks is to locate the piece of the sentence and the click together, while attending to the entire sentence. To explain the preposition effect, we may refer to a Helmholtzian unconscious inference. Our conscious reconstruction of perceiving and understanding the speech stream as it was presented, leaves the click unanalyzed within the reconstruction of the speech. If it is the case that the click is perceived without the reconstruction processes, the unconscious inference is that it occurred earlier than it actually did. If one insists that this is merely an explanation of a well-known "positive time displacement" or prior entry effect, at least it is an *explanation*.

The notion that conscious awareness of serial order can involve reconstruction is not novel. There is a distinguished line of research, stimulated by Husserl's (1917/1990) considerations of the conscious perception of time, and most famously re-introduced by Fraisse (1967, 1974). However, most of the research in this vein involves relatively short intervals or rapid sequences of short and simple stimuli. For example, in demonstrations of metacontrast, a later stimulus will "absorb" an earlier one into an "exploding" or moving single object—indeed, this is a large part of how continuous motion is perceived in cinematic projections of at least 1 every tenth of a second. However, the language sequence cases described involve much longer and more complex prospective and retrospective reconstructions. Thus, we have a potential demonstration that the "psychological moment" is itself determined by the perceptual units required: as they become more complex and hierarchical, the physical size of the "moment" can expand dramatically.

Up to now, I have emphasized evidence for retrospective processing of language, because it is the most dramatic demonstration of the reconstructive nature of our conscious experiences. But as I have mentioned, various researchers have suggested that most processing is prospective, that is predictive templates are generated early during each utterance, and the remaining act of perception is actually based on confirmation of an already formed structure. Certainly, we can experience this with close friends and spouses—we often have a strong expectation of what they are about to say and are just waiting for confirmation of it.

While I think it dubious that comprehension of novel discourses always proceeds in this way, let us suppose for a moment that it does. It would not change the implications of for our proposal that during comprehension, conscious awareness is sometimes retrospective. In that case, later input triggers confirmation of a waiting hypothesis, rather than triggering fresh computational processes. Either way, the conscious awareness of the prior input depends on later input.

This concept returns us to the flagship issue of modularity in perceptual processing and representation, which Fodor famously explored. The corresponding puzzle for present and future research is how the distinct levels/modules of representation are actually integrated into the conscious experience of continuous integrated processing. That is, when I understand the sentence "a sentence is like a (miniature) opera" spoken conversationally, my conscious experience is that I hear and interpret the input as a coherent continuous object that unifies the acoustic input and the representational analysis; this occurs even though detailed examination of the sort I have reviewed here shows that the computational details belie this belief. In Fodor's formulation, the "central processor" is the mental cloaca where inputs and outputs to the different modules can meet: but, by definition, the central processor is relatively slow and woolgathering. So it remains to be spelled out how it could create the introspective belief that we understand sentences synchronously with their presentation. In Fodorian terminology, maybe it will turn out that consciousness itself is made up of the simultaneous output of a number of modules that interconnect with some degree of automaticity. As Fodor might say, stranger things have turned out to be true.

Thus, in this exploration, the study of language may become a theory-rich touchstone for yet another aspect of cognitive science—the nature of conscious experience.

THE *REAL* POVERTY OF THE STIMULUS

I began this discussion noting the significance of "the poverty of the stimulus" for all of cognitive science, as discussed by Fodor (1981).

Now consider the implications for the language-learning child of how sentences are acoustically mangled in normal conversation. There is evidence that child-directed "motherese" is often clearer than normal conversations in many cases (Bernstein-Ratner, 1996; Bernstein-Ratner & Rooney, 2001), but not all (see Van de Weijer, 1998); it may use devices to clarify word boundaries (e.g., Aslin et al., 1996) and it may be that infants prefer motherese when they have a choice (e.g., Fernald, 1985; Cooper et al., 1997). In any case, it is likely that the vast majority of speech that children hear is between adults, or older children, and there are considerable cultural differences in whether motherese is used at all (Lieven, 1994). Furthermore, various studies have shown that the syntactic or phonetic quality of the child's input may bear little relation to the child's emerging language (C. Chomsky, 1986; McColgan, 2011). In any event, well-articulated motherese is not always dominant even in child-directed speech. Consider a transcribed example from a real motherese sentence. First, attempt to understand the following fragment (five words!), taken from an actual utterance by a mother to her child:

(12) [ɨ̃nʷ:ɹɨpə̃m]

Now see the whole utterance that follows; (if you are a phonetician) try sounding out the phonetic version alone to see if you can (suddenly) understand the whole utterance. In the acoustic version, adults cannot understand this sentence excerpt; but it immediately pops into perfect comprehension, with the conscious intuition that the entire utterance was reasonably clearly pronounced, which is immediately heard as in (14).

(13) [o gɹe(t) mamɪ mu ðoz mæɣəzɨ̃ns si jy kʰɨ̃n: gɪɾɨ̃mɨ̃nʷ:ɹɨpə̃m]
(14) Oh great, mummy put those magazines away so you can't get them and rip them

It is amazing enough that adults can understand conversational speech like this. For a child the problem is doubly compounded, since its grammatical knowledge is incomplete, and it has not yet had time to build up complex language patterns. This simple fact vastly increases the poverty of the stimulus problem, since in many cases the child may not be able to even encode the utterance in enough detail to serve as a learning model.

There is an important implication of these analyses for how sophisticated the child's comprehension system must be. Over many years, it has been argued that linguistic processes are *structure dependent* (Chomsky, 1980): rules are characteristically sensitive to hierarchical structure. This part of Universal Grammar has been shown to account for pathways to language in first language acquisition (e.g., Crain & Nakayama, 1987 and many later discussions). Recent attempts have been made to show that serial learning models can converge on such sensitivity but such models fail to generalize realistically, omit structure dependence in fact (Perfors et al., 2006), or focus on simulating structure dependence (Reali & Christansen, 2005; see Berwick et al., 2011, for general discussion). It has been shown that adults can learn serial rules but in so doing they utilize different brain areas than those characteristic of language (Musso et al., 2003; Moro, 2011). In the current "minimalist" treatments of language, hierarchical trees are constructed as sets, that is, without serial order constraints (Chomsky, 2013, 2015). On this view, the surface order in language is imposed by how it interfaces with our systems of input and output: but many actual computation of linguistic rules operate strictly on the hierarchical structures without reference to the serial structure of overt language sequences: thus, the comprehension system is building chunks of hierarchically organized structures which themselves may be internally order-free, corresponding to order free processing of the input.

Consider now, the implications of our idea that during language processing, there are "time free" processing zones that mediate between the serial input, structural analysis and simultaneous consciousness of the serial input and its meaning. Earlier, I suggest that the simplest available phase is the unit in which processing can occur both forward and backward. But this is to say in its strong form, that in certain defined domains, serial order is unconsciously suspended during sentence comprehension—allowing for structural dependencies to take precedence. In brief, within certain domains, even the externalization of language as serial may be ignored during behavior in favor of pure structure dependence.

A moment's thought suggests that this must be so, as part of the solution to how the child manages to comprehend normal conversations and build up linguistic knowledge from them: s/he must be listening for phrasal categories that integrate and organize local word sequences. How else could s/he latch onto meanings and structural regularities so automatically and quickly? So the argument that structure dependence appears spontaneously in children's learning language structure applies perforce to early stages of language processing itself (Christophe et al., 2008; for related discussion of models of how the language learning child might benefit from unsegmented input, see Pearl & Phillips, 2016).

These considerations are consistent with an analysis by synthesis model of language acquisition, proposed in general terms in Bever (1970), developed more specifically in Townsend and Bever (2001), and elaborated in later writings (e.g., Bever, 2008, 2013). On this model, children alternate (logically) between accessing available structures/representational constraints and building generalizations over the language it experiences as represented by those categorical structures. The role of the generalizations is to provide form-meaning pairs for sentences that have not yet been assigned a full grammatical representation. These pairs can then be the input data for further elaboration of grammatical analysis, accessing the categorical structures. The categorical structures are in part innate—unique to language, in part innate as a part of general thought and perceptual processes. The categorical framework itself becomes more elaborate and uniquely adapted to language structure in particular. (See Bever, 2008, for further discussion and examples of this iterative process; see Lidz and Gagliardi, 2015 for a discussion of the interaction of generalizations and structure building during learning; see Bever, 2013 for a general discussion of this model of language acquisition as an instance of intrinsically motivated human problem solving.)

The significant feature of this model is the dynamic integration of probabilistic and categorical information to yield both a repertoire of statistically valid generalizations and a constructed grammatical representation for all the sentences in the language and many of the semi-sentences. While the model has some general support from acquisition data, it is not sufficiently precise to be adequately testable in detail: in part this is because it is a framework for how associative and structural processes can interact, but allows for considerable differences between individuals and the data they experience.

Of course, this is not the first attempt to create a model of language behavior and acquisition that combines both associative and symbolic information. Indeed, the initial flowering of "psycholinguistics" under the leadership of Charles Osgood (Osgood & Sebeok, 1954; Osgood, 1968) was an explicit attempt to show that the then current model of mediated stimulus-response learning could account for the then current phrase structure model of language structure. (Alas, both models were inadequate for their respective goals, but were consonant with each other because the inadequacies corresponded well; see Bever, 1968, 1988 for discussions). In recent years, a class of explicit computational models has appeared that instantiates a dynamic integration of available categorical structures/processes and Bayesian inference algorithms. These models ostensibly connect to Fodor's notion of the language of thought (LoT), the set of symbols and processes that manipulate symbols. The recent models add a Bayesian statistical component to LoT, and recast it as the probabilistic language of thought (pLoT). Researchers in this vein show that many graded kinds of category knowledge can be accounted for as well as apparent category and concept formation. (See Perfors et al., 2006, Goodman & Lassiter 2014; Piantadosi & Jacobs, 2016, for representative discussions among the many articles now appearing on pLoT.) It remains to be seen if such models can actually learn or even render grammatical representations, including processes that involve structure dependent constraints. At the moment these models do not generally address such problems.

This is not to say that no attention is given to how statistically non-categorical input can result in arriving at grammars appropriate to the child's native language. A number of models have also used Bayesian and other statistical techniques of how variable input data may discriminate between candidate grammars. This includes many different target architectures, but all in the general method of using statistically variable input to reinforce or distill out candidate rules or grammars. (For example, see Yang, 2002, 2004; Yang & Roeper 2011; Pearl & Goldwater 2016; Lidz & Gagliardi, 2015.) The critical feature that seems to discriminate these approaches from the emerging pLoT variants of Fodor's LoT is that these approaches presuppose the availability of candidate grammars or rules, both in the child and as the ultimate goal of language learning.

IMPLICATIONS FOR OLD AND NEW RESEARCH

A cautionary note on the issue of how children and adults deal with normal conversational speech: *sometimes* our spoken utterance may be clear enough, with adequate serial cues for a diligent listener to develop immediate representations of what s/he is hearing. This *may* be especially true of instances of so called child-directed "motherese." But what is important in our examples is that this is not always the case, and may not even be true of the majority of cases. If indeed, most of our comprehension has to deal with compressed and cue-poor input, this also calls into question the generalizability of the many studies of carefully pronounced "laboratory speech" that comprise the overwhelming majority of experimental studies, never mind the use of complete word-by-word visual presentation.

The reader will note that I have extrapolated very far ahead of a very small number of facts, but I hope in ways that are amenable to empirical investigation. For example, one can use the backward reconstruction phenomenon as a tool to study what units are the relevant bridges between serial input and structural output. Here is a(n in principle) simple way to do this. Take conversational corpora and analyze the transcripts (which presumably already have interpreted the conversations into complete words, phrases and sentences); pick out candidate phases according to a theory of what phases are relevant [e.g., as postulated in (10)]; test gated increments of each candidate from its beginning for recognition of the input by subjects (that is, start with an initial fragment, then successively longer ones to see when the initial fragment becomes (retrospectively) clearly interpretable; do the corresponding testing starting from the final part of such fragments. The same kind of procedure can be applied to child-directed speech to examine empirically the claim that a great deal of it is also heavily encoded and dependent on both forward and backward processing. No doubt, these are big projects, but the payoff could be even bigger in leading to a theoretical understanding of how serially presented units build up hierarchical structures and meaning in comprehension and language learning, and to information about normal speaking with many practical applications.

Such research programs can be viewed as the latest step in making good on the implications of the original discoveries by Fodor and his colleagues that in ongoing speech comprehension, sentences are automatically segmented into natural units.

AUTHOR'S NOTE

It should be obvious to the reader how much this paper owes to Jerry Fodor. Along with Merrill Garrett, we pioneered click mislocation as a method to demonstrate the active online use of syntax during comprehension: this is the foundation of many subsequent explorations of the initial compositional strategies of comprehension. More personally, my many conversations with Jerry, co-teaching a course in the early 1960s and co-authoring our 1974 book (*The Psychology of Language*), gave me wide-ranging instructions in how to think about the general problems of cognitive science. We did discuss the poverty of the stimulus, both in relation to adult comprehension and language acquisition. But we did not discuss consciousness at all, to my recollection: it was viewed at the time as a slightly embarrassing romantic problem not a scientific one. But as Jerry noted in his 2007 review of Strawson's edited book on consciousness, "[it] is all the rage just now. . . . What everybody worries about most [is] what philosophers have come to call "the hard problem." The hard problem is this: it is widely supposed that the world is made entirely of mere matter, but how could mere matter be conscious? How, in particular, could a couple of pounds of grey tissue have experiences?" In considering this question, I (TGB) follow the general approach in "biolinguistics" to an understanding of the biology and genetics of language: to discover what makes consciousness possible, we first have to determine what consciousness *is*, how it is acquired as a habit from the alleged "blooming buzzing confusion" of infancy, how it is represented, how it works. This chapter is not a solution to all that, but a pointer to a problem that I hope will attract the interest of today's graduate students. Without them, our science will be lost to the world.

I am indebted to Roberto de Almeida and Lila R. Gleitman, editors of the volume where this appeared, for many helpful early criticisms and comments; also to Mahmoud Azaz for conceptual advice, to David Poeppel for reminding me about the broad evidence for what I coin the "computational fractal" in language processing, and especially to Michael Tanenhaus for deeply considered advice. Caitlyn Antal and Valerie Kula were invaluable bibliographic assistants. Other helpful comments are due to Stuart Hameroff, Al Bergesen, Felice Bedford, Virginia Valian, Rudi Troike, Louann Gerken, Billl Idsardi, Gary Dell, Florian Jaeger, Mark Pitts, Maryellen MacDonald, Lisa Pearl, Massimo Piattelli-Palmarini, and Noam Chomsky. Most of all, I am indebted to my colleague Natasha Warner, who not only developed a methodology of collecting natural conversations, but also made available her materials, her bibliographic advice, her perspective, and all the phonetic transcriptions in this chapter.

NOTES

1. For other discussions of reduction in casual speech, see Ernestus (2000), Tucker & Warner (2011), Ernestus & Warner (2011), Dilley & Pitt (2010), Gahl et al. (2012), and chapters in the special 2011 issue of the *Journal of Phonetics*, edited by Ernestus and Warner.
2. Readers interested in the examples discussed in this chapter can email me for a PowerPoint file with sound. tgb@:mail.arizona.edu

REFERENCES

Abrams, K., & Bever, T. G. (1969). Syntactic structure modifies attention during speech perception and recognition. *Quarterly Journal of Experimental Psychology, 21*, 280–290.

Arai, T. (1999). A case study of spontaneous speech in Japanese. In *Proceedings of the International Congress of Phonetic Sciences, San Francisco, 1*, 615–618.

Arai, T., & Warner, N. (1999). Word level timing in spontaneous Japanese speech. Proceedings of the 14th International Congress of Phonetic Sciences, San Francisco, August 1999. 1055–1058.

Aslin, R. N., Woodward, J. Z., LaMendola, N. P., & Bever, T. G. (1996). Models of word segmentation in fluent maternal speech to infants. In J. L. Morgan & K. Demuth (Eds.), *Signal to syntax: Bootstrapping from speech to grammar in early acquisition* (pp. 117–134). Hillsdale, NJ: Erlbaum.

Bernstein-Ratner, N. (1996). From signal to syntax: But what is the nature of the signal. In J. L. Morgan, & K. Demuth (Eds.), *Signal to syntax: Bootstrapping from speech to grammar in early acquisition* (pp. 135–150). Hillsdale, NJ: Erlbaum.

Bernstein-Ratner, N., & Rooney, B. (2001). How accessible is the lexicon in motherese? In J. Weissenborn., & B. Höhle (Eds.), *Approaches to bootstrapping: Phonological, lexical, syntactic and neurophysiological aspects of early language acquisition Vol. I.* (pp.71–78). Philadelphia: Benjamins.

Berwick, R. C., Pietroski, P., Yankama, B., & Chomsky, N. (2011). Poverty of the stimulus revisited. *Cognitive Science, 35*(7), 1207–1242.

Bever, T. G. (1968). Associations to stimulus-response theories of language. In T. R. Dixon & D. L. Horton (Eds.), *Verbal behavior and general behavior theory* (478–494). Upper Saddle River, NJ: Prentice-Hall, Inc.

Bever, T. G., (1970/2013). The cognitive basis for linguistic structures. Reprinted in Laka, I., Sanz, M. and Tanenhaus, M. (Eds.), *Language down the garden path: the cognitive and biological bases for linguistic structures* (pp.1–70). New York, NY: Oxford University Press.

Bever, T. G. (1973). Serial position and response biases do not account for the effect of syntactic structure on the location of brief noises during sentences. *Journal of Psycholinguistic Research. 2*, 287–288.

Bever, T. G. (1988). The psychological reality of grammar: A student's eye view of cognitive science. In W. Hirst (Ed.), *The making of cognitive science.* Cambridge, England: Cambridge University Press.

Bever, T. G. (2008). The canonical form constraint: Language acquisition via a general theory of learning. In Guo et al. (Eds.), *Cross-linguistic approaches to the psychology of language* (pp.475–492). New York, NY: Oxford University Press.

Bever, T. G. (2013/2015). The biolinguistics of language universals: The next years. In M. Sanz, I. Laka, & M. K. Tanenhaus (Eds.), *Language down the garden path: The cognitive and biological basis of linguistic structures* (pp. 235–405). New York NY: Oxford University Press.

Bever, T. G., Lackner, J. R., & Kirk, R. (1969). The underlying structures of sentences are the primary units of immediate speech processing. *Perception & Psychophysics, 5*(4), 225–234.

Bever, T. G., Lackner, J. R., & Stolz, W. (1969). Transitional probability is not a general mechanism for the segmentation of speech. *Journal of Experimental Psychology, 79*(3), 387–394.

Bever, T. G., & Townsend, D. J. (2001). Some sentences on our consciousness of sentences. In R. Dupoux (Ed.), *Language, Brain and Cognitive Development* (pp. 145–155). Cambridge, MA, MIT Press.

Bever, T. G., & Poeppel, D. (2010). Analysis by synthesis: A (re-)emerging program of research for language and vision. *Biolinguistics, 43*(2), 174–200.

Bicknell, K., Florian Jaeger, T., & Tanenhaus, M. K. (2016/1). Now or . . . later: Perceptual data are not immediately forgotten during language processing. *Behavioral and Brain Sciences, 39*, e67. Publisher Cambridge University Press.

Bicknell, K., Tanenhaus, M., and Jaeger, T., Rationally updating uncertainly about previous words. Unpublished manuscript.

Boeckx, C. (2006). *Linguistic minimalism: origins, concepts, methods, and aims.* New York, NY: Oxford University Press.

Brown, M., Dilley L. C., & Tanenhaus, M. K. (2012). Real-time expectations based on context speech rate can cause words to appear or disappear. In N. Miyake, D. Peebles, & R. P. Cooper (Eds.), *Proceedings of the 34th Annual Conference of the Cognitive Science Society* (pp. 1374–1379). Austin, TX: Cognitive Science Society.

Citko, B. (2014). *Phase theory: An introduction.* Cambridge, UK: Cambridge University Press.

Chapin, P. G., Smith, T. S., & Abrahamson, A. A. (1972). Two factors in perceptual segmentation of speech. *Journal of Verbal Learning and Verbal Behavior, 11*(2), 164–173.

Chomsky, C. (1986). Analytic study of the tadoma method: Language abilities of three deaf/blind children. *Journal of Speech, Language and Hearing Research, 29*(3), 332–347.

Chomsky, N. (1959). A review of BF Skinner's verbal behavior. *Language, 35*(1), 26–59.

Chomsky, N. (1965). *Aspects of the theory of syntax.* Oxford, England: MIT Press.

Chomsky, N. (1975). *Reflections on language.* NewYork, NY: Pantheon Books.

Chomsky, N. (1995). *The minimalist program.* Cambridge, MA: Massachusetts Institute of Technology.

Chomsky, N. (1980). *Rules and representations.* New York, NY: Columbia University Press.

Chomsky, N. (2000). *New horizons in the study of language and mind.* Cambridge, England: Cambridge University Press.

Chomsky, N. (2008). On phases. In R. Freidin, C. Peregrín Otero, & M. L. Zubizarreta (Eds.), *Foundational issues in linguistic theory. Essays in honor of Jean-Roger Vergnaud* (pp. 133–166). Cambridge, MA: MIT Press.

Chomsky, N. (2013). Problems of Projection. *Lingua, 130*, 33–49.

Chomsky, N. (2015). Problems of projection: Extensions. In E. Di Domenico, C. Hamann, & S. Matteini (Eds.), *Structures, Strategies and Beyond. Studies in Honour of Adriana Belletti* (pp. 1–16). Amsterdam: John Benjamins.

Christianson, K., Hollingworth, A., Halliwell, J. F., & Ferreira, F. (2001). Thematic roles assigned along the garden path linger. *Cognitive Psychology, 42*, 368–407.

Christianson, K., Williams, C. C., Zacks, R. T., & Ferreira, F. (2006). Younger and older adults' "good-enough" interpretations of garden-path sentences. *Discourse Processes, 42*(2), 205–238.

Christophe, A., Millotte, S., Bernal, S., & Lidz, J. (2008). Bootstrapping lexical and syntactic acquisition. *Language and speech, 51*(1-2), 61–75.

Clark, A. (2013). Whatever next? Predictive brains, situated agents, and the future of cognitive science. *Behavioral and Brain Sciences, 36*(03), 181–204.

Cooper, R. P., Abraham, J., Berman, S., & Staska, M. (1997). The development of infants' preference for motherese. *Infant Behavior and Development, 20*(4), 477–488.

Connine, C. M., Blasko, D. G., & Hall, M. (1991). Effects of subsequent sentence context in auditory word recognition: Temporal and linguistic constraints. *Journal of Memory and Language, 30*, 234–250.

Crain, S., & Nakayama, M. (1987). Structure dependence in grammar formation. *Language, 63*(3), 522–543.

Dalrymple-Alford, E. C. (1976). Response bias and judgments of the location of clicks in sentences. *Perception & Psychophysics, 19*(4), 303–308.

Degen, J., & Tanenhaus, M. K. (2015). Processing Scalar Implicature: A Constraint-Based Approach. *Cognitive Science, 39*, 667–710. doi: 10.1111/cogs.12171.

Dell, G. S., Chang F. (2014). The P-chain: relating sentence production and its disorders to comprehension and acquisition. *Philosophical Transactions of the Royal Society Ser. B, 369*, 20120394. http://dx.doi.org/10.1098/rstb.2012.0394

Dilley, L., & Pitt, M. A. (2010). Altering context speech rate can cause words to appear and disappear. *Psychological Science, 21*(11), 1664–1670.

Ding, N., Melloni, L., Zhang, H., Tian, X., & Poeppel, D. (2016). Cortical tracking of hierarchical linguistic structures in connected speech. *Nature Neuroscience, 19*(1), 158–164.

Ernestus, M. (2000). *Voice assimilation and segment reduction in casual Dutch: A corpus-based study of the phonology-phonetics interface* (Dissertation). Utrecht: LOT.

Ernestus, M., & Warner, N., Eds. (2011). Speech reduction. *Journal of Phonetics* [Special issue], *39*(3), 1–14.

Farmer, T. A., Brown, M., & Tanenhaus, M. K. (2013). Prediction, explanation, and the role of generative models. *Behavioral and Brain Sciences, 36*, 211–212.

Farmer, T. A., Yan, S., Bicknell, K., & Tanenhaus, M. K. (2015). Form-to-expectation matching effects on first-pass eye movement measures during reading. *Journal of Experimental Psychology: Human Perception and Performance, 41*, 958–976.

Ferreira, F., Bailey, K. D., & Ferraro, V. (2002). Good-enough representations in language comprehension. *Current Directions in Psychological Science, 11*(1), 11–15. doi:10.1111/1467-8721.00158.

Ferreira, F., Engelhardt, P. E., & Jones, M. W. (2009). Good enough language processing: A satisficing approach. In N. Taatgen, H. Rijn, J. Nerbonne, & L. Schomaker

(Eds.), *Proceedings of the 31st Annual conference of the Cognitive Science Society*. Austin, TX: Cognitive Science Society.

Ferreira, F., & Henderson, M. (1991). Recovery from misanalyses of garden-path sentences. *Journal of Memory and Language, 31*, 725–745. doi:10.1016/0749-596X(91)90034-H.

Ferreira, F., & Patson, N. D. (2007). The "good enough" approach to language comprehension. *Language and Linguistics Compass, 1*(1-2), 71–83.

Fernald, A. (1985). Four-month-old infants prefer to listen to motherese. *Infant Behavior and Development, 8*(2), 181–195.

Fodor, J. A. (1981). *Representations: Philosophical essays on the foundations of cognitive science*. Cambridge, MA: MIT Press.

Fodor, J. A., & Bever, T. G. (1965). The psychological reality of linguistic segments. *Journal of Verbal Learning & Verbal Behavior, 4*, 414–420.

Fodor, J. A., Bever., T. G., & Garrett, M. F. (1974). The psychology of language: An introduction to psycholinguistics and generative grammar. New York: McGraw-Hill.

Fraisse, P. (1967). *Psychologie du temps*. Paris: Presses universitaires de France.

Fraisse, P. (1974). *Psychologie du rhythme*, Paris: Presses universitaires de France.

Frank, A., & Jaeger, T. F. (2008). Speaking rationally: Uniform information density as an optimal strategy for language production. *In The 30th annual meeting of the Cognitive Science Society (CogSci08)* (pp. 939–944), Washington, DC.

Gahl, S., Yao, Y., & Johnson, K. (2012). Why reduce? Phonological neighborhood density and phonetic reduction in spontaneous speech. *Journal of Memory and Language, 66*(4), 789–806.

Garrett, M. F. (1964). *Structure and sequence in judgments of auditory events* (Unpublished doctoral dissertation). University of Illinois, Chicago.

Garrett, M. F., Bever, T. G., & Fodor, J. A. (1966). The active use of grammar in speech perception. *Perception & Psychophysics, 1*, 30–32.

Goodman, N. D., & Lassiter, D. (2014). Probabilistic semantics and pragmatics: Uncertainty in language and thought. In S. Lappin & C. Fox (Eds.), *Handbook of contemporary semantics*. Hoboken, NJ: Wiley-Blackwell.

Greenberg, S. (1999). Speaking in shorthand—a syllable-centric perspective for understanding pronunciation variation. *Speech Communication, 29*, 159–176.

Greenberg, S., Hollenback, J., & Ellis, D. (1996). Insights into spoken language gleaned from phonetic transcription of the Switchboard corpus. In *Proceedings of ICSLP* (pp. 24–27). Philadelphia: ICSLP.

Husserl, E. (1917). *On the phenomenology of the consciousness of internal time* (1893–1917), Vol. 4, J. B. Brough, Trans. Dordrecht: Kluwer (Original work re-published in 1991).

Jaeger, T. F. (2006). *Redundancy and syntactic reduction in spontaneous speech*. (Unpublished doctoral dissertation), Stanford University, Palo Alto, CA.

Jaeger T. F. (2010. Redundancy and reduction: Speakers manage syntactic information density. Cognitive Psychology, *61*, 23–62.

Johnson, K. (2004). Massive reduction in conversational American English. In K. Yoneyama & K. Maekawa (Eds.), *Spontaneous Speech: Data and Analysis. Proceedings of the 1st Session of the 10th International Symposium* (pp. 29–54). Tokyo, Japan: The National International Institute for Japanese Language.

Johnson, N. F. (1970). The role of chunking and organization in the process of recall. *Psychology of Learning and Motivation, 4*, 171–247.

Just, M. A., & Carpenter, P. A. (1980). A theory of reading: From eye fixations to comprehension. *Psychological Review, 87*(4), 329–354.

Kleinschmidt, D. F., & Jaeger, T. F. (2015). Robust speech perception: Recognize the familiar, generalize to the similar, and adapt to the novel. *Psychological Review, 122*(2), 148–203.

Ladefoged, P., & Broadbent, D. (1960). Perception of sequence in auditory events. *Quarterly Journal of Experimental Psychology, 12*, 162–170.

Levy, R., & Jaeger, T. F. (2007). Speakers optimize information density through syntactic reduction. In B. Schlö˙kopf, J. Platt, & T. Hoffman (Eds.), *Advances in neural information processing systems* (NIPS) 19, 849–856. Cambridge, MA: MIT Press.

Levy, R., Bicknell, K., Slattery, T., & Rayner, K. (2009). Eye movement evidence that readers maintain and act on uncertainty about past linguistic input. *Proceedings of the National Academy of Sciences, 106*(50), 21086–21090.

Levy, R., & Jaeger, T. F. (2007). Speakers optimize information density through syntactic reduction. In B. Scholkopf, J. Platt, & T. Hoffman (Eds.), *Advances in Neural Information Processing Systems 19: Proceedings of the 2006 Conference* (pp. 849–856). Cambridge, MA: MIT Press.

Lewis, S., & Tanenhaus, M. K. (2015). Aligning Grammatical Theories and Language Processing Models. *Journal of Psycholinguistic Research, 44*(1), 27–46.

Lidz, J., & Gagliardi, A. (2015). How nature meets nurture: Universal grammar and statistical learning. *Annual Review of Linguistics, 1*, 333–353.

Lieven, E. V. M. (1994). Crosslinguistic and crosscultural aspects of language addressed to children. In C. Gallaway & B. J. Richards (Eds.), *Input and Interaction in Language Acquisition* (pp. 56–73). Cambridge, England: Cambridge University Press.

MacDonald, M. C. (1999). Distributional information in language comprehension, production, and acquisition: Three puzzles and a moral. In B MacWhinney (Ed.), *The emergence of language* (pp. 177–196). Hillsdale, NJ: Erlbaum.

MacDonald, M. C. (2013). How language production shapes language form and comprehension. *Frontiers in Psychology, 4*, 226.

Marslen-Wilson, W. (1973). Linguistic structure and speech shadowing at very short latencies. *Nature, 244*, 522–523.

Marslen-Wilson, W. (1975). Sentence perception as an interactive parallel process. *Science, 189*, 226–228.

Marslen-Wilson, W., & Zwitserlood, P. (1989). Accessing spoken words: The importance of word onsets. *Journal of Experimental Psychology: Human Perception and Performance, 15*(3), 576–585.

McColgan, K. (2011). *The Relationship between Maternal Language Outcomes* (Unpublished master's thesis). University of Maryland, College Park, MD.

Miller, G. A. (1962). Some psychological studies of grammar. *American Psychologist, 17*(11), 748–762.

Moro, A. (2011). A closer look at the turtle's eyes. *Proceedings of the National Academy of Sciences, 108*(6), 2177–2178.

Musso, M., Moro, A., Glauche, V., Rijntjes, M., Reichenbach, J., Büchel, C., & Weiller, C. (2003). Broca's area and the language instinct. *Nature Neuroscience, 6*(7), 774–781.

Osgood, C. E. (1968). Toward a wedding of insufficiencies. In T. R. Dixon, & D. L. Horton (Eds.), *Verbal learning and general behavior theory* (pp. 495–519). Englewood Cliffs: NJ: Prentice-Hall.

Osgood, C., & Sebeok, T. (1954). Psycholinguistics: A survey of theory and research problems. *Journal of Abnormal and Social Psychology, 49*(4, part 2): 1–203.

Pearl, L. & Goldwater, S. (2016). Statistical learning, inductive bias, and Bayesian inference in language acquisition. In J. Lidz, W. Snyder, & C. Pater (Eds.), *The Oxford Handbook of Developmental Linguistics* (pp. 664–695). New York, NY: Oxford University Press.

Pearl, L. & Phillips, L. (in press). Evaluating language acquisition models: A utility-based look at Bayesian segmentation. In A. Villavicencio & T. Poibeau (Eds.), *Language, cognition and computational models*. Cambridge, England: Cambridge University Press.

Perfors, A., Tenenbaum, J., & Regier, T. (2006). Poverty of the stimulus? A rational approach. In R Sun and N Miyake (Eds.), *Proceedings of the 28th Annual Conference of the Cognitive Science Society* (pp. 663–668). Mahwah, NJ: Erlbaum.

Phillips, C., & Lewis, S. (2013). Derivational order in syntax: Evidence and architectural consequences. *Studies in Linguistics, 6*, 11–47.

Piantadosi, S., & Jacobs, R. (2016). Four problems solved by the probabilistic language of thought. *Current Directions in Psychological Science, 25*(1), 54–59.

Poeppel, D., Idsardi, W., & von Wassenhjove, V. (2007). Speech perception at the interface of neurobiology and linguistics. *Philosophical Transactions of the Royal Society Series B*. doi:10.1098/rstb.

Pollack, I., & Pickett, J. M. (1964). Intelligibility of excerpts from fluent speech: Auditory vs. structural context. *Journal of Verbal Learning and Verbal Behavior, 3*, 79–84.

Reali, F., & Christiansen, M. H. (2005). Uncovering the richness of the stimulus: Structural dependence and indirect statistical evidence. *Cognitive Science, 29*(6), 1007–1028.

Rumelhart, D. E., McClelland, J. L., & PDP Research Group (1986). *Parallel distributed processing: explorations in the microstructure of cognition*, Volumes I, II. Cambridge, MA: MIT Press.

Slattery, T. J., Sturt, P., Christianson, K., Yoshida, M., & Ferreira, F. (2013). Lingering misinterpretations of garden path sentences arise from competing syntactic representations. *Journal of Memory and Language, 69*(2), 104–120.

Spence, C., & Parise, C. (2009). Prior-entry: A review. *Consciousness and Cognition, 19*(1), 364–379.

Swinney, D. (1979). Lexical access during sentence comprehension: (Re) consideration of context effects. *Journal of Verbal Learning and Verbal Behavior, 18*, 645–659.

Titchener, E. B. (1908). *Lectures on the elementary psychology of feeling and attention*. New York, NY: Macmillan.

Townsend, D., & Bever, T. G. (1991). The use of higher-level constraints in monitoring for a change in speaker demonstrates functionally distinct levels of representation in discourse comprehension. *Language and Cognitive Processes, 6*(1), 49–77.

Townsend, D. J., & Bever, T. G. (2001). *Sentence Comprehension: The integration of habits and rules*. Cambridge, MA: MIT Press.

Tucker, B. V., & Warner, N. (2010). What it means to be phonetic or phonological: The case of Romanian devoiced nasal. *Phonology, 27*(2), 1–36.

Van de Ven, M. A. M. (2011). *The role of acoustic detail and context in the comprehension of reduced pronunciation variants* (PhD thesis), Radboud Universiteit Nijmegen, Nijmegen, Netherlands.

Van de Weijer, J. (1998). *Language input for word discovery* (Ph.D. thesis). Nijmegen, Netherlands: Max Planck Series in Psycholinguistics, 9.

Warner, N., Fountain, A., & Tucker, B. V. (2009). Cues to perception of reduced flaps. *Journal of the Acoustical Society of America, 125*(5), 3317–3327.

Wundt, W. (1918). *Grundriss der Psychologie*. Leipzig, Germany: Alfred Kroener Verlag.

Yang, C. (2002). *Knowledge and learning in natural language*. New York, NY: Oxford University Press.

Yang, C. (2004). Universal grammar, statistics or both? *Trends in Cognitive Sciences 8*(10), 451–456.

Yang, C., & Roeper, T. (2011). Minimalism and language acquisition. In C. Boeckx, (Ed.) *The Oxford handbook of linguistic minimalism*. New York, NY: Oxford University Press.

5

Semantics for a Module

ROBERTO G. DE ALMEIDA AND ERNIE LEPORE

Modularity is a hypothesis about a nomological distinction between perceptual, input-driven computations and background knowledge. It hinges on the very nature of representations and processes computed by input systems—and, crucially, on what input systems deliver to higher cognition. Perceptual computations are said to be encapsulated or have only relative—principled—access to background knowledge in the course of running its default algorithms. Moreover, perceptual computations are supposed to deliver "shallow" representations of transduced inputs. This is where we begin, and this is pretty much where *The Modularity of Mind* (Fodor, 1983) left off: the theoretical and empirical research programs were—and still are—to determine the scope of perceptual computations and their degree of autonomy; and, more broadly, to search for the line that divides perception from cognition, hardwired computations from contingent and malleable operations.

Of course, *Modularity* was not only about the encapsulation of some psychological capacities. It advanced an epistemological thesis about the distinction between observation and inference in the acquisition of knowledge—or the fixation of belief. In the present chapter, we are concerned with the psychological, rather than the epistemological, thesis that *Modularity* advanced. We tie together two threads bearing on sentence representation and processing: one is that sentence perception is, to some extent, computationally encapsulated; and the other is that sentence *meaning* is, to some extent, context insensitive, or at least its sensitivity is rule-governed.

These threads come together in the following way: we argue that the output of sentence encapsulation is a minimally and highly constrained, context-sensitive propositional representation of the sentence, built up from sentence constituents. Compatible with the original *Modularity* story, we argue that the output of sentence perception is thus a "shallow" representation—though it is *semantic*. The empirical cases we discuss bear on alleged cases of sentence indeterminacy, and how such cases might (a) be assigned (shallow) semantic representations, (b) interact with context in highly regulated ways, and (c) whether and, if so, how they can be enriched. In the course of our discussion, we will advance and defend a proposal for a semantic level of representation that serves as output of the module and as input to other systems of interpretation, arguing for a form of modularity or encapsulation that is minimally context sensitive provided that the information from context—whatever it may be—is itself determined nomologically, namely, by linguistic principles.

THE CASE FOR MODULARITY

The Modularity of Mind (Fodor, 1983) raised fundamental questions about the architecture of perception and cognition, and, in particular, about linguistic and visual computations—whether they are to some degree encapsulated from background knowledge, beliefs, and expectations. These questions have long been the focus of inquiry in cognitive science, with implications not only for standard cases of sentence parsing and early visual perception but, with language, for related questions such as whether there is a semantics/pragmatics distinction, and the nature of compositionality. In this chapter, we explore these latter topics and their relevance for the general hypothesis of language modularity. In particular, we discuss which type of *semantic* representations might be computed within a linguistic module, or, rather, serve as the *output* of computations performed by the module.

Pertinent to our general goals is the following question (Fodor, 1983, p. 88):

> [W]hich phenomenologically accessible properties of an utterance are such that, on the one hand, their recovery is mandatory, fast, and relevant to the perceptual encoding of the utterance and, on the other, such that their recovery might be achieved by an informationally encapsulated computational mechanism?

Although Fodor does not offer a precise answer to this question, he suggests a research program, which we plan to elaborate on:

> [W]hile there could perhaps be an algorithm for parsing, there surely could not be an algorithm for estimating communicative intentions in anything like their full diversity. Arguments about what an author meant are thus able to be interminable in ways in which arguments about what he said are not. (p. 90)

The research program lies, of course, in determining where the line should be drawn between sentence parsing and the recovery of speaker intentions that go beyond communicating information the sentence conventionally contributes. We contend that somewhere between parsing (viz., syntactic analysis) and the points a speaker intends to get across lies an encapsulated semantic representation—if you like, the specification of the proposition the uttered sentence expresses—which we take to be an excellent candidate for what the language module delivers to higher cognitive systems.

Our case for a *semantic* (propositional) output of the module will be made along the following lines: first, we present general assumptions about the cognitive architecture that underlies our view about the workings of a language module—in particular, about how the module computes sentential (and perhaps even discourse) meaning. We then discuss how a semantic representation might be seen as a "shallow" representation, which the module outputs. We support our view by presenting theoretical and experimental evidence for so-called cases of *indeterminacy*—specifically, sentences which, when tokened, are alleged to only express propositions after enrichment via pragmatic operations or lexical-semantic decomposition. We claim, contra the mainstream, that these discourses are arguably, by default, output by the module as complete, *un*enriched propositions, and that these propositions are only minimally context-sensitive (viz., sensitive *to antecedents* within a discourse), and further, that the recovery of whatever information theorists claim is required to understand their utterances is the product of functions performed by other cognitive systems of interpretation—that is, distinctly not linguistic ones.

MODULARITY AND COGNITIVE ARCHITECTURE

Most discussions on the properties of a language faculty come infused with assumptions about the nature of mental representations and processes—more precisely, what kinds of representations the language system encodes, and how the system performs its functions. We want to make our assumptions on these issues explicit because what we will say about modularity—and, in particular, about the output of the language module—requires a clear understanding of the type of cognitive architecture we assume supports the workings of the language system and its interfaces with other domains of cognition. The need for an explicit commitment to a particular type of cognitive architecture also derives from the kind of thesis we advance here concerning the very nature of the *semantic* output of the language module: in short, we will postulate that a semantic output is not only plausible but perhaps even necessary vis-à-vis a sharp distinction between operations of an encapsulated linguistic system and operations contingent on world knowledge, beliefs, expectations, and so forth. A question, to which we will return, is what sort of semantic system that might be. But we begin with rather orthodox cognitive-architecture commitments.

We are committed to two closely related guiding assumptions on cognitive architecture and modularity: first, we take representations to be *symbolic* insofar as the codes that serve for perceptual and cognitive computations are symbols—simple and complex. We will see that in order to distinguish between semantic representations and other types of knowledge and beliefs we need to adopt a symbolic architecture; what the system outputs is the product of its semantically interpretable computations—namely, a symbolic expression encoding a proposition. A key feature of this representation is that it accounts for the compositionality of whatever linguistic expression that expresses the proposition. Compositionality is a fundamental property of symbolic systems and, as far as we know, of symbolic systems alone. Although we take these symbols to be physically instantiated as patterns of neuronal code or, perhaps, as Gallistel (2018; Gallistel & King, 2010) has proposed, as molecular changes within neurons, the assumption that representations are symbolic is largely independent of hypotheses on the actual form of neuronal implementation. This is so because we take representations and processes to constitute, in principle, an autonomous explanatory level of how cognitive systems work.[1] So much for our first assumption.

Our second assumption is that mental processes—that is, how information is perceived, analyzed, transformed, and so forth—are *computational* and, to a large extent, realized algorithmically. That is, many of the operations that the mind performs—and, in particular, the ones that ought to be taken as *intra*-modular—follow rule-like symbol manipulation. These processes, to a first approximation, are operations performed entirely in a mechanical fashion, akin to logical or mathematical proofs. Now, it is certainly a matter of empirical investigation which cognitive processes can be cast in terms of *algorithms*, which ones follow *heuristic* principles, and which ones are entirely subject to contingencies.[2] Indeed, it is perhaps this second guiding assumption—the extent to which certain processes are algorithmic—that constitutes the main overarching hypothesis bearing on the modularity of mind: that at least some perceptual computations are fixed and operate without interference from background knowledge and beliefs.

Symbolic representations and computational processes are well known guiding assumptions, adopted by a great number of cognitive scientists, in particular, those who subscribe to some version of the Representational/Computational Theory of Mind.[3] It is important, though, to recite them because much of our discussion on the nature of modularity and the semantic system that we take to be internal to—or as an output of—the language module rests on there being *algorithmic* processes, that is, fixed rule-like computations over symbolic representations in the course of analyzing a sentence.[4] We should further qualify these assumptions regarding the nature of representations and processes, mainly because the symbolic-computational view has generated numerous controversies (and misunderstandings), particularly on the semantic nature of "amodal" symbols and whether or not symbolic computations are "meaningless" (e.g., Searle, 1980).

Although we assume that the computations of a module are formal, the symbols the language system computes, according to us, specify, inter alia, truth conditions, and therefore, the system must distinguish *types* of representations in the course of performing its computations. In other words, the computations must be sensitive to *symbol types* and symbolic structures, and so, the relevant semantic distinctions between representations must be coded in the symbols or symbolic structures. Thus, while operations are performed in a mechanical fashion, semantic distinctions ought to be coded symbolically.[5] For example, sentence-parsing computations must be sensitive to distinctions between verb *types*, which in turn must determine the course of processing routines bearing on the *types* of arguments a verb takes and, hence, interpretation should be sensitive to the nature of symbols and how they are structured.[6] We will return to this issue below, in the context of examining so-called indeterminate sentence interpretation.

The upshot of this brief discussion is that a module ought to perform its computations algorithmically, with computations being sensitive to type/token distinctions that are supposed to be encoded in the elementary symbols and symbolic expressions. We will turn now to how computations performed within the module—especially within the hypothetical language module—might be carried out, and what sort of output the module might produce.

THE BOUNDARIES OF LINGUISTIC MODULARITY

Fodor (1983) discussed, at length, criteria for what he called "vertical faculties" (e.g., information encapsulation, domain-specificity, neurological specialization, genetic determination, fast and mandatory computations). So, we won't further exegete here. Our focus, instead, will be primarily on two criteria that bear more directly on our main point: information encapsulation of modular computations, and "shallow" outputs that modules are supposed to produce. Arguments for information encapsulation, simply put, turn on the degree of functional autonomy of a given perceptual domain (e.g., language) with regards to other "vertical" faculties (e.g., vision), general "horizontal" faculties (e.g., long-term memory) and, more broadly, the central system (the "Quineian" holistic space), where encyclopedic and episodic knowledge, beliefs and so forth reside. Arguments for "shallow" outputs turn on the types of representations a module produces: by hypothesis, they do not produce full analyses but, rather, minimally, translations of post-transduced inputs that preserve the nature of the distal stimulus (viz., relevant properties of linguistic structure and lexical denotations). To a certain degree, we assume that whatever the language system computes—the operations on its natural-kind linguistic data—is to a large extent encapsulated. As Fodor puts it,

> . . . data that can bear on the confirmation of perceptual hypotheses includes, in the general case, considerably less than the organism may know. That is, the confirmation function for input systems does not have access to all of the information that the organism internally represents; there are restrictions upon the allocation of internally represented information to input processes. (Fodor, 1983, p. 69)

The question becomes what exactly "the system might know" in order to yield a sufficiently rich representation of the input without being contingent on "all of the information" the organism might internally represent. The standard answer, in the course of the modularity debates in psycholinguistics, has been to focus on syntactic parsing (see, e.g., Townsend & Bever, 2001, for a review). And the research strategy has been to show that syntactic parsing might be unsettled by semantic (i.e., "background") knowledge. Crucially, this strategy rests on the assumption that syntactic analyses are immune to semantic variables—thus, any demonstration of semantic influence on syntactic analyses ought to be a violation of the key encapsulation principle. But while this research strategy has proven fruitful, producing an enormous amount of data (as Fodor says, "that's why we have careers"), it seems to us that it also misses the mark. This is so because what is modular is entirely dependent on the sort of fixed linguistic information the input system might encode. It may turn out that some "semantic" data bearing on structural analyses is encoded in the symbolic rules that govern the workings of the parsing routines, and thus, might have influence on what sort of parsing *choices* a syntactic system might make. To wit, it may be, for example, that thematic/"semantic" roles assigned by verbs to their arguments are part of the database that the parsing system consults in making its choices; it may also be that properties of thematic/"semantic" structure enter into determining what sort of representation the input system might produce. For instance, it has been demonstrated that, traditionally, "semantic" information encoded in different verb classes are affected selectively in cases of brain traumas or diseases (see, e.g., Piñango, 2006; Thompson & Lee, 2009). Data from Alzheimer's patients, in fact, suggest that verb-thematic hierarchy plays a significant role in patients' preferences for how the arguments of a verb map onto syntactic structure (see Manouilidou, de Almeida, Nair, & Schwartz, 2009; Manouilidou & de Almeida, 2009). Patients have no difficulty with canonical sentences in which the noun phrase in subject position is assigned an *Agent* role, as in (1a). However, patients have difficulty understanding the sentence when the subject position is occupied by an *Experiencer*, as in the case of subjects of psychological verbs, as in (1b). Moreover, when the verb assigns the role of *Experiencer* not to the noun phrase in subject position but to the one in object position, as in (1c), patients show a much greater impairment in comprehension. This effect is independent of voice, that is, it is obtained even in the passive form of the same sentences, where the linear order of constituents is inverted.

(1) a. The gardener cultivated the carrots (*Agent, Theme*)
 b. The public admired the statue (*Experiencer, Theme*)
 c. The statue fascinated the public (*Theme, Experiencer*)

It is quite plausible that parsing operations rely on more than the to-be-saturated arguments of a verb and their structural arrangements: decisions might also take into account the role the arguments play in the semantic representation of a sentence. This view, of course, does not commit us to a particular ontology of thematic roles, but simply suggests that semantic information can enter into decisions the parser makes. A parsing model such as Bornkessel and Schlesewsky's (2006), although "incremental," seems to be entirely governed by principles that include algorithmic and heuristic rules for determining structural choices concerning verb-argument relations. This seems to be a "modular"—input-driven—system whose choices are not dependent on background information but on encoded syntactic and semantic principles.

We end this section with a summary of our guiding assumptions, and how they relate to our view that a semantic analysis is the output of the language module. We assume modular systems operate as computations over post-transduced symbolic expressions. Moreover, we assume modular computations are sensitive to semantic distinctions among symbolic expressions, and thus, the input to linguistic analysis could well be guided by encoded—fixed—semantic principles. As we saw, there is a case to be made for semantic representations being the determinants of intra-modular decisions: thematic-role assignment is just one, enriching the nature of the computations that the input system for language computes. In the next two sections, we elaborate, first, on what sort of representation serves as output for the module. We aim to show that the module computes shallow semantic information on the assumption that the input system knows "considerably less" than what the sentence is *about*. We then focus on a particular case: sentences whose propositional contents are alleged to require enrichment in order to explain what their uses can communicate. We aim to show that whatever this sort of enrichment includes is a function of contextual information that goes beyond input analysis.

OUTPUT: A SHALLOW PROPOSITION

The proposal that the language module outputs a type of *semantic* representation suggests that one function of the perceptual system is to analyze utterance *content*. But is the idea of *intra*-modular semantic representations and processes in conflict with sentence encapsulation? Encapsulation, after all, requires that semantics not be served by background knowledge and other systems of interpretation, and this requires a clear distinction between semantic properties that are encapsulated (thus, algorithmic) and other knowledge systems that are not. This is all true, but still, we will defend a view of semantics that is compatible with modularity, where *semantic* representation is recoverable from what is expressed overtly by sentence constituents (viz., lexical, morphological, and functional constituents) and syntactic (and discourse) arrangement. Our proposal, more specifically, is that the symbolic expression the language module outputs carries all the relevant information for (further) elaboration by higher cognitive systems. We take, in short, the proposition that a sentence conveys to be recoverable from its constituents, its structure, and its linguistic relations to other sentences in a discourse. More importantly, we argue that these broader contextual effects—always lurking as a threat to modularity—are either, as a matter of fact, intra-modular (viz., linguistically determined by semantic or syntactic operations inside the module), or are post-modular (i.e., higher cognitive effects on modular outputs).

Determining the nature of the semantic output of the module depends fundamentally on what one takes to be "semantic." In *Modularity*, Fodor often mentions that, among the tasks of an encapsulated language system is that of producing a *logical form*. Although this is not explored in detail, if we take the symbols the module computes to be distinguished by semantic properties, then the logical form that the module outputs has many of the ingredients interpretative processes require in order to perform their basic functions. In other words, if symbolic expressions carry semantic properties that distinguish them from one another, we can assume that much of what the input system does is to produce the semantically relevant representation that symbolic combinations yield.

This view is a bit more explicit in Fodor (1990):

> [W]e are committed to the traditional assumption that the outputs of parsers are representations of utterances that determine their meaning. Roughly, a parser maps wave forms of (sentence-length) utterances onto representations of the proposition that they express. (p. 4)

Significantly, what the parser outputs determines what the sentence means; its (output) representation is mapped onto the proposition that the sentence expresses. Fodor adds to this "a position that is quite appealing":

> Parsing recovers the lexico-syntactical structure of an utterance, but *no* semantic–level representation (no representation that generalizes across synonymous utterances; hence, for example, no model theoretic representation). (p. 8)

Fodor's main reason for keeping semantics out of the module's computational tasks is his belief that, in order to perform any sort of semantic computation "the speaker/hearer would have to know the semantics of their language in the way that they indisputably have to know, say, its lexical inventory." (p. 8)[7]

It is not clear exactly what sort of semantics the speaker/hearer "would have to know" nor is it clear what sort of representations they would need to encode in order to compute sentential meaning while preserving modularity. We assume something akin to lexical denotations (or pointers to lexical denotations; viz., morphemes) and whatever apparatus yields a logical form would suffice. Fodor (1990), in fact, leaves the door open to some form of semantic representation within the module by proposing the following:

(i) We will use 'parsing' to name a process that starts with acoustic specifications and ends with meaning representations.

(ii) We assume that meaning representations are probably distinct from, and probably constructed on the basis of, syntactic structural descriptions; (. . .) (pp. 8–9)

The proposal that parsing "ends with meaning representations" entails some form of representation that might be available to other systems of "interpretation." This is clearer in the model Fodor proposes, encompassing the following serial stages:

(a) acoustics → (b) structural description (syntax) → (c) meaning representation → (d) discourse model → (e) real world model[8]

With the exception of the last stage (e), Fodor leaves it open which operations might be encapsulated. It might be safe to assume that the first transformation, from (a) to (b), follows from the initial transduction of linguistic codes devoid of semantic content: the operations, by hypothesis, are transformations over symbols or symbolic expressions representing the likes of grammatical categories. That is where the standard view of modularity assumes information encapsulation ends, and that is where many studies have suggested there is penetration by meaning representations or even the discourse model. But Fodor's (1990) revised version of modularity traces the line higher, admitting that meaning representation and even a discourse model could be computed by the modular parser. In fact, faced with many studies suggesting that parsing might be influenced by local context (e.g., Crain & Steedman, 1985; Altmann & Steedman, 1988), Fodor assumes the level of discourse representation (d) might provide intra-modular constraints to decisions made at (b), the building of syntactic representations.

This last point is crucial for our understanding of a modular mechanism that is semantically shallow at the output level, while assuming that outputs are representations of the propositional content of input sentences. We take "propositional content" to include denotations of what is overtly specified in the sentence—namely, its lexical or morphological constituents—as well as how these constituents are structurally arranged. Moreover, we also take this propositional content to include the specification of the linguistically active but perhaps phonologically null elements that constitute the sentence. These elements are like the nominal antecedents of pronouns, elliptical verb phrases, and other linguistically specified elements (including cross clausal and sentential mechanisms for establishing discourse anaphora).

Thus far, this amounts to assuming that propositional content is compositional—or, to be more precise, obeys "classical" compositionality—for even in cases where propositional content might be attributed to phonologically null elements, they must be linguistically (i.e., syntactically) licensed. However, there are cases in which the elements called for by phonologically null and overt elements are outside the scope of the sentence proper. One can imagine contexts in which pronouns have their antecedents in the immediate discourse, or cases of indexicals (e.g., "there" and "now"), which pick up their contents from already referenced discourse elements, or even from the visual context. Below, we discuss in more detail experimental evidence for linguistically determined discourse enrichment of sentences.

We assume that these cases can be accommodated by a theory that takes the building of a proposition to be conventionally governed by contextual factors—linguistically determined and part of the local discourse. One case is presupposition. Assuming that what speaker A says to B makes reference to what is in their common ground, one can take pieces of the common ground to aid in the proposition-building process. To be clear, what is presupposed is, to a large extent, linguistically determined. Thus, it might be part of the "discourse model" that Fodor refers to, providing intra-modular constraints on the types of information that constituents of the proposition pick up.

Consider now reference in a visual context. Imagine that upon referring to a particular person on the scene, speaker A says to B: "That is the girl I told you about." Indexicals such as "that," "I" and "you" as well as what has been talked "about" (the girl) constitute elements of discourse that may enter into the propositions that A and B coordinate on during their linguistic exchange. What A and B talk about, or refer to, in the context are not sentential, but rather (local) discourse, constituents that hold a special relation to both the sentence A utters and the information B exploits in building a propositional representation of what A says. A modular output might build the proposition that the sentence expresses taking into account the elements that are within the immediate discourse (discourse referents).

For another case in point consider a discourse like (2):

(2) A man walked in. He sat down.

On one of its readings, 'He' is interpreted as co-varying with 'A man.' (2) is all true just in case some man walked in and sat down. The pronoun resolution for (2) is guided by an implicit organization that knits together information in discourse. On this anaphoric reading, the discourse begins with a description involving 'A man' and proceeds directly to develop a narrative: accordingly, 'He' is interpreted as dependent on 'A man.' This information is entirely encapsulated to this discourse. Confronted with this discourse, without any attendant pointings or other sorts of gestures, speakers know automatically how to interpret it and resolve its pronoun. (For many other examples and for a general defense of the claim that all context sensitivity in resolved in this rule governed conventionalized manner, see Stojnic, Stone & Lepore, 2017).

Our view of Fodor's revised program for a module assumes that the basic representation the module outputs is sensitive to contextual information but always in a rule-governed fashion. The elements upon which B builds propositions corresponding to A's contribution to the conversational record begins with what A says with an utterance and might include what they both take to be common ground as well as other conventional contributions. In keeping with modularity, our proposal, then, is that the output encodes "considerably less" than what an individual takes away from another's utterance. It encodes what is linguistically determined.

We are now ready to turn to cases where the interface between modular computations and semantic/discourse interpretations are in play: resolution of (alleged) indeterminacy.

SENTENCE INDETERMINACY

What happens when the linguistic contribution of an uttered sentence underdetermines what its speaker is able to get across with her utterance? For example, on hearing in isolation an utterance of (3a), we might infer reference to an event in which a man began *reading* a book. But obviously, *reading* is not the only possible inference; why not *eating*? And if nothing is off limits, shouldn't we conclude (3a) is *indeterminate* with regards to the event it refers to, that is, indeterminate with regards to what some man began doing with some book?

(3) a. A man began a book
b. $\exists x$(=man), $\exists y$(=book) (*begin* (x, y))[9]

To be clear, indeterminacy issues from the activity the aspectual verb *begin* scopes over. There is, however, a default interpretation for this type of sentence—that a man began doing something with a book—which is the proposition (3a) expresses, *ceteris paribus*. If we assume that what the module computes is minimally a logical form that captures the proposition that the sentence expresses, then (3a) ought to output something like (3b).

We have an initial observation about the encoding of (3a) as (3b). Clearly, (3b) does not exhaust everything that an utterance of (3a) can get across, nor is it supposed to. What (3b) might specify is how the transduced symbols of the proximal stimulation of (3a) is to be encoded. (3b) is a proposed *symbolic* output of stimulus (3a). (As we mentioned, other symbols might enter into the shallow semantic representation for (3a). For instance, if thematic roles are encoded, they might enter into the representation the module outputs.[10] Crucially, what (3b) might account for is the logical structure of (3a) together with the translation of its lexical/morphological constituents.)

There are linguistic-theoretic and experimental treatments of (3a) that assume it gets enriched even at the linguistic level of analysis by a default operation of *coercion*. The key idea is that a verb such as *begin* (and many others within the same class) require an event complement, the absence of which triggers, roughly, a change in the nature of the internal argument (in (3a), *book*) to make it fit with the requirements of the verb. One proposal, we call *coercion with interpolation*, hypothesizes that the supposed mismatch between the event-taking verb (*begin*) and an entity-type object (*book*) is resolved by, first, extrapolating an event information from the lexical entry *book*; and, second, by interpolating this event information into the semantic composition (deemed *enriched*) of the sentence (see Pustejovsky, 1995, and Jackendoff, 2002, about variants of this process). Another coercion proposal, we call *type-shifting*, makes no direct claims for interpolation, nor does it assume any form of extrapolation of information from the lexical entry: what it proposes is that the entity denoting a book is supposed to change its type to an *event*, to respect the requirements of the event-selecting verb (e.g., Pylkkanen, 2008).

These two proposals agree that (3a)'s being enriched linguistically in the "classical" way does not work, for what's needed is a way to make the arguments "fit" with their selecting verbs. The proposals differ about the source of enrichment and, consequently, about their commitments vis-à-vis the nature of semantics. Type shifting rests on an ontology of semantic *types* that has not been established—one we are not prepared to adopt. The idea that the alleged verb-argument mismatch is resolved by changing the semantic type of the complement strikes us as affirming the consequent. But our main reason for suspicion that enrichment is obtained via type shifting is that it requires postulating at least one of two things: that token nouns are loaded with information about their possible types and their modes of combination with their host verbs; and that semantic principles are informed about these modes of combination.

The assumption that items are informed about their possible semantic types entails that lexical items are polysemous between diverse types they can be coerced into. One has to assume that this is true of *all* lexical forms. The assumption that semantic-type combinations are driven by rules also assumes that the rules ought to be informed about the default types of token items. Either version of this approach to coercion relies on internal analyses of token items to yield appropriate combinations as well as to reject anomalous ones. But how should the linguistic system be informed about such semantic properties (viz., semantic types and their appropriate combinations) without also being informed about putatively holistic world facts—arguably determinants of plausibility? To put it simply, our main point is that to know that *book* can be read on the basis of the linguistic system as an *event performed with a book* requires knowing that the noun *book* allows for such an *event* reading upon the implausibility generated by its *entity* reading. Simply postulating that the *entity→event* shift is demanded by the verb does not work because type shifting relies on analyzing the noun default type before triggering the shifting operation.

Coercion with interpolation runs into its own problems. Most notoriously, it rests on an analytic/synthetic distinction, as pointed out elsewhere (see Fodor & Lepore, 2002; de Almeida & Dwivedi, 2008; de Almeida & Riven, 2012). Semantic interpolation requires a vastly rich, encyclopedic lexicon, whose properties are supposed to provide filler information—what can be interpolated in the resulting semantic composition. For instance, we would need to know a great deal about books in order to find out what is possible or likely for one to begin doing with them, in order to select an appropriate filler to enrich (3a). And besides, the lack of a principled analytic/synthetic distinction leaves us wondering what sorts of properties—the ones which are supposed to constitute the lexical-semantic information—are to be regarded as part of the "semantic lexicon" and which are not. Moreover, there is no evidence that nouns such as *book* are actually constituted by properties or features, let alone that this process takes place at a linguistic level of analysis, a level independent from general knowledge.

One issue common to both the coercion proposals we have sketched is that they assume *entity* arguments don't fit with verbs such as *begin*. A linguistic test used to support the alleged oddness of the *begin-entity* combination rests on showing that so-called *event* nominals do not require coercion, such as in (4).

(4) The general began the war

This argument, however, has little validity, for sentences with *begin-event* forms might also require enrichment. Simply put, if *x begins entity* calls for an event interpretation, so does *x begins event*: (4) is neither synonymous with (5a), nor does it entail (5a). Rather, (4) can be roughly paraphrased by something like (5b) because one can begin a war without fighting it (see de Almeida & Riven, 2012, for further discussion on this issue; see also Cappelen & Lepore, 2005, for a defense of slippery slope arguments of this sort).

(5) a. The general began fighting the war
    b. The general caused the war to begin

An alternative to coercion assumes that (3a) remains linguistically "indeterminate" with respect to what sorts of enrichments its tokenings might admit of. Crucially, this view assumes enrichment is beyond the linguistic level of analysis—it comes from post-linguistic processes. Such processes are most likely abductive, for they take into account what might be contextually appropriate, what might be most probable, etc. Essentially, this view, assumes what the module outputs is based on what the sentence expresses (at least in a discourse) without lexical decomposition (*a la* coercion with interpolation) or type-shifting. Moreover, this view assumes the enrichment of (3a) is linguistically motivated by the *syntax* of VP. Key here is that verbs such as *begin* are represented by an argument structure that specifies a syntactic position for the filler event, as in (6).

(6) $[_{VP}\ [_{V0}\ \textit{began}\ [_{V0}\ e\ [_{OBJ}\ \text{NP}]]]]$

There are many distributional arguments for the *linguistic* reality this gap within VP (see de Almeida & Dwivedi, 2008; and de Almeida & Riven, 2012). For instance, it is within VP in the second clause of (7) where verb ellipsis is realized—that is, where the second-clause *reading* is syntactically determined to re-appear.

(7) I started reading with my contacts but finished $[[_{VP}\ [_{V0}\ e][_{PP}$ with my glasses$]]]$

What is important about this proposal is that the gap (*e*) might serve as the trigger for the inferences that would ultimately enrich (3a). It may be that the proposition (3a) expresses, then, allows for the gap that we suggest occurs in VP of a verb such as *begin* (but see note 9). The key point we want to register is that whichever form this representation takes, (a) it does not specify *how* (3a) is enriched (i.e., it does not determine a default content for (3a)); and (b) it provides a linguistically-motived basis for enrichment without committing to a type-shifting analysis of the complement. In both cases, the syntactic gap analysis provides a linguistically-motivated source for ulterior enrichment thus avoiding the problems that afflict the different views of coercion.

We now turn to experimental evidence, first in support of coercion views, and then, against coercion.

Experimental work supporting coercion is slim. Earlier studies (e.g., McElree, Traxler, Pickering, Seely, & Jackendoff, 2001) have shown that (8a) takes longer to process at post-verbal positions when compared to (8b). This extra time was assumed to be due to the process of semantic interpolation.

(8) a. The secretary began the memo before it was due
b. The secretary typed the memo before it was due

Obtaining longer reading times at post-verbal positions need not constitute support for "interpolation." Alleged indeterminate sentences such as (3a) or (8a) differ syntactically from fully determined ones such as (8b) (see de Almeida & Dwivedi, 2008, for linguistic analyses). Thus, longer RTs could be due to structural differences between them. Besides, results obtained by McElree et al. (2001) could not be replicated by de Almeida (2004, Experiment 1), employing a similar experimental paradigm and conditions. And while Pickering et al. (2005) have attempted to replicate McElree et al.'s results, most effects were statistically weak or reflected relatively late processes (e.g., second-pass reading), compatible with post-parsing enrichment. Replicability is of the essence for establishing a given phenomenon. But even if those results were to be consistently replicated, they only suggest there are differences between sentence types, without exactly accounting for what yields those differences; more specifically, they cannot be taken to support "interpolation" or "type-shifting" forms of coercion directly.

Similarly, experiments involving ERPs (event-related potentials) have shown processing differences between sentences such as (8a) and (8b), but without determining how these sentences differ (see, e.g., Kuperberg et al., 2010; Baggio et al., 2010). MEG (magnetoencephalography) experiments have also suggested that processing sentences such as (8a) and (8b) yield different magnetic patterns, but they too have not accounted for the source of the difference.

Most studies that claim support for coercion have in fact served two purposes: either they have been specifically designed to show that these sentences behave differently (thus, supposedly supporting some form of coercion), or they have focused on determining the anatomical source of the difference, on the assumption that coercion is necessarily at play. On both accounts, they are compatible with the view that indeterminacy is attributable to sentential structural properties. At the very least, they have shown that differences in processing are manifestations of structural differences. At most, they have shown that these sentences call for different enrichment processes, coercion or something else. More directly related to our concerns, a view that takes indeterminate sentences to be initially analyzed based on their constituents and syntactic form—thus, initially without enrichment—stands as the default.

Experiments using brain imaging—in particular, functional magnetic resonance imaging (fMRI)—can be further illuminating with regards to the source of enrichment, on the assumption that different anatomical sites might be engaged in processing different kinds of stimuli. There are, however, a few caveats regarding the use of fMRI to determine the nature of linguistic and nonlinguistic processes involved in indeterminacy resolution. First, accounting for differences between sentence types in terms of anatomical resources involved requires having a clear understanding of "where" or even "how" different kinds of syntactic, semantic, and pragmatic processes take place in the cortex or even in subcortical areas of the brain. Lacking such a clear understanding of the mapping of language (and post-linguistic processes) leaves us with a fair amount of speculation. This is akin to finding reading-time, ERP, or MEG differences without knowing the source of these differences. Second, even if we were to have a firm foundation upon which to build our neuroanatomical hypotheses, it is quite possible that similar networks might be deployed to achieve functionally different ends. While this is certainly a strong argument against a strict physicalist explanation, it is also a call for keeping the spotlight on the very theories that underlie the anatomical predictions. And third, there are numerous constraints on the analysis of fMRI data, which relies for the most part on set parameters of what is to be considered "activated" in the course of processing a given stimulus. At the voxel (unit of activation) level, this means determining a significance parameter; at the cluster level this means determining a particular number of contiguous voxels (the "regions") while leaving lower quantities at bay (the heap paradox comes to mind: why 100 voxels and not 99?); and, overall, establishing activation levels often requires leaving unreported networks that do not reach a given threshold but which, nonetheless might be engaged in processing the stimuli. Despite these general constraints on the use and interpretation of fMRI data, this technique can be used to complement both linguistic analyses of indeterminate sentences as well as studies employing behavioral and electrophysiological techniques.

Thus far, the main anatomical sites involved in the resolution of alleged indeterminacy (or in the *attempt* to resolve it) have been elusive. MEG studies (e.g., Pylkkanen & McElree, 2007) have suggested that the main area involved in interpreting sentences such as (8a) compared to (8b) is the ventro-medial prefrontal cortex (vmPFC). This area was activated following an initial bilateral temporal activation, though the estimate that the vmPFC is the main "site" of coercion is, at this juncture, highly speculative given the involvement of other areas. Also, the advantage that MEG has over fMRI in terms of temporal resolution, it lacks in spatial resolution. Employing fMRI, Husband et al. (2011) found no evidence of vmPFC activation but greater activation at the left inferior frontal gyrus (IFG), suggesting that this region "supports the application of coercion processes" and "the detection of semantic type mismatch between the verb and its complement" (pp. 3260–3261). While these results are consistent with the idea that indeterminacy might involve a structural-gap detection, the claims that Husband et al. make go far beyond that. For them, activation of the IFG suggests "the mismatch and its repair only affect semantic composition and do not recruit other processes for repair or rejection" (p. 3262). Their idea is that semantic composition incorporates mechanisms of detection of anomaly and repair, though it is not clear on what grounds semantic anomaly is detected, or how repair is obtained. The only way to assume this is happening is to assume that—as we discussed above—the semantic composition system is informed about world contingencies. This seems to be the position they take: "Assuming event meanings for nouns are also stored in the lexicon (Pustejovsky, 1995), IFG may function to select and retrieve the noun's event-related meaning" (Husband et al, 2011, p. 3262). But of course, this cannot be achieved unless there is an account of the analytic/synthetic distinction for lexical-semantic representations—which, as far as we know, nobody has.

The first caveat above—regarding the lack of clear neuroanatomical parameters for linguistic and post-linguistic processes—requires us to investigate phenomena that are poorly understood by taking a broad stance. The most parsimonious approach is to map out the process, typically reporting its neuronal correlates, by contrasting several variables. For instance, contrasting sentences such as those in (9), representing a wide spectrum of normal and abnormal constructions allow us to dissociate indeterminate sentences such as (9a) from sentence types such as those that are determinate (9b) or semantically/pragmatically anomalous (9c), or even syntactically anomalous (9d). Underlying this approach is the assumption that differences and similarities in terms of regions, activation levels, or even number of activated voxels obtained between these sentences are indicative of the nature of the resources involved in the processes of parsing and interpretation.

(9) a. The author started the book.
b. The author wrote the book.
c. The author drank the book.
d. The author yawned the book.

In the fMRI study conducted by de Almeida, Riven, Manouilidou, Lungu, Dwivedi, Jarema, and Gillon (2016), employing sentences such as those in (9), the neuronal correlates of indeterminacy resolution were found to be somewhat different from those in previous studies.[11] Indeterminate sentences such as (9a) were found to activate a wide network, in particular, the left and right IFG, both

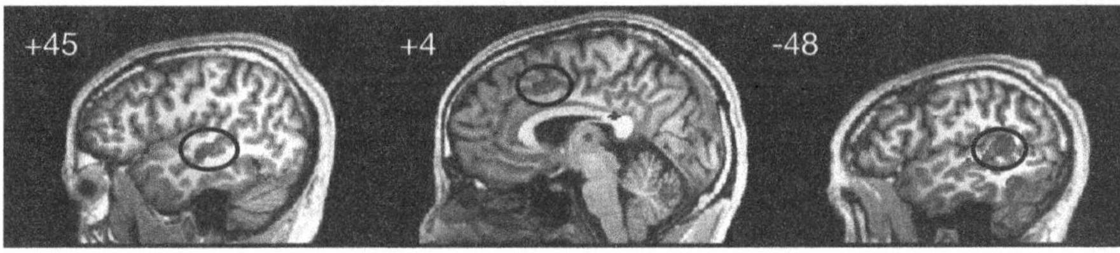

Figure 5.1 Partial results from de Almeida et al.'s (2016) fMRI study. Areas within ellipsis represent some the main regions activated in the contrast between "indeterminate" (such as (9a)) and "determinate" (9b) sentence types. Activation maps represent (a) the right hemisphere, superior temporal gyrus (Talairach +45), (b) medial right hemisphere (+4), with activation of the anterior cingulate cortex (ACC), and (c) the left hemisphere, superior temporal gyrus (–48) regions. For color figures and more details, see http://journal.frontiersin.org/article/10.3389/fnhum.2016.00614/full.

temporal lobes and the anterior cingulate cortex (ACC). While other sentences also showed above set threshold activation in so-called "language areas" (left superior temporal lobe and L-IFG), indeterminate sentences surpassed other sentences in all those regions. Figure 5.1 shows data for the contrast between indeterminate and determinate (control) sentences—(9a) and (9b), respectively.

In addition, as Figure 5.2 shows, the number of voxels activated for indeterminate sentences by far surpasses those activated for other sentences in (9)—even in cases of blatant semantic and syntactic violations, such as in (9a) and (9b).

While these data do not completely rule out coercion, they point to a different perspective, one compatible with the one we proposed: greater activation beyond traditional linguistic areas for indeterminate sentences allied to overall greater

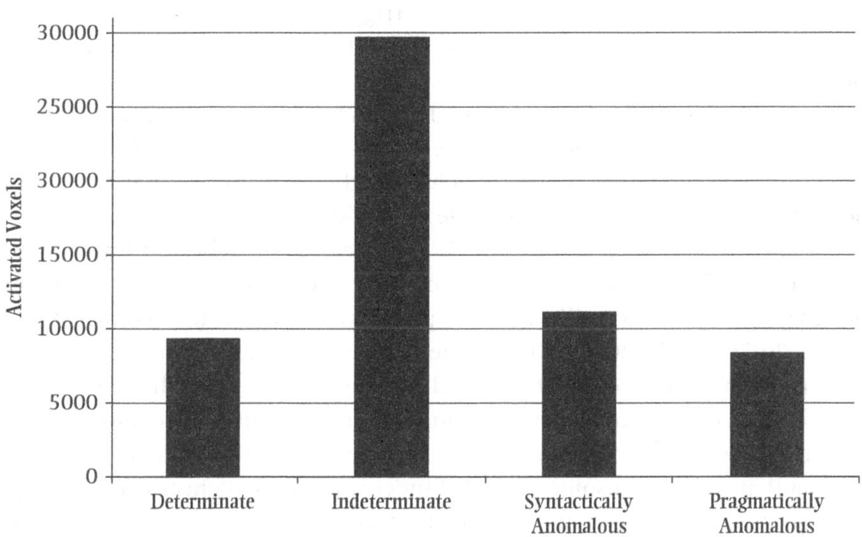

Figure 5.2 Number of whole-brain significantly activated voxels for sentences in (8). From de Almeida et al. (2016).

number of activated voxels bilaterally suggest that indeterminate sentences trigger a search for a resolution, consistent with a state of uncertainty—more so than with a default intra-linguistic semantic coercion.

MODULARITY AND CONTEXT SENSITIVITY

So-called indeterminate sentences are supposed to constitute a challenge to modularity: if they are resolved during initial parsing, they ought to be resolved based on knowledge that traditionally lies outside the module. The lack of an analytic/synthetic distinction defers the resolution of indeterminacy to post-parsing mechanisms of interpretation. We have assumed that the output of the module is something akin to a proposition, but one *un*enriched by local lexical-semantic processes. In our proposal, however, syntactic (and discourse) triggers work to signal higher interpretive processes where enrichment might be due. And if the syntactic analysis presented in section (5) holds, the trigger is within the VP. The widespread activations that indeterminate sentences cause suggest that there is at least an attempt to resolve indeterminacy.

In principle, cases such as *A man began a book* appear to be well resolved or enriched because they come embedded in utterance contexts. Nobody relatively sane addresses you with this sentence without having first established a frame of reference or presuppositions, outside of a common ground. Few experimental studies have attempted to manipulate the role of context in processing indeterminate sentences (de Almeida, 2004; Traxler et al., 2005), and the results have been inconsistent. We have assumed that so-called indeterminate sentences are indeterminate only in isolation, that no enrichment takes place by default, not at least by coercion. But we have also speculated that these sentences harbor a syntactic position that might serve as a "trigger" for processes of enrichment down the stream. We have also suggested that there are ruled governed (that is, conventionally determined) resolutions for some (much?) of what goes under the general rubric of "indeterminacy," for example, as in the cases of pronoun resolution in a discourse, as in (2) above. This is entirely consonant with Fodor's (1990) revised modularity model, which takes the scope of modularity to be not the sentence but, more broadly, what he called "discourse model."

This discourse model, to be clear, is also local, for it relies on linguistically determined links among sentence and clauses, and various discourse elements, such as pronouns, tenses, elliptical verb phrases and the like (cf., Stojnic, Stone & Lepore, 2017; Stojnic, 2016a, 2016b). The very use of indefinite article is taken to presuppose the introduction of a novel discourse referent. By calling for "a man," "a book," etc., one grounds their interpretation to elements not yet established in the prior discourse. Perhaps, more directly related to our immediate concerns is the idea that the VPs of some so-called indeterminate sentences carry a trigger, as in (6), above. We take it that the role of this trigger, in the absence of a supporting context, is to generate inferences—some abductive—that will attempt to put an end to any appearance of indeterminacy. But uttering such sentences within a discourse, allows the trigger to operate locally, picking out elements that have been either clearly established or hinted, and thus part of the propositions that the preceding context generates.

Effects of a preceding discourse on the processing of an alleged indeterminate sentence have been investigated in only two studies, with results somewhat consistent. De Almeida (2004, Experiment 2) found that a "context" such as (10a) facilitated the processing of a sentence such as (10b)—a contextually preferred sentence (following norms)—compared to less appropriate sentences (10c) and (10d) which took equally longer to process at the complement NP (*the memo*) than (10a). While this does not constitute facilitation of (10c), the relevant findings here are that (i) both (10c) and (10d) were less contextually appropriate and (ii) there was no extra cost associated with indeterminacy when context provided a relevant (local) filler for the indeterminate sentence (say, *working* on the memo).

(10) a. The secretary would always be sure to work ahead of schedule. She was asked to work on a memo.
b. The secretary typed the memo long before it was due.
c. The secretary began the memo long before it was due.
d. The secretary read the memo long before it was due.

An eye-tracking study by Traxler et al. (2005) was closer to obtaining a real facilitation effect of an indeterminate sentence by its local discourse. It is perhaps in their Experiment 3 where we can find clearer results.[12] They presented "contexts" such as (11a) or (11b), which were followed by "target" sentences such as (12a) or (12b) in a factorial design.

(11) a. The student read a book in his dorm room.
b. The student started a book in his dorm room.

(12) a. Before he read the book about the opium trade, he checked his email.
b. Before he started the book about the opium trade, he checked his email.

While they found differences in reading times between "context" sentences (11), they found no differences between "target" sentences (12). We have seen that indeterminate sentences in isolation can produce longer reading times—though not consistently so. That's the case of their "context" sentences, which precede their targets. Also, we have seen that the cost associated with indeterminate sentences compared to controls can be accounted for by differences in syntactic structure, as in (6). Thus, here again, coercion cannot be the only explanation. More importantly, the null effects they obtained in the "target" sentences in (12) can be seen as an effect of attenuation of target by context. First, it is expected that (11a) primes (12a) by virtue of repetition of the VP *read a/the book*. The same can be said of the pair (11b) and (12b). When "context" and "target" types are crossed, however, having "read a book" in the context, as in (11a), does not speed up reading of the indeterminate (12b) any more than having "begin a book" facilitates "begin the book."

To summarize, processing an indeterminate sentence in a biasing discourse—one that provides a potential filler event—facilitates resolution. Key here is that sentences—or propositions they yield—are sensitive to information within the "discourse model." That does not constitute a violation of modularity, for the information that so-called indeterminate sentences seek are within the local context and do not depend on analytic lexical decompositions. Put somewhat differently, they do not violate modularity because the resolutions are entirely predictable (because these discourse resolutions are entirely conventionally (linguistically) governed).

It is important to highlight that what we are here calling enrichment goes beyond the local discourse in the relevant sense. Sentences in discourses that dictate resolutions are not enriched. Rather, the effects of prior discourse in the enrichment of indeterminate sentences unfolds across the discourse (in a rule governed fashion). A study from Riven and de Almeida (2017) might be taken to support this view. Participants heard biasing contexts such as (13a) and, either immediately after the critical clause *Lisa began the book*, or 25 seconds after it (with intervening neutral discourse), they were presented visually with one of the probe sentences (13b)–(13d). Participants were asked to press a button indicating whether probe sentences were identical to segments they heard.

(13) a. Lisa had been looking forward to the new Grisham novel ever since it came out. She had finally managed to set aside some time this weekend and made sure to make her home library nice and cozy. First thing Saturday morning, Lisa curled up on the sofa in her library with a blanket and a fresh cup of coffee. With everything in place, *Lisa began the book.* [Immediate probe point; discourse continues for 25 seconds]

b. Lisa began the book (identical/indeterminate)

c. Lisa began reading the book (biased foil)

d. Lisa began writing the book (non-biased foil)

This procedure is similar to one employed by Sachs (1967, 1974) showing the effect of propositional (or "gist") encoding of sentences in memory, with quick loss of verbatim representation. Crucial in our manipulation, however, was the effect that the context would have on participants' acceptance of the biased foil as if it was the original sentence. Here, contrary to previous studies, there is nothing in the context providing a clear event for enriching the indeterminate sentence other than suggestions that Lisa was about to *read* a book. In the case of (13a), the context is much closer to "hinting" about what is happening than providing a filler event. As can be seen in Figure 5.3, results show a clear effect of enrichment of the indeterminate sentence over time.

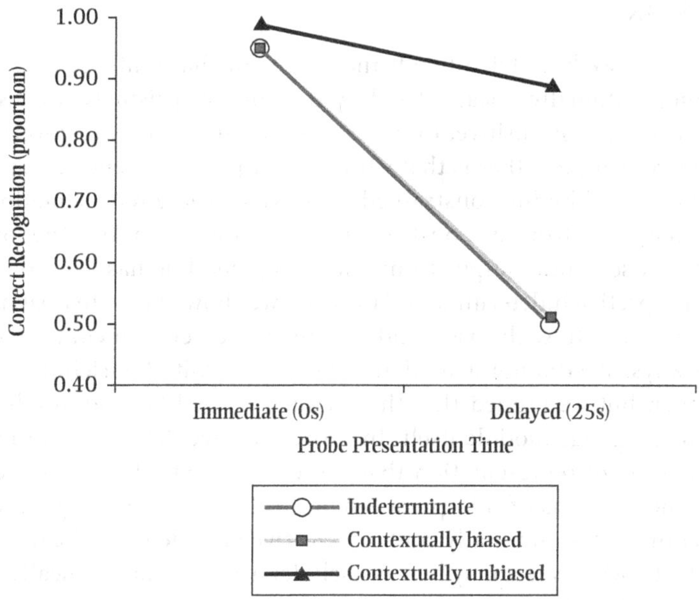

Figure 5.3 Proportion of correct recognition for sentences (13b)–(13d) following contexts such as (13a) at two probe points: immediate and delayed by 25s, with intervening neutral discourse. From Riven & de Almeida (2017).

The biased foil (13c) is accepted as much as the original sentence heard in context. Confidence ratings collected after each trial, in fact, show that subjects are more confident that (13c) is the sentence they heard than they are of the true stimulus (13b). But these effects only obtain at the later probe point, not at the early one.

Overall, the results suggest that a sufficiently rich context might create a false memory—an effect of enrichment of the proposition—that is not driven by the local "discourse model" but one that comes from what Fodor referred to as "real world model." The line between the two, as we suggested, is thin, but one that makes a crucial distinction between encapsulated and unencapsulated processes: the former relies on linguistically determined enrichment, the latter not. While the discourse model provides a local source for antecedents of determinate noun phrases, pronouns, and the like, the "real world" hints on what is possibly the best way for enriching a sentence but without providing the actual information.

To summarize these results: the suggestion is that operations between sentence-level representations and local context are obtained within module, that is *local* enrichment is modular, for it is driven by linguistic processes. It is in this sense that sentences can be said to be mildly context-sensitive. In particular, local context or *co-text* provides the fillers that linguistic elements (syntactic gaps, pronouns, etc.) call for.

CONCLUSION

In this chapter, we have tried to advance the view that sentence perception is largely computationally encapsulated; and, more surprisingly, that sentence *meaning* is context insensitive, or at least its sensitivity is rule-governed. The way these two work together is that while the output of sentence encapsulation is a minimally and highly constrained, context-sensitive representation of the sentence composed from its constituents, it remains semantic. The long-term challenge to a semantic output from a language module has been the alleged cases of interpretive indeterminacy. However, we showed how to assign semantic representations to such cases, and that they interact with context in highly regulated ways. We did not deny that such cases admit of enrichment of some sort or other, but we argued that these issues go well beyond anything concerning the language module itself. In short, we have defended a proposal for a semantic level of representation that serves as output of the module and as input to other systems of interpretation, arguing for a form of modularity or encapsulation that is minimally context sensitive provided that the information from context—whatever it may be—is itself determined nomologically, namely, by linguistic principles.

AUTHORS' NOTE

Research for this article was supported by a grant from the Natural Sciences and Engineering Research Council of Canada (NSERC). We are thankful to Caitlyn Antal for comments on an earlier draft.

NOTES

1. We endorse the view that representations and processes are autonomous *qua an explanatory level*, but also that a full account of a cognitive system cannot dispense with a proper characterization of the biological implementation of its functions. See, e.g., the tri-level hypothesis of Pylyshyn (1984); and Gallistel (2018, this volume) for specific implementation proposals.
2. In common usage, *algorithms* are used for computational procedures that are guaranteed to produce a certain outcome, while *heuristic* rules are incomplete computational procedures. As noted by several authors (e.g., Pylyshyn, 1984; Haugeland, 1981), all computations involve algorithms and the distinction amounts to the end result—whether or not it is guaranteed to produce a given outcome. We will assume, in the present discussion, that algorithms—say, semantic ones—computed by the module are deterministic, while procedures on modular *outputs* (e.g., computing something akin to implicatures) are generally heuristic. These two options, of course, do not exhaust the range of possible cognitive mechanisms—including the possibility that some mechanisms might be entirely contingent on the individual's belief states.
3. By contrast one could postulate a connectionist type of system, with representations being nodes in a network and processes being activation patterns over those

nodes. Such a system could in principle be "modular" as long as the operations it performs are *encapsulated*, that is, are not subject to influences from other systems, and, in particular, from background knowledge (see below). In this case, the nodes at the "encapsulated" part of the system would have to be severed from feedback from higher-up nodes, in particular, those at the hypothetically *un*encapsulated part of the network. But this contrasts sharply with what connectionist networks stand for: that patterns of activation at lower levels are in large part constrained by patterns of activation at higher levels. Also, even if this could be fixed, the system would not operate *algorithmically,* nor would it be *compositional,* thus, lacking key architectural features to which our proposal adheres.

4. This certainly does not entail that modules (or "input analyzers," as Fodor reluctantly calls them) are the only systems to operate algorithmically, but they are the ones that compute algorithms on *post-transduced* symbols and, so, autonomously. Moreover, this does not entail that modules operate only algorithmically. It is possible to conceive of modular operations that are heuristic, as long as the choices it makes in the course of its computations are internal to the module; that is, encapsulated from general knowledge.
5. We take Pylyshyn (1984) to be rather clear about this: "formal symbolic structures mirror all relevant semantic distinctions to which the system is supposed to respond" (p. 74). See Pylyshyn (1984) for extensive discussion on symbols and their interpretation.
6. See, e.g., de Almeida and Manouilidou (2015) for a review on verb argument structure and on the content of arguments.
7. One reason Fodor keeps semantics out of the module is that semantic descriptions often appeal to lexical decomposition; and semantic theories that do so patently have to rely on an analytic/synthetic distinction. Fodor's rejection of this distinction implies that the module is open to all the possible beliefs the speaker/hearer might have encoded, which, of course, is exactly what modularity denies. But as we will show, there is a sense in which semantic representation need not invoke semantic decomposition, and thus, can constitute the level of representation the module outputs.
8. The final "stage" in this model, *the real world model*, is distinguished from *the discourse model* on the assumption that one ought to construct a representation of the (linguistic) discourse before checking it against the hearer's knowledge or an "aggregate picture representation of how things are" (p. 9). The discourse model is the wider-scope linguistic representation of the sentence, which prevails even when it conflicts with real-world knowledge. We assume that information contained within the discourse model can be conceived as being intra-modular while the real world model cannot.
9. Notice that on one analysis (Davidson, 1967) of (3a), the verb *begin* introduces a variable—say, *w*—which in (3a) ranges over not the action/event *x* began doing with *y*, but *begin* itself, thus yielding something like $\exists w$ (*begin* (x, y, *w*)).
10. A view similar to this one has been proposed by Parsons (1990; see also Pietroski, 2015).
11. It is important to note that de Almeida et al. (2016) employed a different method, materials, and analyses, thus it did not constitute an attempt to replicate Husband et al. (2011) also because the data collection predates the publication of this latter study.

12. Traxler et al.'s (2005) findings, however, are difficult to interpret given the inconsistent results between and within experiments—both in terms of region, where effects are found and in terms of eye-tracking measures that yield the effects. Moreover, many of their statistical analyses—including some that are taken to support their views—are "tendencies," not statistically significant results. And although their results are offered in support to coercion, they can also be claimed to support the perspective we take.

REFERENCES

Altmann, G., & Steedman, M. (1988). Interaction with context during human sentence processing. *Cognition, 30*(3), 191–238.

Baggio, G., Choma, T., Van Lambalgen, M., & Hagoort, P. (2010). Coercion and compositionality. *Journal of Cognitive Neuroscience, 22*(9), 2131–2140.

Bornkessel, I., & Schlesewsky, M. (2006). The extended argument dependency model: A neurocognitive approach to sentence comprehension across languages. *Psychological Review, 113*(4), 787–821.

Cappelen, H., & Lepore, E. (2005). *Insensitive semantics: A defense of semantic minimalism and speech act pluralism*. Oxford, England: Wiley-Blackwell.

Crain, S., & Steedman, M. (1985). On not being led up the garden path: The use of context by the psychological parser. In D. R. Dowty, L. Karttunen, & M. Zwicky (Eds.), *Natural language parsing* (pp.320–358). Cambridge, England: Cambridge University Press.

Davidson, D. (1967). The logical form of action sentences. In D. Davidson (Ed.), *Essays on actions and events* (pp. 105–122). Oxford, England: Oxford University Press.

de Almeida, R. G. (2004). The effect of context on the processing of type-shifting verbs. *Brain and Language, 90*(1-3), 249–261.

de Almeida, R. G., & Dwivedi, V. (2008). Coercion without lexical decomposition: Type-shifting effects revisited. *Canadian Journal of Linguistics, 53*(2), 301–326.

de Almeida, R. G., & Manouilidou, C. (2015). The study of verbs in cognitive science. In R. G. de Almeida & C. Manouilidou (Eds.), *Cognitive science perspectives on verb representation and processing* (pp. 3–39). New York, NY: Springer.

de Almeida, R. G., & Riven, L. (2012). Indeterminacy and coercion effects: Minimal representations with pragmatic enrichment. In A. M. Di Sciullo (Ed.), *Towards a biolinguistic understanding of grammar: Essays on interfaces* (pp. 277–302). Amsterdam, Netherlands: John Benjamins.

de Almeida, R. G., Riven, L., Manouilidou, C., Lungu, O., Dwivedi, V. D., Jarema, G., & Gillon, B. A. (2016). The neuronal correlated of indeterminate sentence in interpretation: An fMRI study. *Frontiers in Human Neuroscience, 10:614.* doi: 10.3389/fnhum.2016.00614.

Fodor, J. A. (1983). *The modularity of mind*. Cambridge, MA: MIT Press.

Fodor, J. A. (1990). *On the modularity of parsing: A review.* Unpublished manuscript, Rutgers University.

Fodor, J. A., & Lepore, E. (2002). The emptiness of the lexicon: Reflections on Pustejovsky. In J. A. Fodor & E. Lepore (Eds.), *The compositionality papers* (pp. 89–119). Oxford, England: Oxford University Press.

Gallistel, C. R. (2018). The neurobiological bases for the computational theory of mind. In R. G. de Almeida & L. Gleitman (Eds.), *On Concepts, Modules, and Language*. Oxford, England: Oxford University Press.

Gallistel, C. R., & King, A. P. (2010). *Memory and the computational brain: Why cognitive science will transform neuroscience*. New York, NY: Wiley-Blackwell.

Haugeland, J. (1981). Semantic engines: Introduction to mind design. In J. Haugeland (Ed.), *Mind design: Philosophy, psychology, artificial intelligence* (pp. 34–50). Cambridge, MA: MIT Press.

Husband, E. M., Kelly, L. A., & Zhu, D. C. (2011). Using complement coercion to understand the neural basis of semantic composition: Evidence from an fMRI study. *Journal of Cognitive Neuroscience, 23*(11), 3254–3266.

Jackendoff, R. (2002). *Foundations of language: Brain, meaning, grammar, evolution*. Oxford, England: Oxford University Press.

Kuperberg, G. R., Choi, A., Cohn, N., Paczynski, M., & Jackendoff, R. (2010). Electrophysiological correlated of complement coercion. *Journal of Cognitive Neuroscience, 22*(12), 2685–2701.

Manouilidou, C., & de Almeida, R. G. (2009). Canonicity in argument realization and verb semantic deficits in Alzheimer's disease. In S. Featherston & S. Winkler (Eds.), *The fruits of empirical linguistics* (pp. 123–149). Berlin: Mouton de Gruyter.

Manouilidou, C., de Almeida, R. G., Schwartz, G., & Nair, N. V. (2009). Thematic roles in Alzheimer's disease: Hierarchy violations in psychological predicates. *Journal of Neurolinguistics, 22*(2), 167–186.

McElree, B., Traxler, M., Pickering, M., Seely, R., & Jackendoff, R. (2001). Reading time evidence for enriched composition. *Cognition, 78*(1), B17–B25.

Parsons, T. (1990). *Events in the semantics of English: A study in subatomic semantics*. Cambridge, MA: MIT Press.

Pickering, M. J., McElree, B., & Traxler, M. J. (2005). The difficulty of coercion: A response to de Almeida. *Brain and Language, 93*(1), 1–9.

Pietroski, P. M. (2015). Lexicalizing and combining. In R. G. de Almeida & C. Manouilidou (Eds.), *Cognitive science perspectives on verb representation and processing* (pp. 43–66). New York, NY: Springer.

Piñango, M. (2006). Thematic roles as event structure relations. In I. Bornkessel, M. Schlesewsky, & A. Friederici (Eds.), *Semantic Role Universals and Argument Linking: Theoretical, Typological, and Psycholinguistic Perspectives*. Berlin, Germany: Mouton.

Pustejovsky, J. (1995). *The generative lexicon*. Cambridge, MA: MIT Press.

Pylkkänen, L. (2008). Mismatching meanings in brain and behavior. *Language and Linguistics Compass, 2*(4), 712–738.

Pylkkänen, L., & McElree, B. (2007). An MEG study of silent meaning. *Journal of Cognitive Neuroscience, 19*(11), 1905–1921.

Pylyshyn, Z. W. (1984). *Computation and cognition: Toward a foundation for cognitive science*. Cambridge, MA: MIT Press.

Riven, L., & de Almeida, R. G. (2017). *Context breeds enriched interpretations of indeterminate sentences*. Manuscript submitted for publication.

Searle, J. R. (1980). Minds, brains, and programs. *Behavioral and Brain Sciences, 3*(03), 417–424.

Stojnić, U. (2016a). One's modus ponens: Modality, coherence and logic, *Philosophy and Phenomenological Research*, doi:10.1111/phpr.12307.

Stojnić, U. (2016b). *Context-sensitivity in a coherent discourse*. Unpublished PhD Dissertation. Rutgers University, Piscataway, NJ.

Stojnić, U., Stone, M., & Lepore, E. (2014). *Discourse and Logical Form: Pronouns, Attention and Coherence*. Linguistics and Philosophy, doi:10.1007/s10988-017-9207-x.

Thompson, C. K., & Lee, M. (2009). Psych verb production and comprehension in agrammatic Broca's aphasia. *Journal of Neurolinguistics*, *22*, 354–369.

Townsend, D. J., & Bever, T. G. (2001). *Sentence comprehension*. Cambridge, MA: MIT Press.

Traxler, M. J., McElree, B., Williams, R. S., & Pickering, M. J. (2005). Context effects in coercion: Evidence from eye movements. *Journal of Memory and Language*, *53*(1), 1–25.

6

Center-Embedded Sentences

What's Pronounceable Is Comprehensible

JANET DEAN FODOR, STEFANIE NICKELS, AND ESTHER SCHOTT

Doubly center-embedded relative clause constructions (henceforth 2CE-RC), with the structure shown in (1), are notoriously difficult to process. This is so for classic examples as in (2), whose difficulty seems disproportionate to their brevity, and equally for longer examples such as (3), tested in a much-cited experiment by Gibson & Thomas (1999).

(1) [NP1 [NP2 [NP3 VP1] VP2] VP3]

(2) The girl the man the cat scratched kicked died.
The rat the cat the dog chased killed ate the malt. (from Chomsky & Miller, 1963, p. 286)

(3) The ancient manuscript that the graduate student who the new card catalog had confused a great deal was studying in the library was missing a page.

Fodor, Bever and Garrett (1974) made a virtue of this unwieldy construction, by using 2CE-RC sentences as their experimental materials in a number of studies of how the parsing mechanism extracts cues from surface sentences in order to establish their deep structure. Their foundational work in experimental psycholinguistics was achieved at a time when tools for stimulus presentation and response measurement were primitive: DMDX didn't yet exist; event-related potentials (ERPs) hadn't even been dreamed of; some responses were timed with stop-watches. Making fine distinctions of syntactic processing difficulty with the blunt instruments to hand could be tricky and frustrating. But by working with a sentence type so difficult that comprehension often failed, Fodor, Bever and

Garrett were able to expand the scale of response measures so that performance differences of interest could be observed.[1]

The 2CE-RC construction has three well-established peculiarities.

1. First is its unusually difficult comprehension. Such sentences have been deemed incomprehensible, unacceptable, even ungrammatical. Intuitively, the increment of processing cost due to embedding one object-gap RC inside another one is much greater than the cost of embedding the same RC inside a main clause.
2. Second is an observation by Bever (1988), who first noted the ameliorating effect of using a pronoun as NP3, as in *The girl the man I scratched kicked died.*
3. Third is that a 2CE-RC sentence may be perceived, wrongly, as equally or more grammatical if VP2 is omitted, as in *The girl the man the cat scratched died*, which may be judged acceptable. This is the "missing-VP illusion." (References to experimental data in several languages are in Experiment 2.)

Many explanations have been offered over the years since Miller & Chomsky (1963) first drew attention to this recalcitrant construction type. We summarize a handful of them in Box 6.1.

Along with these different accounts of the source of the difficulty, there are corresponding proposals about how the difficulty can be minimized, thus acknowledging the considerable range of variation in the severity of the problem that is observed across examples. Hudson (1996) ran a series of informal experiments in which students had to recall a spoken sentence; he reports error rates for 2CE-RC constructions ranging widely, from 7% for sentence (4) to 81% for sentence (5), though matched for number (if not length or frequency) of words.

(4) The shells that the children we watched collected were piled up.
(5) People that politicians who journalists interview criticize can't defend themselves well.

We will argue for a significant role of prosodic phrasing in creating the difficulty of the 2CE-RC construction, and correspondingly a role for prosodic phrasing in facilitating its processing. Specifically, we propose that there is an alignment problem at the syntax-prosody interface, consisting of a mismatch between the heavily nested syntactic structure and the flat structure required by prosodic phrasing.[2] We predict as a corollary that if the prosody can be made natural, the syntax will be computable without the usual extreme difficulty. Of course such sentences will never be very easy to parse and comprehend. They contain two relative clauses, each of which modifies a subject and contains an object "gap," properties well-known to increase processing difficulty; prosody cannot eliminate these complexities. But our data suggest that the difficulty of double center-embedding per se *can* be tamed by cooperative prosody. We

Box 6.1

A Sample of Proposed Explanations for the Processing Difficulty of 2CE-RC Sentences

- The parser cannot recursively call the same sub-routine (Miller & Isard 1964).
- A three-NP sequence with no relative pronouns is misparsed as coordination (Blumenthal 1966).
- Exponential increase in number of potential grammatical relationships (Fodor & Garrett 1967).[a]
- The parser cannot assign both subject and object roles to NP2 (Bever 1970).
- The Sausage Machine parser can't correctly "chunk" the word string (Frazier & Fodor 1978).
- "Disappearing" syntactic nodes in complex tree structures (Frazier 1985).
- Syntactic prediction locality theory (SPLT, Gibson & Thomas 1999).

[a] "Given one embedding, two nouns must be assigned to each of two verbs as subject and two nouns must be assigned to each of two verbs as object. Hence, we have four possible analyses of N1 N2 VI V2 into NVO assuming no noun is both subject and ob[ject] of the same verb. However, given two embeddings, three nouns must be assigned [to] each of three verbs as subject and three nouns must be assigned to each of three verbs as object. Still assuming no noun may be assigned as both subject and object of the same verb, we have 18 possible analyses of the double-embedded case (if the final verb is intransitive, there are two possible analyses for the single embedding and 12 for two embeddings)" (Fodor & Garrett 1967, pp. 296).

present examples showing that selective shrinking and lengthening of phrases can coax the prosodic processor into creating rhythmic packages that do fit well with the nested syntactic tree structure. Short inner phrases help with that, while short outer ones hinder. The appropriate prosody is difficult to achieve, for reasons that will be explained, and typical syntactic phrase lengths in 2CE-RC sentences do not cooperate in this regard, which may be why this prosodic phenomenon has not been widely recognized. We will show that the prosodic approach offers explanations for all three distinctive peculiarities of the 2CE-RC construction listed: the near-incomprehensibility of most standard examples; the pronoun effect; and (perhaps) the missing-VP effect.[3]

A FACILITATIVE PROSODIC PHRASING

Suppose a speaker wishes to tell a friend "The girl the man the cat scratched kicked died." The syntactic structure of this 2CE-RC sentence is sketched in Figure 6.1, with some details omitted so as to focus attention on the main configurational relations.[4]

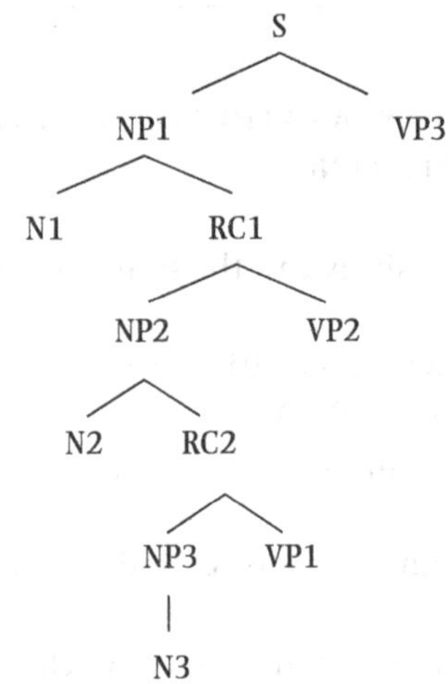

Figure 6.1 Syntactic tree structure (simplified) for the 2CE-RC construction.

How could the would-be speaker set about assigning a prosodic structure to this syntactic tree? The sentence is too long, even with these short constituents, to be expressed in a single prosodic phrase,[5] so it needs to be snipped apart at natural syntactic breaks, presumably starting with the major break between the subject and predicate of the sentence. It turns out that a critical issue is how many units to divide the structure into: 2 units or 3 or 4 or more. As often noted, 2CE-RC sentences are frequently pronounced with a "list intonation," which amounts to dividing the word string into 6 prosodic phrases, each NP and VP a unit to itself. This is not helpful; in fact it is a clear mark of failure to comprehend. Thus the challenge is posed: not dividing the word sequence prosodically is impossible, but dividing it into too many pieces obscures the syntactic structure.

An optimal division must satisfy two criteria: it must do as little damage as possible to the syntactic tree, while also satisfying prosodic constraints. Doing least damage to the syntactic tree structure means cutting the tree not arbitrarily but at natural syntactic joints. In other words, the prosodic units should be aligned with syntactic phrases, as far as is possible. However, the constraints that apply at the syntax-prosody interface are a heterogeneous set, and they include eurhythmic constraints on optimal phrase length and balance which may compete with alignment constraints (Box 6.2). These are presented in Optimality Theory as "soft" constraints, which apply except where they are out-ranked by some more prominent constraint in the language in question.

Cutting the word string at the highest syntactic level, between the matrix subject and its verb phrase, yields (6). (In all examples, || indicates a prosodic phrase boundary.)

(6) The girl the man the cat scratched kicked || died.

Box 6.2

Some Constraints on Prosodic Phrasing

A. **Relation to syntax/semantics**

Edge alignment (Align$_R$ XP): "The right edge of any XP in syntactic structure must be aligned with the right edge of a MaP in prosodic structure" (Selkirk 2000: 232)

Wrap: "Each syntactic XP must be contained in a phonological phrase" (Truckenbrodt, 1995: 10)

B. **Prosodic phrase length constraints**

Binary Minimum: "A major phrase must consist of at least two minor/accentual phrases." (Selkirk, 2000, p. 244)

Binary Maximum: "A major phrase may consist of at most two minor/accentual phrases." (Selkirk, 2000, p. 244)

Uniformity: "A string is ideally parsed into same length units." (Ghini, 1993, p. 56; see also the Balance principle of Gee & Grosjean, 1983)

A note on reading the examples: It is most illuminating to read them aloud, or at least to sound them out in one's head. They should be pronounced with a prosodic break everywhere where shown by || and nowhere else.

Although it fits the syntactic structure, prosodic phrasing (6) flagrantly violates the Uniformity/Balance principle. There are 9 words, divided into 8 for the first prosodic phrase and 1 for the second prosodic phrase. Counting stressed syllables is more appropriate for (English) prosody than counting words, but still there is an imbalance of 5 + 1.

For the 2-phrase prosody to be successful, it needs the encouragement of balanced phrase lengths, as in (7).

(7) The girl the man I love met || died of cholera in 1960.
Balanced aligned prosody: 7 + 5 words; 4 + 4 stressed syllables

Although this example is longer than (6), remarkably the 2CE-RC construction now sounds very much like a normal sentence.

However, a prosodically balanced example like (7) is rare. The sentence has both RCs within the prosodic phrase that encompasses its matrix subject NP, which is followed and balanced by a long matrix VP. Squeezing 2 RCs into the space of a single prosodic phrase is quite an art, so it is not likely to occur often in normal language use. The stressless pronoun in the inner relative clause (RC2) in (7) provides almost the only way to achieve it.[6] It allows the 7-word subject, containing 2 relative clauses, to be pronounced with only 4 stressed syllables. Otherwise there would have to be at least 5 stressed syllables in the subject, as in examples (8) and (9), and this is usually judged to be too much; it oversteps the maximum length limit for an (intermediate) prosodic phrase.

(8) The girl the man Jill loves met || died of cholera in 1960.
Balanced aligned prosody: 7 + 5 words; 5 + 4 stressed syllables

(9) Girls men Jill loves met || died of cholera in 1960.
Balanced aligned prosody: 5 + 5 words; 5 + 4 stressed syllables

To summarize so far: Except with a pronominal NP3, a 2-chunk prosody compatible with the syntax is hard to achieve, since an NP containing two RCs is not usually as short as a prosodic phrase needs to be (in English). For a more stable solution, therefore, we need to snip the syntactic tree structure again, creating a 3-phrase prosody.

A cut at the next level down in the syntactic tree would be between NP1 and the RC1 that modifies it (see Figure 6.1), creating a sequence of three prosodic phrases: NP1 || RC1 || VP3. This clearly should be helpful in easing the crush in the overstuffed matrix clause subject in examples (8) and (9). However, once again the constituent lengths have to cooperate. Separating off RC1as a prosodic phrase does not by itself ameliorate syntactic processing, as can be seen in (10), where the phrase lengths are seriously imbalanced.[7]

(10) The girl || that the young man I love met in Barcelona || died.
Unbalanced aligned prosody 2 + 9 + 1 words; 1 + 6 + 1 stressed syllables

However, the same syntactic cut with cooperating phrase lengths, as in (11), does permit fairly painless processing. Note that the outer phrases (NP1 and VP3) are longer in (11) than in (10), and they balance a central RC1 which is about as short as it can be.

(11) The elegant woman || that the man I love met || moved to Barcelona.
Balanced aligned prosody: 3 + 6 + 3 words; 2 + 3 + 3 stressed syllables

The striking difference in naturalness between (10) and (11) underscores the importance of phrase lengths in making 2CE-RC constructions pronounceable. Indeed, with encouraging phrase lengths as in (12), the 3-phrase prosody works quite well even with a non-pronominal inner subject, suggesting that this prosodic pattern is indeed more stable and realistic than the 2-phrase prosody we considered earlier.

(12) The elegant woman || that the man Jill loves met || moved to Barcelona.
Balanced aligned prosody: 3 + 6 + 3 words; 2 + 4 + 3 stressed syllables

Taking stock at this point: We have found a successful recipe for creating a 2CE-RC structure that is recognizable, more or less, as a normal English sentence. The trick is to adjust the lengths of the lexical/syntactic phrases so that they are also acceptable as prosodic phrases. To the best of our knowledge this is a novel observation, though it is prefigured in large part by the Sausage Machine account of the processing difficulty of 2CE-RC sentences (Frazier & Fodor, 1978, pp.306–312); see Fodor (2013) on how the Sausage Machine's PPP (Preliminary Phrase Packager) morphed into a Prosodic Phrase Processor, as here. It is especially interesting that compatibility between syntactic phrasing and prosodic phrasing is not achieved, as might have been expected, by ensuring that all six syntactic units have the length of a typical prosodic phrase. Instead, the successful strategy packs most of the syntactic structure inside a single prosodic phrase, cramming NP2 NP3 VP1 and VP2 together without any breaks between them.

What we have arrived at so far is that 2CE-RC sentences are relatively easily parsed if their phrase lengths permit a prosodic division of the word string into weight-balanced units NP1 || RC1 || VP3, achieved by lengthening NP1 and VP3, and shortening RC1. However, there are practical limits on how short RC1 can be. In order to accommodate more typical sentences in which RC1 is more substantial than in (12), we could apply the snipping procedure once more, to break up that complex constituent. The next natural cutting point in the syntactic tree is indeed inside RC1, between its complex subject and its VP (see Figure 6.1).[8]

For example (12) as it stands, this is not a success; the resulting (13) is prosodically very unnatural. To satisfy the optimal length constraints on prosodic phrasing, we need to lengthen VP2, as in (14), to achieve prosodic balance inside RC1.

(13) The elegant woman || that the man Jill loves || met || moved to Barcelona.
Unbalanced aligned prosody: 3 + 5 + 1 + 3 words; 2 + 3 + 1 + 3 stressed syllables

(14) The elegant woman || that the man Jill loves || met on a cruise ship || moved to Barcelona.
Balanced aligned prosody: 3 + 5 + 5 + 3; 2 + 3 + 3 + 3 stressed syllables

However, though intended to appease the prosodic processor, this extra cut, dividing the sentence into a sequence of four balanced prosodic phrases, is not obviously an improvement for the syntactic processor. According to our intuitions and those of other English speakers we have consulted, sentence (14) feels as if it is beginning to break up into a list-like structure, reminiscent of the familiar unhelpful 6-phrase pronunciation of (2) and (3). Thus the additional prosodic break in (14), though it would have been expected to contribute by relieving the crush inside RC1, seems to be a move in the wrong direction from the perspective of syntactic processing. Dividing the word string at its joints is good but this division goes a step too far.[9] Therefore the 3-phrase prosody NP1 || RC1 || VP3 may be the best truce between syntax and prosody that can be achieved. Our goal is to understand why this is so. But at least, the fact that this prosody imposes such stringent constraints on phrase lengths does explain why it is so rarely encountered.

Box 6.3

Summary of Intuitive Judgments of Processing Difficulty in Relation to Prosodic Phrasing

Division of 2CE-RC sentence structure into 2 syntactically aligned prosodic phrases (NP1 NP2 NP3 VP1 VP2 || VP3) is very difficult to achieve, but when phrase lengths permit it, it is helpful for comprehension.

Division of the sentence structure into 3 syntactically aligned prosodic phrases (NP1 || NP2 NP3 VP1 VP2 || VP3) is difficult but *can* be achieved if the inner constituents are short and the outer ones are long. It greatly facilitates parsing and comprehension.

Division into 4 syntactically aligned prosodic phrases, by breaking VP2 out of the upper relative clause (NP1 || NP2 NP3 VP1 || VP2 || VP3), is less acceptable prosodically and less helpful for parsing than the 3-phrase prosody. It shares some of the unnaturalness of the common but unhelpful 6-phrase "list intonation" pronunciation (NP1 || NP2 || NP3 || VP1 || VP2 || VP3).

To summarize: We have observed here a struggle in 2CE-RC sentences between balanced prosodic weight and prosody-syntax alignment. Depending on the lexical content of a particular sentence, there may or may not be a good way of reconciling these conflicting concerns. Box 6.3 summarizes the intuitions we have already presented informally. In section "Elicited Prosody Experiments" we report two experiments which corroborate these intuitions. In section "Explanation" we offer a theoretical explanation.

ELICITED PROSODY EXPERIMENTS

We report two experiments here, each described in more detail later on, to assess the predicted facilitating effect of the 3-phrase prosody. In Experiment 1 (Fodor & Nickels, 2011) participants read sentences first silently for comprehension, then aloud for recording, followed by judgments of pronounceability and comprehensibility. A familiarization procedure was employed in hope that it would increase the percentage of successfully parsed items. In Experiment 2 (Schott, 2012; Schott & Fodor, 2013) the "missing-VP2 illusion" described earlier was employed as a more objective measure of successful syntactic parsing. Participants read the sentences first silently, then aloud for recording, followed by a yes/no answer to the question "Is something missing from this sentence?"

In both experiments, we manipulated phrase lengths in order to compare sentence versions designed to be susceptible to the helpful 3-phrase prosody and versions which were designed to resist that prosody. We refer to the former as ENCouraging, and the latter as DISCouraging. In both cases RC1 was introduced by *that* and RC2 was not.

Experiment 1 (Rating Task with Familiarization)

MATERIALS

Experiment 1 manipulated both the length and the "weight" of the six phrases in a sentence, and compared the 2CE-RC structure with items with a single RC embedding. Items were constructed as follows; examples of each type are in Box 6.4.

Box 6.4

EXAMPLES OF EACH TYPE OF EXPERIMENT 1 MATERIALS

2CE-RC(length)

ENC: The rusty old ceiling pipes that the plumber my dad trained fixed continue to leak occasionally.

DISC: The pipes that the unlicensed plumber the new janitor reluctantly assisted tried to rcpair burst.

2CE-RC(weight)

ENC: The soufflé that the waitress the boss hired brought disintegrated.

DISC: The drink that the hostess the nightclub employed stirred spilled.

1CE-RC

ENC: The elderly Austrian woman that the retired boxer danced with just died in an automobile accident.

DISC: The woman that the recently retired middle-weight boxer had danced with on a South-American cruise died.

2CE-RC(G&T)

DISC: The prayer that the monk the religious fanatic had persecuted relentlessly was chanting every day was echoing in the empty church.

4 types of filler items

1. If Barbara wasn't crying because she lost her excellent exam notes, what was her problem?
2. The engineers continued, even though they knew it was hopeless, to try to repair the damaged bridge support.
3. Bertram told the physiotherapist that whenever he tries to exercise his leg muscles start to cramp.
4. Professor Thompson knew the internationally famous literary critic giving the speech was a fraud.

2CE-RC(length): 4 pairs of 2CE-RC sentences, with phrase length manipulation. Paired items had the same total number of words, plus or minus one. They had similar though not identical semantic content, but differed in their distribution of phrase lengths. To ENCourage the 3-phrase prosody, the outer constituents NP1 and VP3 were long and RC1 was quite short (by relative clause standards), with the result that these three constituents were more or less equal in length. In their DISCouraging counterparts, the outer constituents were too short to be phrased alone,[10] while the RC1 was too long to be phrased as a single unit.

2CE-RC(weight): 4 pairs of 2CE-RC sentences, with lexical "weight" manipulation. In contrast to 2CE-RC(length) sentences, each of the 6 phrases were matched in word count across the ENC/DISC items in a pair.[11] Paired items had roughly similar semantic content, but they differed in the predictability (corpus frequency, default status) of their content words, to either ENCourage or DISCourage the 3-phrase prosody, on the hypothesis that less predictable words would be less susceptible to phonetic reduction and thus would create prosodically "weightier" phrases. The mean lexical frequencies for the ENC and DISC sentences in a pair were matched.

1CE-RC: 4 pairs of sentences with the structure NP1 [NP2 VP2] VP3, in which a single-level RC modifies the subject of the main clause. Paired items had, again, the same number of words, plus or minus one, similar though not identical semantic content, but differed in the distribution of phrase lengths. To ENCourage the 3-phrase prosody the outer constituents were long and the RC was short; to DISCourage the 3-phrase prosody the outer constituents were short and the RC was long. In their overall length and their phrase length distributions these sentences were comparable to the 2CE-RC(length) items, although their syntactic structure was shallower.

2CE-RC(G&T): 4 typical 2CE-RC items from a previous study (Gibson & Thomas, 1999), with uniformly long constituents, as in sentence (3). We regarded these phrase lengths as DISCouraging the 3-phrase prosody.

16 assorted filler items, of 4 subtypes that differed in structure but contained multiple clauses and mild parsing challenges: the *if not because* construction, parenthetical adverbial clauses, early/late closure garden paths, NP/clausal complement garden paths.

ENC and DISC examples of each item type are shown in Box 6.4.

PARTICIPANTS AND PROCEDURE

Twenty-eight native English speaking participants (9 male) recruited at CUNY Graduate Center were tested individually. Their task was to judge the pronounceability and comprehensibility of sentences that were displayed visually on a computer screen. On the assumption that even in their most ENCouraging versions these materials would be too challenging for many people to process, we employed a familiarization procedure with the aim of increasing the overall level of comprehensibility and thus avoiding floor effects that could obscure judgment differences between item types. Each sentence (including fillers) was built up in 5 successive steps, as illustrated in (15) and (16) for the ENC and DISC versions respectively of 2CE-RC(length), and in (17) for a filler item.

(15) **2CE-RC(length), ENC version**

My dad trained a plumber.
Here is the plumber my dad trained.
The plumber my dad trained fixed the rusty old ceiling pipes.
Here are the rusty old ceiling pipes that the plumber my dad trained fixed.
The rusty old ceiling pipes that the plumber my dad trained fixed continue to leak occasionally.

(16) **2CE-RC(length), DISC version**

The new janitor reluctantly assisted an unlicensed plumber.
Here is the unlicensed plumber the new janitor reluctantly assisted.
The unlicensed plumber the new janitor reluctantly assisted tried to repair the pipes.
Here are the pipes that the unlicensed plumber the new janitor reluctantly assisted tried to repair.
The pipes that the unlicensed plumber the new janitor reluctantly assisted tried to repair burst.

(17) **Filler sentence**

The bridge support was damaged.
The engineers were trying to repair it.
The engineers were trying to repair the damaged bridge support.
They continued to try, even though they knew it was hopeless.
The engineers continued, even though they knew it was hopeless, to try to repair the damaged bridge support.

The 5 sentences in a set were displayed successively, each one on a single line, across the middle of the screen. The participant was instructed to read each sentence silently first for comprehension, then aloud for recording, and then to press an arrow key to remove that sentence and bring up the next one in the set. The first four sentences in a set were in white font against a dark background; the fifth one was in yellow font, and the participant knew s/he would have to judge the yellow sentence on two 5-point scales (5 = best) that appeared in succession on the screen: How easy was it to pronounce? How easy was it to understand?

PREDICTIONS

ENC versions were expected to be rated higher on the pronounceability scale than their DISC counterparts. With regard to the prosody with which they were pronounced, we anticipated that ENC versions would more often exhibit the optimal NP1 || RC1 || VP3 prosodic structure, while DISC versions would be divided into more chunks, creating a less natural and more "list-like" prosody for the sentence. On the hypothesis that a more natural prosodic phrasing would facilitate construction of the correct syntactic structure, ENC versions were expected to be rated higher on the comprehensibility scale than their DISC counterparts.

RESULTS: PRONOUNCEABILITY JUDGMENTS AND EVALUATION OF PRODUCED PROSODY

Participants' ratings of pronounceability are shown in Figure 6.2.

A one-way repeated-measure ANOVA including all eight different conditions revealed significant differences among them ($F(7, 189) = 26.11, p < .001$). Pairwise contrasts were computed to reveal which conditions differed specifically, only a selection of which will be reported here for reasons of space. The ENC versions of the 2RC-CE(length) items were rated as significantly easier to pronounce than their DISC versions ($F(1, 27) = 25.35, p < .001$), as predicted. The ratings for the 2CE-RC(G&T) items did not differ reliably from those for the DISC versions of the 2CE-RC(length) items ($F(1, 27) = 1.73, p = .199$); however, this may not be a fair comparison since the G&T sentences were longer, by 4.75 words on average, than the 2RC-CE(length) DISC items that we constructed. The ratings

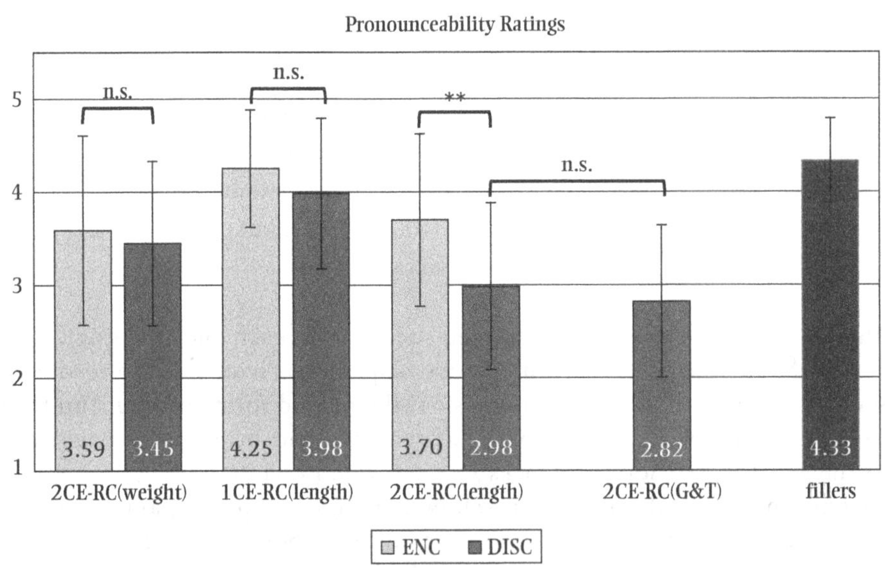

Figure 6.2 Mean scores on the pronounceability judgment scale, by stimulus type for $n = 28$ subjects. Whiskers indicate the standard deviation. ** indicates $\alpha < .001$.

for the ENC versions of the 2CE-RC(length) sentences were significantly lower than those for the ENC versions of the 1CE-RC items ($F(1, 27) = 11.21$, $p < .01$) and than those for the fillers ($F(1, 27) = 20.90$, $p < .001$), showing that even with favorable phrase lengths there remained some difficulty in finding an appropriate pronunciation of the nested 2CE-RC structure. For the 2CE-RC(weight) items, the ENC and DISC versions differed numerically in the direction expected but the effect was small and not statistically reliable ($F(1, 27) = 0.66$, $p = .424$). For the 1CE-RC (single level relative clause) items, which had phrase length patterns quite similar to those of the 2CE-RC items, ENC versions were also judged to be easier to pronounce than the DISC versions, but this difference showed only a weak trend toward statistical significance ($F(1, 27) = 2.55$, $p = .122$).

These self-reports of pronounceability by participants are corroborated by judgments of the appropriateness of the produced prosodic contours, by two trained judges (graduate students of linguistics) who were unaware of the design of the experiment. They judged only the doubly center-embedded sentences (i.e., 2CE-RC(weight), 2CE-RC(length) and 2CE-RC(G&T); see Figure 6.3). Their judgments, on a scale of 1 to 5 (5 = fully natural), were very similar to the pronounceability ratings by participants. The differences between the five conditions were confirmed by an ANOVA ($F(4, 104) = 27.84$, $p < .001$). Follow-up pairwise comparisons showed significantly better ratings for the 2CE-RC(length) in their ENC versions than in their DISC versions ($F(1, 26) = 34.54$, $p < .001$). The 2CE-RC(G&T) sentences did not differ from the 2CE-RC(length) DISC items (F(1,

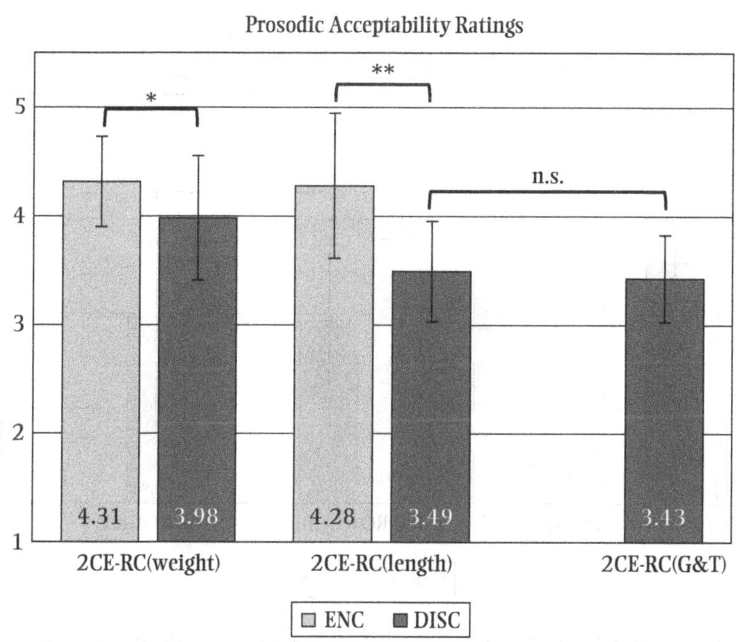

Figure 6.3 Mean appropriateness ratings of produced prosody, by trained judges. Whiskers indicate standard deviations. * indicates α < .05; ** indicates α < .001.

26) = .58, *p* = .45). In contrast to the pronounceability judgments by participants, the judges' ratings of prosodic appropriateness for the ENC and DISC versions of the 2CE-RC(weight) sentences showed a significant difference in favor of the ENC version ($F(1, 26) = 7.34, p < .02$).

In short: the ENC phrase length manipulation did make the 2CE-RC(length) sentences easier for readers to pronounce, and expert judges evaluated the overall prosodic contour of the ENC versions as more appropriate than that of the DISC versions for both the length and weight manipulations.

RESULTS: COMPREHENSIBILITY JUDGMENTS

The comprehensibility ratings by participants (see Figure 6.4) also showed significant differences among the eight tested conditions ($F(7, 189) = 33.95, p > .001$). Specifically, among the 2RC-CE(length) items the ENC versions were judged to be easier to understand than the DISC versions ($F(1, 27) = 30.98, p < .001$), as predicted. However, the ENC(weight) items showed no comprehensibility advantage over their DISC versions ($F(1, 27) = 1.75, p = .197$), possibly because the unfamiliarity of some of the words they contained was their most prominent property.

It is noteworthy that even the ENC versions of the 2CE-RC(length) items were judged to be less comprehensible than the filler items ($F(1, 27) = 22.40, p <.001$) and than the ENC versions of the 1CE-RC(length) items ($F(1, 27) = 22.68, p <.001$). Thus, it cannot be claimed that the prosodic facilitation achieved by ENC phrase lengths eliminated all processing difficulty from the doubly

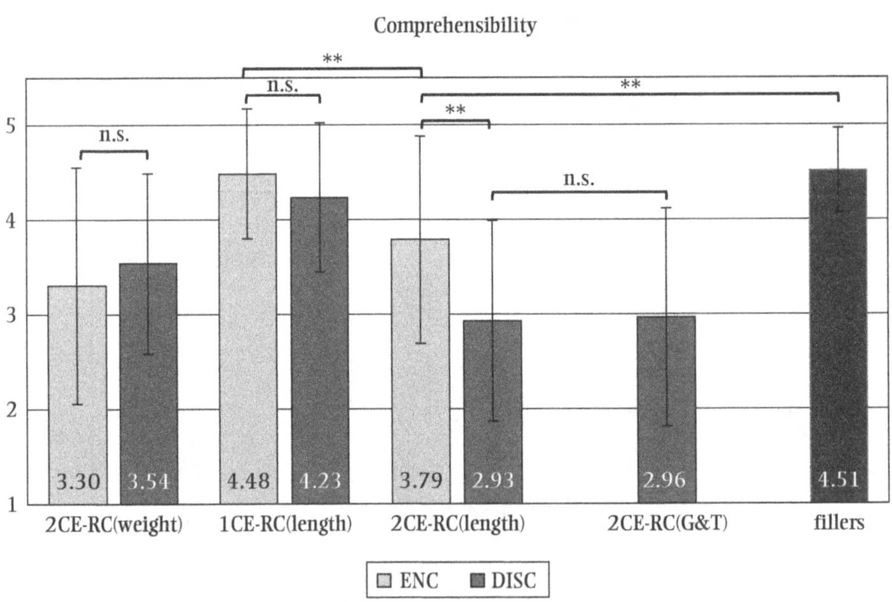

Figure 6.4 Mean scores on the comprehensibility judgment scale, by stimulus type. Higher scores indicate higher judged comprehensibility. Whiskers indicate standard deviations. ** indicates α < .001.

center-embedded construction. As anticipated, the comprehensibility ratings of the 2CE-RC(G&T) items did not differ from those for the 2CE-DISC(length) items ($F(1, 27) = .07, p = .787$).

RESULTS: PRODUCED PROSODIC PHRASING

A closer look at the produced prosody of these sentence types supports the hypothesis, originally formulated on the basis of intuitive judgments, that the prosodic phrasing most conducive to comprehensibility consists of three prosodic phrases, with the entire RC1 as the middle one: NP1 || NP2 NP3 VP1 VP2 || VP3. The two expert judges, discussed earlier, established the locations of prosodic boundaries, coding them either as typical phonological ("intermediate") phrase boundaries (level 3 in the ToBI transcription system) or as stronger (ToBI level 4). Sentence prosody patterns were then classified into 3 types, focusing on the latter part of the sentence, the sequence of three VPs, where the greatest differences were observed. The contrast of primary interest was between pronunciations in which VP2 was integrated into the RC1 prosodic phrase, and pronunciations in which VP2 was preceded by a boundary separating it from the preceding sequence of NP2 NP3 VP1. This "separated VP2" prosody was expected to be more common for DISC than for ENC items. Since it precludes the 3-phrase grouping believed to be facilitative of syntactic parsing, it was predicted to be associated with low comprehensibility ratings. The third prosodic pattern encoded was one that lacked a prosodic boundary before VP3. This was an occasional pronunciation, which the judges had regarded as very unnatural, and which we deemed to indicate complete failure to comprehend the VP sequence.

Figure 6.5 shows the 2CE-RC(length) items, in their ENC and DISC versions, and also the 2CE-RC(G&T) items, which had patterned like the DISC versions of the 2CE-RC(length) items in the judgment data reported earlier.

A chi-square test confirmed that the frequencies of the observed phrasing patterns differed significantly between conditions ($\chi^2(4, N = 216) = 22.54, p < .001$). Specifically, an integrated VP2 (i.e., absence of any prosodic break immediately preceding VP2) was more common for ENC than for the DISC and G&T item types ($p < .05$). In fact, the DISC and G&T items exhibited a strong bias toward the separated VP2 pattern ($p < .05$). The pattern with no break before VP3 was rare in all conditions. These data provide some support for the hypothesis that the ENC pattern of phrase lengths which improves comprehensibility does so by facilitating the inclusion of VP2 within the central prosodic phrase of a 3-phrase prosodic contour.

Further evidence for this conclusion comes from the relationship between the integrated-VP2 prosody and the distribution of comprehensibility scores, as shown in Figure 6.6.

For the ENC versions there is an overwhelming tendency to rate the integrated VP2 items as highly comprehensible (scale positions 4 and 5). Even for the DISC versions, there is a peak at scale point 4 for the integrated VP2 items. By contrast, for both ENC and DISC the separated VP2 items mostly cluster at the mid-scale

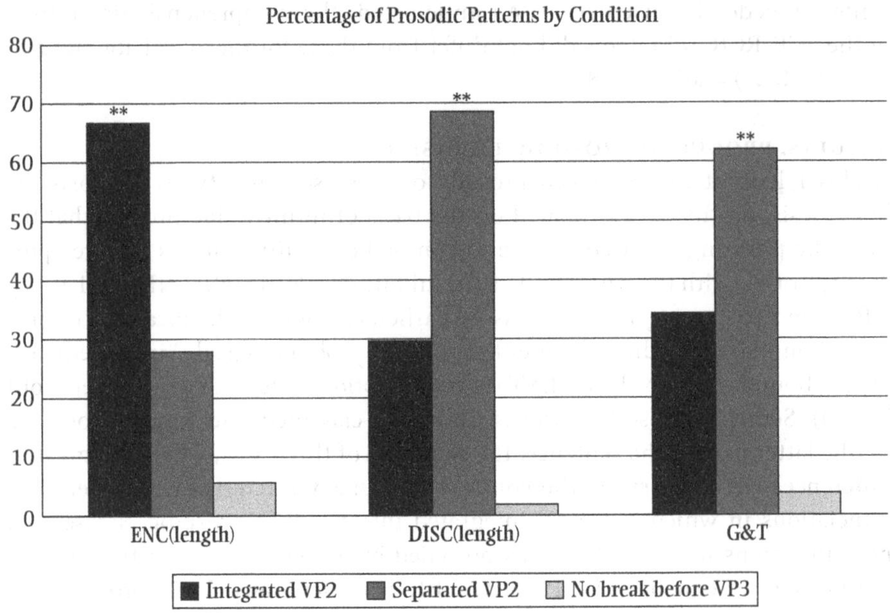

Figure 6.5 Prosodic phrasing patterns in productions of 3 sentence types. Asterisks indicate significant differences from equally distributed frequencies across sentence types (α < .001).

values. It is of interest that the separated VP2 items don't peak at the bottom end of the comprehensibility scale. This suggests that a list-like prosody with distinct separated phrases may improve comprehension of the individual phrases even if it obscures the syntactic/semantic relations among them.

The fact that our phrase length manipulations were not 100% successful in inducing the prosody we had intended (see Figure 6.5) provides the opportunity to

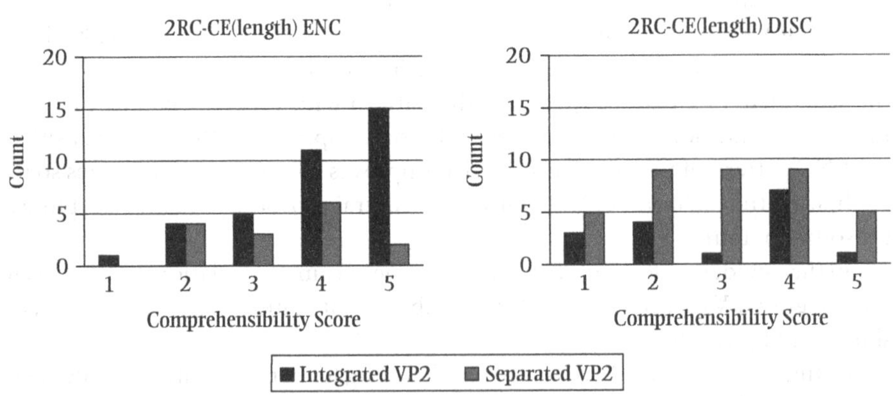

Figure 6.6 Frequency of doubly center-embedded sentences phrased with integrated VP2 and separated VP2 and associated comprehensibility scores, for ENC items (*left panel*) and DISC items (*right panel*). Higher scores indicate higher judged comprehensibility.

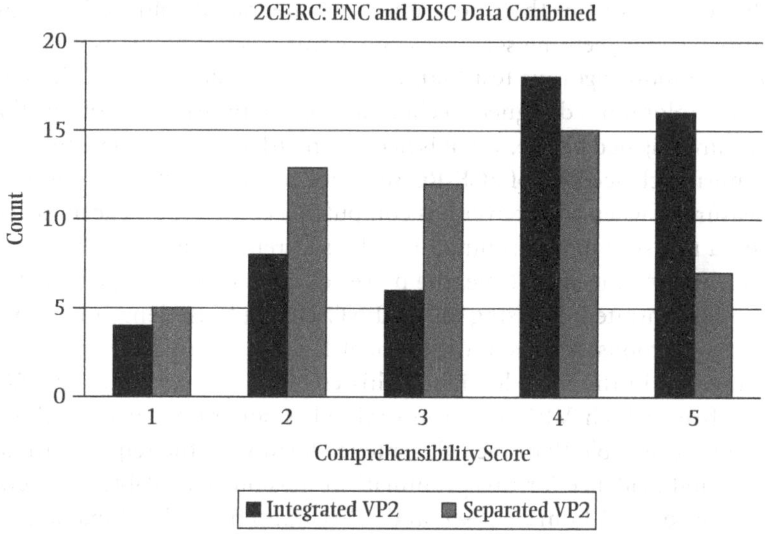

Figure 6.7 Distribution of comprehensibility ratings for items pronounced with integrated VP2 and separated VP2, for ENC and DISC items combined.

distinguish between effects of prosody on comprehensibility and effects of phrase lengths on comprehensibility. Thus, in Figure 6.7 the ENC and DISC data are pooled, and the comprehensibility scores are sorted only by prosodic phrasing of VP2.

It is apparent, at least numerically, that a prosodic pattern with an integrated VP2 is generally conducive to comprehensibility, regardless of whether the stimuli had been intended to be prosodically ENCouraging or DISCouraging. On average, integrated VP2 phrasing was associated with higher comprehensibility scores (mean 3.65), in contrast to sentences where VP2 was phrased separately (mean 3.16). However, this numerical difference did not reach statistical significance in a mixed model regression with prosodic phrasing as the predictor, compared with an intercept-only model ($\chi^2(1) = 2.79, p=.095$). The indecisiveness of this result may be due to the very low power of this experiment (which had only 4 ENC/DISC sentence pairs in the 2CE-RC(length) category); but the trend observed here suggests that the effect of prosodic phrasing on comprehension may emerge more strongly in larger-scale studies.

Overall, the findings of Experiment 1 are compatible with our general hypothesis that phrase lengths can impact prosodic contour assignment and that the appropriateness of the prosodic contour can improve comprehensibility, especially for the nested 2CE-RC structure which otherwise can be exceptionally difficult to process.

Experiment 2: In Search of the "Missing VP Illusion"

PURPOSE

Experiment 1 provided some support for the claim that a prosodic manipulation, induced indirectly by the distribution of phrase lengths, can transform

2CE-RC sentences from their notoriously unintegratable phrase-list sequences into fairly normal-seeming sentences. Since this has not previously been noted, it deserves a more rigorous test than Experiment 1 could provide. That experiment was exploratory, designed to check a range of informal intuitions that had not been investigated before. What is needed in order to consolidate the prosody/syntax mismatch account of 2CE-RC sentence processing difficulty is an objective measure of when a perceiver has computed a complete and accurate syntactic tree—a measure more probing than the self-reports of comprehensibility in Experiment 1. At the same time, the power of the follow-up experiment should be more concentrated, focusing on the ENC/DISC phrase-length manipulation which showed robust results in Experiment 1.

As observed in the introduction of this chapter, an ungrammatical 2CE-RC sentence from which VP2 has been omitted is sometimes perceived as more (or no less) acceptable than the full sentence form with the required three VPs. Experimental evidence for this grammatical illusion is available in several languages (Gibson & Thomas, 1999, and Christiansen & MacDonald, 2009, for English; Gimenes, Rigalleau & Gaonac'h, 2009, for French; Bader, 2011, for German). This could provide the objective test we need, to assess accurate processing.[12] Acceptance of a 2CE-RC sentence lacking VP2 would be a rather clear indication that a veridical syntactic tree structure had not been established. If it had, the absence of a VP2 should have been apparent (or else, if VP3 were pressed into service as VP2 in on-line processing of some examples, the absence of VP3 should have been apparent).

Our hypothesis for Experiment 2 was thus that the absence of VP2 should be easier to detect in ENC items than in DISC items, because the correct syntactic tree structure is more likely to be computed for the ENC items. The participants' task was to read the sentence first silently for comprehension, then aloud for recording, and then judge: "Is something missing from this sentence?"

MATERIALS

There were 16 ENC/DISC sentence pairs, structured in the same way as the 2CE-RC(length) items of Experiment 1, presented either in full or with VP2 missing, counterbalanced across 4 lists. One quartet of target items is shown in Box 6.5. There were 25 filler sentences, which were similar to those of Experiment 1 except that an obligatory word or phrase was omitted from 12 of them. Examples of filler sentences with and without a missing element are also shown in Box 6.5.

Note that for the ENC and DISC target items, overall sentence lengths were matched between Complete and Missing versions by inserting at the beginning of the Missing sentences a number of words equal to those deleted from the Complete version. These words were set off from the main body of the sentence by a comma, to ensure that they did not intrude on the prosodic phrasing of the body of the sentence that followed. This was unlike the materials of Gibson and Thomas (1999) in which the Missing versions were systematically

Box 6.5

Examples of Experiment 2 Materials

ENC Complete: The rusty old ceiling pipes that the plumber my dad trained fixed continue to leak occasionally.
ENC Missing: Admittedly, the rusty old ceiling pipes that the plumber my dad trained continue to leak occasionally.
DISC Complete: The pipes that the unlicensed plumber the new janitor kindly assisted tried to repair still leak.
DISC Missing: To no-one's surprise, the pipes that the unlicensed plumber the new janitor kindly assisted still leak.
Filler Complete: Professor Thompson knew the internationally famous literary critic giving the speech was a fraud.
Filler Missing: If Barbara wasn't crying because she lost her excellent exam notes, what was problem?

shorter than the Complete versions, which could have contributed to their positive evaluation.

PARTICIPANTS AND PROCEDURE

Participants were recruited through an invitation posted on websites. A web-based platform (www.limesurvey.org) was used for conducting the experiment. Participants were asked to read each visually displayed sentence silently first and then aloud while recording themselves. After the recording, they were asked to judge whether something was missing from the sentence. A practice session prior to the experiment provided examples. This was a more challenging task for participants than in Experiment 1 because we could not employ the familiarization protocol with this task. Building up a sentence stepwise despite the absence of one constituent from the final version was not feasible. However, just as for Experiment 1, we expected greater difficulty in assigning a natural-sounding prosodic contour to DISC versions than to ENC versions, and consequently a lower probability of computing the correct phrase structure of DISC versions. This could lead to greater incidence of the missing-VP illusion, i.e., more false acceptances for Missing DISC versions than for Missing ENC versions.

RESULTS

Among the total of 49 participants who successfully completed the questionnaire there was a strongly bimodal distribution in which 23 participants (47%) rejected the majority of the Complete 2CE-RC sentences, ENC as well as DISC, as having "something missing." This was unexpected. Standard reactions to 2CE-RC sentences such as (2) and (3) suggest that, if anything, the Complete items might have been thought to have too many, rather than too few, constituents. These "CE-rejecters" may have been responding to some property that was incidental

Table 6.1. Percentage of "Something Missing" Judgments for Complete and Missing Sentences, by Encouraging and Discouraging Phrase Lengths

	Complete	Missing
ENCouraging	10.4%	94.8%
DISCouraging	20.8%	88.5%

to the purpose of the experiment, such as the absence of an overt complementizer *that* heading the inner relative clause (in all the 2CE-RC items). Or they may have regarded all doubly center-embedded items as unacceptable and used the "something is missing" response as the only means provided for rejecting them. Since our methodology was unable to assess the grounds for this across-the-board rejection of 2CE-RC sentences, we set the data from those participants aside. Balancing the participant numbers across materials lists resulted in a total of 24 "discriminating" participants in the final analyses.

We computed the accuracy of detecting when a constituent was missing from the sentence. Unexpectedly, accuracy was very high in all conditions. There was no striking "missing VP illusion." Table 6.1 summarizes the findings. Clearly, the Missing items were not being treated like the Complete items.

In other respects, the differences are in the directions predicted. For the Complete items, results from simple mixed-effects modeling with only phrase lengths as predictor show that the ENCouraging phrase lengths yielded significantly less rejection than the DISCouraging phrase lengths ($z = 2.29, p = .02$); this is in accord with the comprehensibility ratings of Experiment 1. The absence of VP2 (the *Missing* column in Table 6.1) was more detectable in 2CE-RC sentences which had phrase lengths ENCouraging helpful prosody than when phrase lengths DISCouraged that prosody, but this difference was non-significant ($z = 1.42, p = .15$).

The reason for this failure to robustly replicate the missing-VP2 effect remains unclear. The explicitness of the question posed to participants may have made them more alert to absent constituents than would a general judgment of grammaticality or acceptability. Also, it is possible that the web-based methodology, without ability to monitor participant behavior, allowed the "discriminating" participants to study the sentences at length before responding. Another possibility of some interest is that reading the sentences aloud taps deeper syntactic processing than silent reading as in previous experiments on the missing VP effect. Replicating this study in a laboratory context, as we are currently planning, may help to disentangle these potential explanations.

Trained judges coded the prosody with which the sentences were spoken, as in Experiment 1. Phrase length manipulations predicted prosodic contours significantly, especially in the NP region. We tallied the percentage of items produced with a break after NP1 and nowhere else in the NP sequence, which we

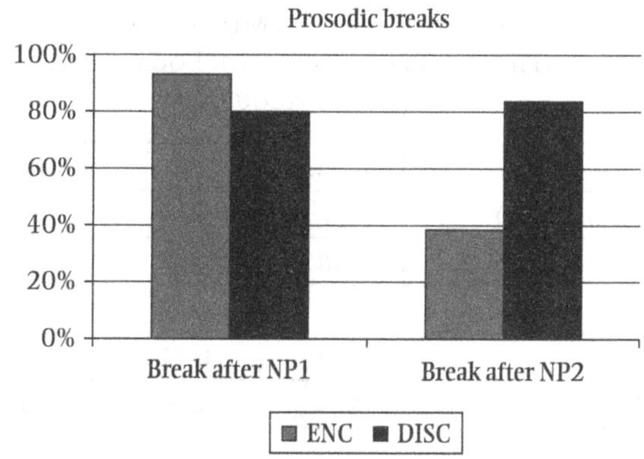

Figure 6.8 Percentage of prosodic breaks in the NP sequence, after NP1 only, and after NP2 (with or without a break after NP1), for ENC and DISC phrase lengths.

regard as optimal for syntactic parsing. We also determined the percentage of items pronounced with a break after NP2 (with or without another break in the NP sequence), which we regard as unhelpful for syntactic parsing. As shown in Figure 6.8, NP sequences in ENC items were pronounced with more breaks after NP1 only ($z = 4.5$, $p < 0.001$), and with fewer breaks after NP2 ($z = -6.1$, $p < 0.001$), compared to DISC items. There were also more DISC items than ENC items with breaks after *both* NP1 and NP2, in the manner of "list intonation" (34.7% for ENC; 67.0% for DISC; $\chi^2 = 38.1$, $p < .001$). Differences in the VP region did not reach significance, contrary to Experiment 1. In discussion that follows we suggest that this reflects the procedural difference between the two experiments: with and without stimulus familiarization.

Next, participants' judgments of sentences as complete or "missing something" were evaluated in relation to the prosody with which they had been produced, pooling ENC and DISC items; see Table 6.2. In this analysis, produced prosodies were grouped by presence/absence of a break after NP2, since this was the strongest factor observed.

When participants produced a sentence with a break after NP2, their judgment of the grammatical status of the sentence was less accurate. Results from extended mixed-effects modeling, with phrase length and prosodic break location as predictors, showed that the best predictor of judgment accuracy was the

Table 6.2. Relation between NP-region Prosody and Percentage of "Something Missing" Judgments (Regardless of ENC/DISC Phrase Length Patterns)

Produced prosody	Complete	Missing
No break after NP2	6.2%	92.9%
Break after NP2	20.7%	91.7%

Table 6.3. Results from Extended Mixed-Effects Modeling with Phrase Lengths and Prosodic Break Locations as Predictors of Judgment Accuracy

Contrast	Estimate	SE	z-Value	p-Value
ENC vs DISC (Complete)	1.11	0.49	1.61	.106
ENC vs DISC (Missing VP2)	0.68	0.48	0.90	.371
Break after NP1	0.10	0.53	0.19	.853
Break after NP2	**−1.54**	**0.77**	**−2.00**	**.046**
Break before VP3	−0.23	0.55	0.43	.672

produced prosody, regardless of whether the sentence was associated with phrase lengths pre-classified as ENC or DISC; see Table 6.3.

DISCUSSION

Experiment 2 did not replicate the missing-VP effect which has been found previously (with materials that we regard as having DISCouraging phrase lengths, as in (3) from Gibson and Thomas). But the results do confirm the link observed in Experiment 1 between phrase lengths and produced prosody, even in the absence of pre-familiarization with the materials. And its major finding bolsters the hypothesis, which was only marginally supported in the smaller-scale Experiment 1, that the produced prosody is what mediates causally between the phrase length manipulations and the ease of syntactic processing.

It is of interest that the most relevant aspect of the prosodic phrasing in Experiment 2 was the location of a boundary within the NP sequence. In Experiment 1, the most significant prosodic indicator was in the VP region: whether or not VP2 was included in a central prosodic "package" consisting of NP2 NP3 VP1 VP2. But Experiment 1 employed the familiarization technique, which provided readers with an opportunity to practice pronouncing the 2CE-RC structure, so difficulties might arise only toward the end of building up the central prosodic unit. Specifically: the complete sequence of three NPs was already present in the penultimate sentence in the sequence (line 4 in (15) and (16)), whereas the complete VP sequence was not presented until the final sentence, which was to be judged. By contrast, in Experiment 2, although items were read silently before recording there was no systematic rehearsal of the complex structure, so prosodic and parsing difficulties could arise earlier, during the NP sequence, specifically at the point of deciding whether or not to group NP1 and NP2 together prosodically. Grouping them would increase the probability of a break following NP2, which would preclude the optimal prosodic phrasing NP1 || NP2 NP3 VP1 VP2 ||VP3. To summarize: In Experiment 1 with familiarization, the prosodic grouping of the central constituents NP2 NP3 VP1 VP2 would most likely be blocked by inability to squeeze VP2 in at the end, whereas in Experiment 2 without familiarization it would most likely be blocked by premature grouping of NP1 and NP2 together at the beginning.

EXPLANATIONS

Our proposal has been that comprehension requires accurate building of syntactic tree structure, which must also be compatible with interface constraints. What remains to be explained is why the interface constraints apparently favor (for English) a 3-phrase prosody in which the upper RC is not divided. We will sketch here one account that lays the responsibility on constraints on prosodic phrasing, and one that invokes constraints on syntactic structure.

A possible prosodic account points to the fact that the 3-phrase prosody can provide maximum satisfaction of a cohesional constraint along the lines of Truckenbrodt's Wrap constraint (Box 6.2), which requires each syntactic phrase to be completely contained within a prosodic phrase. If this is applicable to the three clauses of the 2CE-RC structure,[13] the only way to satisfy it fully would be for the whole sentence to constitute a single prosodic phrase, but that is impractical in terms of typical phrase lengths as we have observed. The 3-phrase prosody does the next best thing: it sacrifices Wrap in the main clause, but complies with it for the two embedded clauses. Alternate ways of dividing the word string would not help in this regard. For instance, dividing the string at the middle of the lower RC would violate Wrap at all three clausal levels. This line of explanation is worth pursuing. For example, the decline in judged acceptability of the 4-phrase prosody in (14) might be attributable to the additional Wrap violation for RC1, which is split in two.

A possible syntactic explanation harks back to very early work on the syntax-prosody interface (Chomsky & Halle, 1968), in which an explanation was sought for the characteristic prosodic phrasing of recursive right-branching constructions as in (18).

(18) This is the cat || that chased the rat || that ate the cheese . . .

English generally aligns prosodic boundaries with the ends of syntactic phrases (Right-alignment with XP; see Box 6.2), but on the assumption that a restrictive RC modifies a noun (or N-bar) inside NP (see note 4) there are no right XP brackets at the prosodic boundary locations in (18). Since the sentence must be divided into shorter chunks in some fashion, one way to account for these pre-RC boundaries would be to assume that limits on maximum phrase length outweigh the XP-alignment principle (i.e., they would rank higher in an Optimality Theory framework; see Selkirk, 2000). But Chomsky and Halle and subsequently Langendoen (1975) took a different tack. They maintained that the prosodic phrasing in (18) is *fully* aligned with syntactic phrasing. Where the prosody cannot match the syntax due to phrase length constraints, the syntactic structure has to be brought into line with the prosody. This is effected by means of "readjustment rules" which rearrange the constituents in the syntactic tree structure.

Typically, readjustment rules move a constituent A out of a larger constituent B, and adjoin A as a sister of B. In (18), for example, string-vacuous application of a relative clause extraposition rule could extract each RC from the noun phrase in which it originates and raise it to become a sister of that noun phrase.

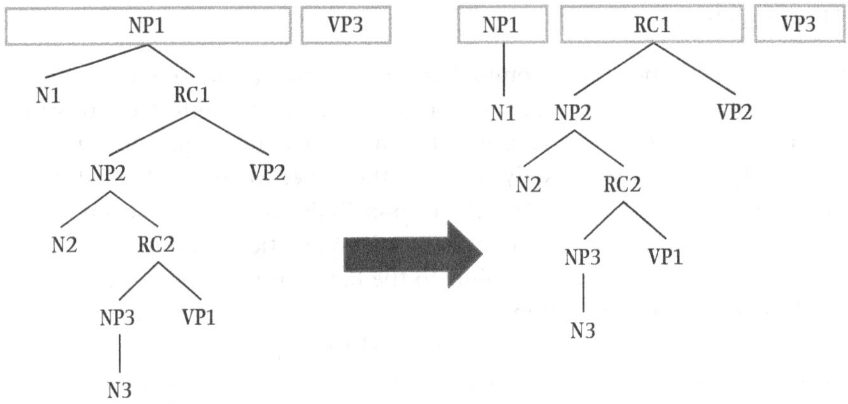

Figure 6.9 String-vacuous extraposition of RC1 in the 2CE-RC construction.

As a result, there would indeed be a syntactic right-edge bracket for the prosodic phrasing to align with preceding each RC in (18). A similar analysis of the center-embedded RC construction (Figure 6.1) would lift RC1 out of the noun phrase in which it originates as a modifier of NP1 and attach it as a sister to that noun phrase,[14] permitting a prosodic boundary between the two constituents and thus licensing the 3-phrase prosody (Figure 6.9).

Is string-vacuous RC-extraposition legitimate? Certainly, *non*-vacuous RC-extraposition is possible, as in examples (19) and (20) with interpretations as indicated.

(19) The children e_i were weeping RC_i[who the principal had scolded].

(20) Nobody e_i puts anything e_j into this sink RC_j[that would block it] RC_i[who wants to go on being a friend of mine].

String-vacuous movement is by its nature difficult to demonstrate. It has even been argued that it should be banned from linguistic theory altogether (den Dikken & Lahne, 2013). However, Wagner (2010) has developed a positive argument in favor of vacuous extraposition of RCs, noting that when the head noun is an idiom chunk (e.g., *Mary praised the headway that John made*), extraposing the RC away from the noun is unacceptable and so is inserting a prosodic boundary between the head noun and the adjacent RC. This parallelism may suggest that a boundary following a head noun is a sign that vacuous RC-extraposition has occurred.

A benefit of the vacuous extraposition analysis is that it offers a straightforward syntactic explanation for the degraded acceptability of the 4-phrase prosody for 2CE-RC sentences. The 4-phrase prosody, with its separated VP2 (i.e., a boundary between the subject and predicate of the upper RC, as in (14)), would have to come about by vacuous extraposition of VP2 out of the relative clause RC1 in which it originated. But if we may judge on the basis of *non*-vacuous extraposition, that would be syntactically illegitimate: a finite VP cannot be

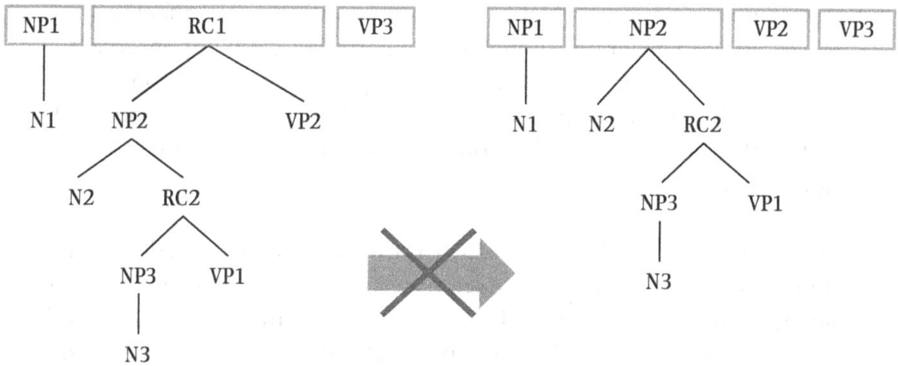

Figure 6.10 Illegitimate string-vacuous extraposition of VP2 in the 2CE-RC construction.

extraposed (Figure 6.10). Example (21), with the interpretation as indicated, is strongly ungrammatical.

(21) *The children $_{RC}$[who the principal e_i yesterday morning] $_{VPi}$ [had scolded] were weeping.

As a result, there is no legitimate readjusted syntactic structure for the 4-phrase prosodic analysis to align with. No matter how tempting the parser may find that prosodic phrasing on-line in order to break up over-heavy prosodic phrases, it does not aid sentence comprehension because it implies an illicit syntactic structure.[15]

CONCLUSION

The experiments we have reported here are just a beginning. They need to be followed up by more substantial studies that can provide insight into the linguistic and psycholinguistic mechanisms at work. But there are already promising indications that a significant factor in the near-unparsability of many doubly center-embedded sentences (at least in English) is the radical misfit between their strongly hierarchical syntactic structure and the prosodic phrasing induced by typical phrase lengths. It can be argued that this prosodic approach to syntactic center-embedding contributes to the explanation of all three well-known peculiarities of 2CE-RC processing:

1. Unusually difficult comprehension. The heavily nested 2CE-RC construction conflicts more radically with the "flattening" tendency of prosodic phrasing (Myrberg, 2013) than other multi-clausal syntactic constructions do.
2. Improvement if NP3 is a pronoun. A pronoun is short and usually unstressed, so it helps to slim down the lower RC, leaving more room to include NP2 and VP2 within the middle prosodic package.

3. Apparent improvement if VP2 is absent (the "missing-VP effect"). This is well-attested generally, though our Experiment 2 did not observe it. When the central phrase of the optimal 3-phrase prosody is too tightly packed to be able to include VP2, or if the NP sequence is mis-phrased so that there is no attachment point for VP2, the parser is better off ignoring it.

This does not mean that other factors such as working memory overload make no contribution to the difficulty of these sentences. Clearly, they contain a great deal of material to be remembered, and unlike right-branching constructions their word order does not allow subjects and predicates that belong together semantically to be composed together without delay. However, the current findings suggest that in addition to this complexity of semantic relations there is a purely syntactic disability in building a coherent tree structure for a 2CE-RC sentence when phrase lengths and hence prosodic structure are fighting against it.

Center-embedded constructions may thus once again become a tool for psycholinguistic research, providing a rich source of data in this case for elucidating details of the syntax-prosody interface, in English and also cross-linguistically. We are particularly interested in exploring the idea that when put under extreme pressure by this construction, the interface negotiations favor different solutions in different languages. There is a growing body of work in which the various interface constraints formulated by phonologists are being tested in psycholinguistic performance. It has become evident that phrase length constraints can be at least as powerful as alignment constraints and sometimes outrank them. It is well-established, beginning with Lehiste (1972) and by many studies since, that the overt prosody of spoken sentences has an impact on syntactic parsing: it can facilitate the comprehension of unambiguous sentences, and bias the interpretation of ambiguous ones. That similar effects can be induced by manipulating the lengths of phrases in visually presented sentences read silently (or aloud) has also been amply demonstrated in recent years, by Hirose (2003) for Japanese, Lovrić (2003) for Croatian, Fernández et al. (2003) for Spanish and English, among others. Thus the present finding that the comprehensibility of doubly center-embedded sentences can be improved by such means fits very naturally into this broader research paradigm.

NOTES

1. "Self-embedded sentences . . . exhibit features that are relevant to testing the significance of certain types of surface clues to deep structure configurations. We have employed them in the present experiments because, with iteration of the self-embedding operation, Ss have difficulty in understanding them. This provides an opportunity for the presumed facilitatory effects of surface structure clues to be revealed more strongly than in the case of sentences which Ss find easy to understand." (Fodor & Garrett, 1967, p. 291)

2. That prosodic structure must be flat was entailed by the Strict Layer Hypothesis of Selkirk (1981 and elsewhere), which forbade recursion in prosodic structure: one prosodic unit could be embedded in another only if they were at qualitatively different levels of the prosodic hierarchy. More recently this constraint has been recast as a violable condition which may be outweighed by other constraints on prosodic structure (Selkirk 1995; Truckenbrodt, 1999; Myrberg 2013). Wagner (2010) presents robust evidence for recursive prosodic phrasing in English coordination constructions, and Féry and Schubö (2010) demonstrate it in RC constructions in German, but it is not the most common pattern in our data for English 2CE-RC. A possible reason is offered in section "Explanation."
3. Fodor & Garrett (1967) compared 2CE-RC stimuli (such as *The pen the author the editor liked used was new*) pronounced with neutral prosody and pronounced with 'expressive' prosody (details not specified) and found little benefit from the latter, compared with the benefit of presence versus absence of relative pronouns in the RCs.
4. Current syntactic analyses of relative clause structure differ with respect to exactly how and where the RC is embedded. For simplicity here, Figure 6.1 does not show DP structure dominating the NPs. Also, the RC is shown beneath NP as a sister to a lexical Noun node (which might be dominated by an N-bar node) rather than as sister to a maximal projection (noun phrase, NP) as in the familiar shorthand representation of 2CE-RC structure in (1). We will continue to use that shorthand for convenience in what follows, and hope that the variant notations create no confusion.
5. We use the term 'prosodic phrase' to denote a unit lower in the prosodic hierarchy than a full intonational phrase (IPh). These units are referred to in the linguistics literature in various terms: *intermediate phrase (ip)*, *major phrase (MaP)*, *phonological phrase* or *p-phrase*. Although RCs are clausal units, they do not commonly constitute IPhs, at least in English (see, e.g., Göbbel, 2013, p.136). Non-restrictive RCs do, but they require a full relative pronoun such as *which* or *who*, and are precluded in our materials which have only a *that* or null complementizer.
6. Non-prosodic explanations of the pronoun advantage have been proposed by Bever (1988) and Gibson and Thomas (1999). More generally, Bever (1970) noted an improvement in processing when the three NPs are varied in form.
7. The examples from this point onward all have an overt complementizer (sometimes termed a relative pronoun) *that* at the beginning of RC1. This is because in all of these examples there is a prosodic boundary at that position, and an overt *that* is preferred after a prosodic boundary (Fox & Thompson, 2007). This lengthens RC1 by one word but does not add to the stressed syllable count. In these examples we have not inserted *that* to introduce RC2, to avoid giving the impression that there should be a prosodic boundary there.
8. There are many other ways of creating a sequence of 4 prosodic units out of the 6 phrases of sentence (12) (e.g., *The elegant woman || that the man Jill || loves met || moved to Barcelona*), but they all align improperly with the syntax and are considered extremely unnatural; see Fodor (2013) for discussion in terms resembling Selkirk's Sense Unit Constraint.
9. Some English speakers may be able to control two degrees of boundary strength (see Liberman, 2013). That could allow a 4-phrase pattern such as *The elegant*

woman || *that the man Jill loves* | *met on a cruise ship* || *moved to Barcelona*, with the break between the subject and predicate inside RC1 weaker than the breaks surrounding RC1. We encountered this rarely in our experiments, but there may be individual variation here, such that speakers who are particularly attuned to prosody are better able to deploy this pronunciation than linguistically naive speakers. Individual differences certainly deserve attention in future research.

10. Short phrases consisting of even a single word can be prosodically acceptable if heavily stressed. Such pronunciations presuppose a rich discourse context with prominent contrasts. However, this observation is of interest because, although we have not tested it yet, it suggests that the prosodic weights of the constituents are more relevant to 2CE-RC parsing than measures of lexical/syntactic length.
11. In all 2CE-RC(weight) sentences, both ENCouraging and DISCouraging, every NP consisted of a definite determiner and a single noun (or a proper name in NP3 position), and every VP consisted of a single verb (sometimes with a particle/preposition), see example in Box 6.4.
12. We chose not to evaluate processing accuracy by means of post-sentence comprehension questions, in case reading the question might interfere with the participant's phonological memory of the target sentence.
13. As defined in Truckenbrodt (1995; though see also Truckenbrodt, 2005), Wrap does not apply to adjuncts (which would include RCs) or to complete sentences (which would include the highest-level clause in a 2CE-RC construction), so this line of explanation may need to be based on some more general prosodic wrapping constraint.
14. Where the extraposed RC1 attaches in the tree structure needs to be established. Figure 6.9 is not precise in this regard. Attachment to the highest syntactic node, as sister to NP1 on its left and to VP3 on its right, would be illegitimate in theories that insist on binary syntactic branching. However, if the extraposition occurs at the level of PF (phonological form), it might not be subject to a binarity constraint. See Callahan (2013) for fuller discussion of the syntax of extraposed RC constructions in subject position.
15. Given that VP2 cannot extrapose, would it be permissable to skip down further and perform readjustment within NP2 by vacuously extraposing RC2? (In our materials that would be impossible due to absence of an overt relative pronoun/complementizer in RC2, but the question can be raised more generally.) RC2 would still be trapped inside RC1, so this would yield a multi-level phrasing NP1 || NP2 | RC2 | VP2 || VP3. This has been reported for German (Féry & Schubö, 2010), and it will be worth probing for in English, though it may be blocked if English tolerates a break between a restrictive RC and the head it modifies (i.e., vacuous RC extraposition if Wagner's argument is correct) only when it is motivated by length considerations or focus (unlike German, which favors a break before restrictive RCs generally; Augurzky, 2006).

REFERENCES

Augurzky, P. (2006). Attaching relative clauses in German: The role of implicit and explicit prosody in sentence processing. *MPI Series in Human Cognitive and Brain Sciences*, *77*, 264. https://books.google.com/books?isbn=3936816514

Bever, T. G. (1970). The cognitive basis for linguistic structures. In R. Hayes (Ed.), *Cognition and language development* (pp. 277–360). New York: Wiley & Sons, Inc.

Bever, T. G. (1988). The psychological reality of grammar: A student's eye view of cognitive science. In W. Hirst (Ed.), *The making of cognitive science: A festschrift for George A. Miller* (pp. 112–142). Cambridge University Press.

Blumenthal, A. (1966). Observations with self-embedded sentences. *Psychonomic Science 6*, 203–206.

Callahan, T. (2013). *Within-DP relative clause extraposition*. Unpublished ms, The Graduate Center, City University of New York.

Chomsky, N., & Halle, M. (1968). *The Sound Pattern of English*. New York: Harper & Row. Paperback edition 1991, MIT Press.

Chomsky, N., & Miller, G. A. (1963). Introduction to the formal analysis of natural languages. In R. D. Luce, R. R. Bush, & E. Galanter (Eds.), *Handbook of Mathematical Psychology* Vol. 2 (pp. 269–321). Hoboken, NJ: John Wiley and Sons.

Christiansen, M. H., & MacDonald, M. C. (2009). A usage-based approach to recursion in sentence processing. *Language Learning, 59*, 126–161.

den Dikken, M., & Lahne, A. (2013). The locality of syntactic dependencies. In Marcel den Dikken (Ed.), *The Cambridge Handbook of Generative Syntax* (ch. 11). Cambridge, England: Cambridge University Press.

Fernández, E. M., Bradley, D., Igoa, J. M., & Teira, C. (2003). *Prosodic phrasing in the RC-attachment ambiguity: Effects of language, RC-length, and position*. Paper presented at the 9th Annual Conference on Architectures and Mechanisms for Language Processing (AMLaP), Glasgow, Scotland.

Féry, C., & Schubö, F. (2010). Hierarchical prosodic structures in the intonation of center-embedded relative clauses. *The Linguistic Review, 27*, 293–317.

Fodor, J. A., Bever, T. G., & Garrett, M. F. (1974). *The psychology of language: An introduction to psycholinguistics and generative grammar.* New York, NY: McGraw-Hill.

Fodor, J. A., & Garrett, M. F. (1967). Some syntactic determinants of sentential complexity. *Perception & Psychophysics, 2*(7), 289–296.

Fodor, J. D. (2013). Pronouncing and comprehending center-embedded sentences. In M. Sanz, I. Laka, & M. K. Tanenhaus (Eds.), *Language down the garden path: The cognitive and biological basis of linguistic structures*, Oxford Studies in Biolinguistics, (ch. 9). New York, NY: Oxford University Press.

Fodor, J. D., & Nickels, S. (2011). *Center-embedded sentences: Phrase length, prosody and comprehension.* Poster presented at AMLaP, September 2007, Paris, France.

Fox, B. A., & Thompson, S. A. (2007). Relative clauses in English conversation: Relativizers, frequency and the notion of construction. *Studies in Language 31*, 293–326.

Frazier, L., & Fodor, J. D. (1978). The sausage machine: A new two-stage parsing model. *Cognition, 6*, 291–325.

Frazier, L. (1985). Syntactic complexity. In D. R. Dowty, L. Karttunen, & A. M. Zwicky (Eds.), *Natural language parsing: Psychological, computational, and theoretical perspectives* (pp.129–189). Cambridge, England: Cambridge University Press.

Gee, J. P., & Grosjean, F. (1983). Performance structures: A psycholinguistic and linguistic appraisal. *Cognitive Psychology, 15*, 411–458.

Ghini, M. (1993). Phi-formation in Italian: A new proposal. *Toronto Working Papers in Linguistics, 12*, 41–78.

Gibson, E. & Thomas, J. (1999). Memory limitations and structural forgetting: The perception of complex ungrammatical sentences as grammatical. *Language and Cognitive Processes, 14*, 225–248.

Gimenes, M., Rigalleau, F., & Gaonac'h, D. (2009). When a missing verb makes a French sentence more acceptable. *Language and Cognitive Processes, 24*, 440–449.

Hirose, Y. (2003). Recycling prosodic boundaries. *Journal of Psycholinguistic Research, 32*, 167–195.

Hudson, R. (1996). The difficulty of (so-called) self-embedded structures. *University College London Working Papers in Linguistics, 8*, 283–314.

Liberman, M. (2013). English prosodic phrasing. *Language Log is a collaborative language blog maintained by Mark Liberman, a phonetician at the University of Pennsylvania*. http://languagelog.ldc.upenn.edu/nll/?p=6810.

Lovrić, N. (2003). Implicit prosody in silent reading: Relative clause attachment in Croatian (Unpublished Ph.D. dissertation), The Graduate Center, CUNY.

Miller, G. A., & Chomsky, N. (1963). Finitary models of language users. In R. Duncan Luce, R. R. Bush, & E. Galanter (Eds.), *Handbook of Mathematical Psychology*, Volume 2 (pp.419–491). Hoboken, NJ: John Wiley and Sons.

Miller, G. A., & Isard, S. (1964). Free recall of self-embedded English sentences. *Information and Control, 7*, 292–303.

Myrberg, S. (2013). Sisterhood in prosodic branching, *Phonology, 30*, 73–124.

Schott, E. (2012). The influence of prosody on comprehending complex English sentences (Unpublished Diplomarbeit), Ludwig-Maximilians University, Munich, Germany.

Schott, E., & Fodor, J. D. (2013). *Prosody induced by phrase lengths facilitates syntactic processing in reading*. Poster presented at AMLaP September 2013, Marseille, France.

Selkirk, E. O. (1981). On the nature of phonological representation. In J. Anderson, J. Laver, & T. Meyers (Eds.), *The cognitive representation of speech* (pp. 379–388). Amsterdam: North Holland.

Selkirk, E. (1995). The prosodic structure of function words. *University of Massachusetts occasional papers: Papers in optimality theory, 18*, 439–469.

Selkirk, E. (2000). The interaction of constraints on prosodic phrasing. In M. Horne (Ed.), *Prosody: theory and experiment* (pp. 231–261). Dordrecht, Netherlands: Kluwer Academic Publishers.

Truckenbrodt, H. (1995). On the relation between syntactic phrases and phonological phrases. *Linguistic Inquiry, 30*, 219–255.

Truckenbrodt, H. (2005). A short report on intonation phrase boundaries in German. *Linguistische Berichte, 203*, 273–296.

7

Getting to the Root of the Matter

Acquisition of Turkish Morphology

NATALIE BATMANIAN AND KARIN STROMSWOLD

INTRODUCTION

Children's initial use of grammatical morphemes varies depending on the language they are learning. In languages with few inflectional morphemes, such as English and Swedish, children typically fail to use grammatical morphemes during the early stages of acquisition and as shown in (1) they often produce lexical roots.

(1) a. Papa have it. (Eve, 1;6, Brown, 1973)
    b. Cromer wear glasses. (Eve, 2;0, Brown, 1973)

Based on such acquisition data, some theorists (e.g., Bloom, 1970; Brown, 1973; Lebeaux, 1988; Radford, 1990; Clahsen, Penke, & Parodi, 1994; Brown, 1996) argue that children attain competence with lexical categories (e.g., nouns and verbs) before they do with grammatical categories (e.g., tense inflections). Others suggest that children learn language one lexical item at a time and that initially they do not have category-wide rules (e.g., Tomasello, 1992; 2000). Yet others (e.g., Poeppel & Wexler, 1993; Deprez & Pierce, 1994) argue that at all stages of acquisition, children's linguistic abilities are fundamentally the same as those of adults.

The unique linguistic properties of Turkish allow us to delve into the complex nature of language acquisition. Using Turkish allows us to see if children produce lexical roots in early language. It also allows us to determine if the complexity and frequency of grammatical morphemes dictates their order acquisition.

If linguistic properties dictate how children acquire language, then early child speech should be different in Turkish than it is in languages like English. For example, in English, lexical morphemes often occur in root form (Brent & Siskind, 2001), and children use lexical roots. Therefore, the absence of root form in Turkish should manifest itself in the absence of lexical roots in child speech. If however, the processes of language acquisition are the same across languages, then there should be cross-linguistic similarities in the ways children acquire language, we would expect to see lexical roots in the early language of both English and Turkish speaking children.

We argue that the first stage in acquisition children focus on lexical roots to isolate and learn their meaning of lexical roots because they carry more of the information about communicative intent and their meaning is easier to glean from non-linguistic cues. In order to isolate lexical roots, children must first learn to distinguish them from grammatical morphemes.

One way for children to distinguish lexical roots from grammatical morphemes is to use phonological cues. Selkirk (1996) argues that phonological contrasts in English between lexical roots and grammatical morphemes may make syntactic information accessible to children. We propose that a phonological property in Turkish called vowel harmony allows children to distinguish lexical roots from grammatical morphemes.

At the second stage of acquisition, children need to discover the different semantic meanings that are mapped onto a grammatical morpheme. In inflectional languages acquisition is dependent on the child's ability to decipher the multiple meanings that are mapped onto single grammatical morphemes. For example, in Italian, the morpheme *–erà* in *mang-erà* "eat-Fut 3rd Sg" and *-iò* in mang-*iò* "eat-Past 3rd Sg" convey information for both tense and subject agreement.[1] In inflectional languages, grammatical morphemes generally convey more than one piece of semantic information. Contrastingly, in Turkish the correspondence between a grammatical morpheme and its semantic meaning is one-to-one.

In line with Slobin (1973), we hypothesizes that when the correspondence between morphological form and grammatical meaning is one-to-one, morphemes are likely to be acquired earlier, than when the correspondence between form and meaning is one-to-many.

The unique linguistic properties of Turkish allow us to test this hypothesis.

In Turkish, almost all grammatical morphemes follow the one-morpheme-one-meaning rule (e.g., *gel-di-ler* "come-Past-3rd Pl," *gel-ecek-ler* "come-Fut-3rd Pl"). However, even in Turkish, some morphemes have multiple meanings mapped onto them (e.g., optative mood marker and the evidential mood marker). According to the one-to-one mapping hypothesis these morphemes should be acquired later than others.

Another aspect of the acquisition of grammatical morphemes that has been debated is the effect of their frequency in adult language. In English, it has been shown that the confounded semantic and syntactic complexity of grammatical morphemes plays a role in the order of their acquisition and that their frequency does not affect it (e.g., Brown, 1973; Pinker, 1981; Stromswold, 1999). Since we are able to tease apart the semantic and morpho-syntactic properties of grammatical morphemes in Turkish, we may also be able to examine the role of frequency in the acquisition of morphemes. We conjecture that if two morphemes have the similar semantic and morpho-syntactic properties, then their frequency in adult language will play a role in the order of their acquisition.

In the next section, we will discuss the relevant linguistic properties of Turkish followed by the data to date on the acquisition of Turkish. In this paper, we report two studies. The first one examines the grammatical morphemes in adult input that are relevant to the proposed hypotheses. The second study examines child spontaneous speech to determine the evidence to support the hypotheses.

Isolating Roots with Vowel Harmony: Phonology of Turkish

In Turkish, each lexical root can occur with a large number of bound derivational and grammatical morphemes. The richness in derivational and inflectional morphology allows a word root to be expressed in thousands, or even millions of forms (Hankamer, 1989). For example, when a verb root is suffixed with 3 grammatical morphemes, 11,313 word formations can be derived (Hakkani-Tür, Oflazer, & Tür, 2002).

We hypothesize that to learn lexical roots children will isolate them. An ability to isolate lexical roots suggests that children have an abstract understanding of lexical roots and that they entertain hypotheses about what their phonological form may be (Pinker, 1984). One might argue that, Turkish-speaking children use a simple phonologically based algorithm to strip off inflectional morphemes to discover the lexical roots. For example, many verb roots in Turkish are CVC and some of these verb occur in root form in adult speech (see Study 1), one could imagine that, working from right-to-left, Turkish-speaking children successively strip off syllables and consonants until they arrive at the initial CVC. If children rely on this simple phonological mechanism, then we would expect errors with word roots that are not CVC. For example, children may misanalyze word roots that end in consonant clusters combining the last consonant with a grammatical morpheme.

We propose that vowel harmony in Turkish can help children differentiate lexical morphemes from grammatical morphemes, because across many forms of a verb present in the child's input, the verb root stays phonologically intact (see *gel* in (3c)), whereas the grammatical morphemes change their phonological form (compare *-di* in (3c) to *-du* in (3d) and *-me* in (3c) -to *ma* in (3d)). Vowel harmony is a subtle but reliable cue for language learners in correctly identifying lexical roots (see Incorrect Use of Inflections).

In Turkish, vowels within a word share certain phonological properties: the same feature specification for backness and, if vowels are high, feature specification for rounding. This phenomenon is called vowel harmony (Kornfilt, 1997;

p. 398). Vowel harmony extends from word root to inflections (see (2a) through (2f)). In other words, while the vowels in the lexical root stay intact, the vowels in the inflections change to match the vowels in the lexical root. For example, as shown in (2a) and (2b), the accusative case marker can appear as *–u* or *–ü* depending on the feature specification for rounding present in the root.[2] For a few exceptional verbs vowel harmony is triggered backward from inflectional suffix to verb root (see Discussion on how exceptions affect the acquisition of grammatical morphemes).

(2)						
a.	*Bülbül*	'nightingale'	*bülbül-ü*	'nightingale-Acc'	*bülbül-e*	'nightingale-Dat'
b.	*Okul*	'school'	*okul-u*	'school-Acc'	*okul-a*	'school-Dat'
c.	*Gelmek*	'to come'	*gel-di*	'come-Past'	*gel-me*	'come-Neg'
d.	*Koşmak*	'to run'	*koş-du*	'run-Past'	*koş-ma*	
e.	*Su*	'water'	*su-suz*	'water-less'		
f.	*Ev*	'house'	*ev-siz*	'house-less'		

One-to-One Mapping Between Form and Meaning: Morphosyntax Turkish

Turkish is considered to be an agglutinative language because for almost all morphological inflections there is a one-to-one correspondence between morphological form and semantic meaning, and inflections are suffixed onto lexical roots in a fixed order (Aksu-Koç & Slobin, 1985; Kornfilt, 1997). The six nominal case-markers indicate grammatical relations and thematic roles, and word order is used to convey discourse and pragmatic information (Erguvanlı, 1984; Kornfilt, 1994; Küntay & Slobin, 1996). A Turkish noun may be suffixed with up to 3 inflections, which, if present, must appear in the following order: plural-possessive-case.

Verbal inflections express notions of tense, aspect, mood, modality, person and number. Modulo pragmatic restrictions, a verb root may be suffixed with up to 12 inflections. These, if present, must appear in the following order: verb root-reflexive-reciprocal-causative-passive-abilitative-negative-necessitative-tense-conditional-question-person-agreement (Aksu-Koç & Slobin, 1985). There are 4 tense inflections, which also combine for aspectual information; 4 voice inflections—passive, reflexive, causative and reciprocal; 6 mood inflections—conditional, necessitative, abilitative, negative, interrogative, and evidential/past. In addition, there are four agreement paradigms: the first has the widest distribution, and it is found on all simple tense forms; the second is the agreement paradigm used with the Past tense inflection *–di*. The third is the paradigm for the optative mood, which marks agreement and meaning onto one suffix. The fourth paradigm is restricted to the imperative (Kornfilt, 1997).

Note that the optative and evidential mood markers are the exception to the one-morpheme-one meaning rule in Turkish because it is not possible to discern the semantic component and the subject agreement component. The optative mood marker is a subjunct infinitive and common in adult speech (e.g., *gel-sin* "come- Opt 3rd Sg"). The evidential mood marker conveys that the verb is in past tense, but unlike the simple past tense *-di,* it also conveys that the given information is second hand (e.g., *Ayşe dün gel-miş* "Ayşe. yesterday come-Past" "(It is said that) Ayşe came yesterday").

Data to Date

Despite of the fact that Turkish-speaking children's task of distinguishing morphological information from lexical information seems daunting, some of the research on Turkish acquisition suggests that children acquire the morphology of Turkish very early and error-free (Aksu-Koç & Slobin, 1985). Although the limited number of studies in Turkish have concentrated on children's multimorphemic utterances, some researchers have noted that Turkish-speaking children go through a period, prior to age 2, in which they produce ungrammatical lexical roots (Çapan, 1988; Ekmekçi, 1982; Ketrez & Aksu-Koç, 1999). Additionally, in their study of early language development in Turkish, Aksu-Koç (1988) and Ketrez & Aksu-Koç (1999) have documented that the past tense inflection and the present tense inflection are one of the first morphemes to be acquired before other tense inflection and before the evidential/past inflection.

One vexing problem in research with spontaneous speech data is deciding whether a child is using a certain structure or morpheme productively. When an English-speaking child says "baby cry," the bare utterance could either mean that the child omitted the progressive tense maker or that she has not acquired it yet. In the same situation, a Turkish child might say *bebek ağlı-yor* "baby cry-Prog" (the baby is crying). However, it is not clear whether the child has analyzed the different grammatical components of the word *ağlıyor* or whether it is an unanalyzed unit. Thus, just as a bare stems in English cannot be taken as evidence for lack of knowledge for grammatical morphemes, an inflected word uttered by a Turkish-speaking child cannot be taken as evidence for productivity with the grammatical morphemes present in the utterance. This signifies that words and sentences that are memorized units should not be used as evidence that a child has acquired a morphosyntactic form because the child always hears inflected words. For a child to use a form productively, she must have some understanding of the structure and use of the form and its components. Determining productivity is especially problematic for utterances that are frequent in either parental input or child speech. It is also difficult to determine productivity in an agglutinative language in which words do not appear in isolation and rules of affixation are regular.

We studied adult spontaneous speech directed to children to determine the role of input in the acquisition of grammatical morphemes. With pre-established criteria for productivity, we analyzed longitudinal and cross-sectional spontaneous speech from Turkish-speaking children to determine whether, as hypothesized, children will isolate lexical roots and whether they will acquire some grammatical morphemes sooner than others.

STUDY 1: FREQUENCY OF LEXICAL ROOTS AND MORPHOLOGICALLY COMPLEX WORDS IN ADULT SPEECH TO CHILDREN

We analyzed the grammatical morphemes in the speech of a Turkish-speaking mother to determine what aspects of adult input affected child language. We looked at verbs and nouns that are in root form to determine how their presence in adult speech influences their presence in child speech. In other words, if lexical roots are frequent in adult speech, then it would not be surprising to see them in high percentages in child speech. On the other hand, if lexical roots are not frequent in adult speech, then an explanation is needed for the kind of mechanism that allows children to isolate lexical roots. We also determined the types of grammatical morphemes and their frequency to determine whether frequency influences the ordering of their acquisition.

Participants

Spontaneous speech data were collected monthly from a Turkish-speaking mother and her two children. The older child Ali was taped longitudinally for approximately a year. The sessions, which were recorded during playtime between the mother and an individual child, lasted approximately 30 minutes. Data from 11 transcripts were collapsed into one file because analyses of three transcripts when Ali was 2;1 (2 years; 1 month), 2;5 and 2;8 revealed no differences in the mother's use of inflections.

Analyses

First, we examined the percentage of nouns and verbs that were inflected with grammatical morphemes. We analyzed 76 verb types in total, with 2317 tokens, but we did not include the existential predicative verbs *var* and *yok* and the negation predicate *değil* because they are not suffixed with the full range of verbal inflections. We also examined how frequently tense inflection, mood inflections, voice markers and agreement inflections were used. This analysis was conducted on inflections attached to 36 verbs for which there were more than 25 tokens each.

Results

LEXICAL ROOTS

The example in (3) is a speech excerpt from the mother that illustrates a common occurrence in Turkish, noun ellipsis (i.e., the pronoun for first person plural). The utterance also contains a direct object *oyun* without any casemarking. Recall that nominative case in null in Turkish. As shown in Table 7.1, half of the nouns in the mother's utterances were in root form (i.e., they were not casemarked), 32% were with one inflection and 16% were with two inflections.

Table 7.1. Percentage of Verbs with 0, 1, or 2 Inflections Mother of Ali & Elif

	Ns	Vs
O-infl	52	25(12)*
1-infl	33	34
2+-infl	15	40
Tokens	2189	2879

(3) M: Sen-in-le oyun oyna-yalım hadi.
You-Gen-Instr game play-Opt 1st Pl come-on
"Come on, let's play a game together"

As shown in Table 7.1, of the 76 types of verbs that occurred in the transcripts, 75% of the mother's verb tokens were inflected: 34% of verbs were with one inflection and 40% were with two or more inflections. One quarter of her verbs were in root form (i.e. the imperative mode for second person singular), though this was not true of her verb types. For example, three verbs (*bak* "look," *de* "say," and *al* "take") accounted for half of the 591 verb tokens uttered in root form. Without these three verbs, only 12% of verbs were in root form. In short, while a few verbs were used in root form very frequently, some were used in root form occasionally, and others were never used in root form.[3]

GRAMMATICAL INFLECTIONS

Tense markers: Overall, the inflections that Ali's mother uses most frequently are the past tense inflection *–di* (10% of the time), the present tense inflection *–yor* (10% of the time), and the optative mood marker (9% of the time). Other tense inflections occurred less frequently: the future tense inflection *–acak* (4% of the time); the past tense/evidential inflection *–miş*, and the habitual inflection *–ir* (each 3% of the time). Of the inflections that occurred most frequently, some appeared on more verb types than others. For example, the present tense inflection *–yor* appeared on 92% of the 36 verbs, optative inflections appeared on 75% of them, and the past tense inflection *–di* appeared on 72% of them.

Agreement markers: Unsurprisingly, some subject agreement markers occurred more frequently than others: second person singular (7% of the time); first person singular (6% of the time); first person plural (5% of the time); and the second person plural marker (1% of the time). The third person plural marker, however, occurred less than 1% of the time. Most interestingly, the third person singular, which is not overt, occurred 16% of the time. In other words, 16% of the time the verbs did not have an agreement marker and were suffixed only with tense and/or mood markers.

Voice and mood markers: The infinitival maker *–mek* and the negation marker *–me* occurred 2% of the time. The nominal marker, the passive marker, the causative marker occurred 1% of the time.

To summarize, whereas half of the nouns that Ali's mother utters are in root form, verbs in root form do not occur frequently except for a few particular verbs. An interesting finding in adult language is that despite their agglutinative nature, verbs are frequently suffixed only with one tense inflection. As was previously mentioned, third person singular, which is a null morpheme, is frequent in adult speech. In other words, frequently verb roots appear inflected only with one particle. The tense inflections that occur very frequently in the mother's speech are the past and present tense inflections. Other tense markers appear less frequently in adult speech. The inflections for first and second person also occur frequently though less so than tense inflections. The optative mood (subjunct infinitive) marker appears as frequently as the past and present tense markers. Recall that the optative mood inflection is an exception to the one-form-one-meaning rule in Turkish (see Morphosyntax and Phonology of Turkish). The evidential mood marker *–mış* is also an exception to this rule; however, it occurs less frequently in adult speech than the optative mood marker.

STUDY 2: LEXICAL ROOTS AND GRAMMATICAL MORPHEMES IN CHILD SPEECH

In the second study, we only examine the verb roots and verbal inflections in child speech because half of the nouns uttered by the mother are in root form. We compare the verbs roots in child speech to those in adult speech. Second, we examine child speech for the acquisition of grammatical morphemes. We have hypothesized we do not expect the morphemes that have more than one meaning mapped onto them to occur earlier in child speech than others. We also hypothesized that morphemes that are frequent in the mother's speech are the ones that will occur in child speech. Thus, we expect the past and present tense inflections to be one of the first morphemes to occur in child speech we also expect the optative marker to be acquired later than grammatical morphemes.

Participants

Three monolingual Turkish-speaking children were studied. Ali was audiotaped talking to his mother approximately every two weeks from the ages 2;1 to 2;8. Taping sessions lasted approximately 30 minutes. For this study, we analyzed speech samples collected at ages 2;1, 2;3, 2;4, 2;5, 2;6, 2;7 and 2;8. The same procedure was used to tape Ali's younger sister Elif. For this study, we analyzed Elif's speech samples at ages 1;8 and 2;1. Both of Ali and Elif's parents are monolingual Turkish speakers, and the siblings were growing up in a monolingual Turkish environment in the United States. The third subject is Erel, whose data are part of the Childes database (MacWhinney & Snow, 1985). Erel was recruited from the

Istanbul metropolitan area in Turkey as part of a cross-linguistic study. His interactions with an experimenter were audiotaped at 2;0 and 2;4 (Slobin & Bever, 1982).

Coding

To determine multimorphemic productivity, we had two measures: first, an overall measure to determine percent of inflected verbs and total number of grammatical morphemes, and second, a measure to determine productivity with individual lexical roots and individual grammatical morphemes.

MULTIMORPHEMIC VERBS

In each transcript, we calculated the percentage of all verbs inflected with zero, one, or two inflections.

We also determined productive use of frequently used lexical verbs. We selected ten of the most frequent verbs in each individual participant's transcripts. We analyzed these verbs to determine the earliest age by which the children in our study inflected them with one, or two inflections. If a child produces a word or morpheme in contrasting morphosyntactic environments, then this suggests that the child may be using the morpheme productively (Allen, 1996). Thus, a lexical root was considered productive when the child combined it with at least two different grammatical morphemes (e.g., walk + ed, walk + ing).

GRAMMATICAL MORPHEMES

For each transcript, we determined the total number of different types of verbal inflections that were used. We also determined the age by which the most frequent verbal inflections appeared on two or more lexical stems (e.g., walk + ed, talk + ed).

Based on the assumption that children are likely to omit or overgeneralize newly acquired grammatical inflections, we examined multiword and one word utterances (context permitting) to determine the relative frequency of children's omission errors (e.g., the Turkish equivalent of "she eating" instead of "she is eating") and substitution errors (e.g., the Turkish equivalent of "she are going" instead of "she is going").

Results

PERCENTAGE OF VERBS INFLECTED WITH GRAMMATICAL MORPHEMES

We first determined the percentage of all verbs inflected with grammatical morphemes. Both Ali and Elif produced some multimorphemic words in their speech but many verbs were in root form. Erel, on the other hand, had a higher percentage of verbs inflected with grammatical morphemes.

As shown in Table 7.2, between the ages 2;1 and 2;5, 40% of Ali's verbs had no inflections. Between the ages 2;6 and 2;8 a lower percentage of verbs bore no

Table 7.2. PERCENTAGE OF VERBS WITH 0, 1, OR 2 INFLECTIONS

	Ali		Elif		Erel	
	2;1–2;5	2;6–2;8	1;8	2;1	2;0	2;4
	T1	T2	T1	T2	T1	T2
O-infl	40	21	58	42	9	7
1-infl	60	59	36	46	63	58
2+-infl	0	18	6	12	28	8
Tokens	81	191	60	95	34	68

inflection and about 20% were inflected with two grammatical morphemes. At age 1;8, 60% of Elif's verbs bore no inflection. At age 2;0, only 10% of Erel's verbs bore no inflection and about 30% bore two or more inflections. Recall that in their mother's utterances only one fourth of the verbs are in root form and that half of the tokens are from three verb types.

AGE OF PRODUCTIVE USE OF VERB ROOTS

Despite having 20 to 30 verb types in their vocabulary, the children had few tokens for the majority of the verbs. Thus as shown in Table 7.3, we selected each child's ten most frequent verbs and determined the earliest age by which the child inflected these stems with one, two, three or more distinct inflections (e.g., Open + Past, Open + Prog, Open+Future).

Ali and Elif inflected few verbs productively. As shown in Table 7.4, at Time 1, Ali inflected most of his verbs with only one inflection type: the past tense inflection *-di*. The second inflection type that appears at Time 1 is the present tense inflection *-yor*, although it appears in later transcripts when Ali was close to 2;5. As shown in Table 7.4, at T1, Elif had not yet used verbal inflections productively. At Time 2, she uses a few of her verbs productively. The third participant Erel productively inflected most of the frequent verbs in his vocabulary.

Table 7.3. NUMBER OF TOKENS FOR EACH CHILD'S MOST FREQUENT VERBS AT T1 AND T2 COMBINED

Ali's Verbs	Token	Elif''s Verbs	Token	Erel's Verbs	Token
Ye (eat)	27	*Bak* (look)	36	*Yap* (do)	28
Al (take)	27	*Yap* (make)	10	*Git* (go)	16
Git (go)	24	*Ol* (be)	6	*Oyna* (play)	13
Bak (look)	23	*Ağla* (cry)	4	*Düş* (fall)	12
Bit (finish)	17	*Ver* (give)	4	*Ol* (be)	9
Aç (open)	17	*Oku* (read)	4	*Gel* (come)	7
Gel (come)	14	*Tut* (hold)	3	*Çık* (exit)	9
Düş (fall)	10	*Düş* (fall)	4	*Koy* (put)	5
De (say)	10	*Kalk* (get up)	3	*Sür* (drive)	5
Yat (recline)	9	*Kal* (stay)	2	*Ye* (eat)	4

Table 7.4. Number of Inflection Types Used by Each Child at T1 and T2 with Their Most Frequent Verbs

Ali's	Time 1	Time 2	Elif''s	Time 1	Time 2	Erel's	Time 1	Time 2
Verbs	2;1–2;5	2;6–2;8	Verbs	1;8	2;1	Verbs	2;0	2;4
Ye (eat)	0	2	*Bak* (look)	0	1	*Yap* (do)	5	11
Al (take)	2	3	*Yap* (make)	1	3	*Git* (go)	4	6
Git (go)	2	3	*Ol* (be)		2	*Oyna* (play)	1	3
Bak (look)	0	1	*Ağla* (cry)	1	1	*Düş* (fall)	2	5
Bit (finish)	1	1	*Ver* (give)	1	0	*Ol* (be)		3
Aç (open)	2	4	*Oku* (read)	0	1	*Gel* (come)	0	4
Gel (come)	1	1	*Tut* (hold)		0	*Çık* (exit)		5
Düş (fall)	1	0	*Düş* (fall)		1	*Koy* (put)	2	5
De (say)	1	3	*Kalk* (get up)	0		*Sür* (drive)	2	
Yat (recline)	0	2	*Kal* (stay)		1	*Ye* (eat)	3	4

TOTAL NUMBER OF INFLECTION TYPES

The first measure of productivity with grammatical morphemes was the total number of different types of verbal (out of a total of 40) inflections each child produced in each transcript. Elif and Ali used a handful of verbal inflections productively. Similar to the previous findings, Erel was more productive than the other two children and used a majority of the verbal inflections we studied for this analysis.

Overall, between 2;1 and 2;5, Ali used two different verbal inflections (past and present tense inflections). Between 2;6 and 2;8, there was an increase in the types of verbal inflections he used, acquiring three additional verbal inflections. Similar to her brother at 1;8, Elif had two verbal inflections. At 2;0, Elif had five verbal inflections. At 2;0, Erel had productively used most tense inflections (the future tense being the notable exception). At 2;4, he had used 16 verbal inflections, including all tense and agreement markers and negation.

PRODUCTIVE USE OF INFLECTIONS

As a second measure of productivity with grammatical morphemes, we determined the age by which verbal inflections appeared on two or more lexical stems (e.g., Walk + Past, Eat + Past). The six verbal inflections that we analyzed were past tense (*-di*), habitual (aorist) tense (*-ir*), present tense (*-yor*), future tense (*-acak*), the optative mood inflection, subject agreement, and negation (*-me*). We chose these inflections because they are the basic inflections for verbs and they are frequent in the adult input.

Between 2;1 and 2;5, Ali had used the past tense inflection *-di* with several verbs. By 2;8, Ali used the present inflection *-yor*, the future inflection *-ecek* and first person singular and plural inflections for number agreement with two or more verbal stems.

We also studied past and present tense inflections suffixed to the 14 verbs that were common to Ali's and his mother's speech. First, we wanted to determine whether inflections Ali used correlated with those of his mother. Second, we wanted to determine whether Ali used the past tense inflection arbitrarily or whether he used it primarily in past tense context. The 14 verbs that appeared in Ali's speech in past tense and in present tense, and were also found in his mother speech did not correlate significantly for either tense (past tense $r = 0.16$, $t\,(1,13) = 0.60$, $p > .10$, present tense $r = 0.30$, $t\,(1,13) = 1.16$, $p > .10$).[4]

At 2;0, the only inflections that Elif used on two or more verbal stems were the past and present tense inflections.

Whereas Ali and Elif began to use grammatical morphemes during the period they were studied, Erel was already productively using most of the six verbal inflections. At 2;0, he had productively used all the tense inflections (except the future tense inflection), proper agreement morphology and negation.

It should be noted that despite of being frequent in adult speech the children in our study did not inflect their verbs with the optative mood inflection nor did they inflect it with the evidential/past tense inflection. Additionally, in Ali and Elif's speech agreement, markers were practically absent despite being relatively frequent in adult speech.

OMISSION AND SUBSTITUTION OF INFLECTIONS

The morphological errors we considered were errors of omission and substitution. As with other aspects of their morphosyntactic development, Ali and Elif demonstrated similar patterns in the quantity and type of morphological errors that they made. Erel made fewer errors than the other children but his error patterns were similar to theirs.

As shown in Table 7.5, almost all of Ali's initial morphological errors were those of omission (i.e. lack of verbal and nominal inflections, rendering his utterances ungrammatical).[5] The remainder was substitution errors (i.e., use of verbal inflections that are unrepresentative of adult usage). Of all of his morphological errors, about 70% involved nominal inflections and 30% involved verbal inflections.

Even at the two word stage, Ali used erroneously uninflected verbs. The speech excerpt in (4) is a typical example of an omission error. Ali utters the verb *uyu* "sleep" which is missing a tense inflection. Discourse context suggests that Ali intended to use a different tense inflection than the past tense inflection his mother used. Thus, at 2;3 it is possible that Ali knows enough about the meaning

Table 7.5. Percentage of Morphological Errors and their Relative Distribution by Type

	Ali		Elif		Erel	
	T1	T2	T1	T2	T1	T2
	2;1–2;5	2;7	1;8	2;1	2;0	2;4
Morphological error (%)	19	10	67	24	7	7
Errors of omission (%)	92	78	79	74	73	62
Errors of substitution (%)	7	22	21	26	27	38
Total no. of utterances	315	90	70	156	130	177

of particular Turkish tense inflections to know that *-du* is inappropriate (hence his omission), but not enough to select the correct tense inflection.

(4) M: Uyu-du mu?
Sleep-Past Inter
'Did it sleep?'

A (2;3): *Bu uyu.
This sleep
'This (will) sleep'

M: Uyu-y-acak.
Sleep-Fut
'(It) will sleep.'

We studied the most frequent 14 verbs common to Ali's and his mother's vocabulary, to compare Ali's and his mother's choice of verb roots. Although some verbs appear in root form in his mother's speech, as shown in Table 7.6, there are few direct parallels between Ali's choice of verbs in root form and those of his mother. We did not find a significant correlation between the verbs that Ali used in root form and the verbs his mother used in root form ($r = 0.37$, t (1, 13) = 1.40, $p > .10$). For example, in addition to the verb *uç* "fly," he produced *ağla* "cry," and *düş* "fall." None of these verbs were intended as imperatives.

Some of the errors the children made provide empirical evidence for use of vowel harmony in isolating verb roots. The children made these errors with inflections that do not obey the rules of vowel harmony. Typically vowel harmony does not change the lexical root, but when roots are suffixed with the present tense inflection, some changes occur. For example, the verb root *gel* "come" changes to *gel-i-yor* "(s/he/it) come-Epen-Pres" when suffixed with the present tense inflection, due to the presence of an epenthetic vowel, but it remains unchanged when marked with the past tense inflection *gel-di* "(s/he/it) come-Past." In his earliest

Table 7.6. 14 Verbs Common to Ali and his Mother's Speech that Appear in Root Form and with Inflection

Verbs	Total Token		0-infl (%)		Past tense (%)		Present tense (%)	
	Ali	Mother	Ali	Mother	Ali	Mother	Ali	Mother
Bak (look)	24	282	96	55	4	1	0	13
Aç (open)	17	59	71	34	0	0	12	5
Uyu (sleep	5	114	60	7	20	53	0	0
At (throw)	10	26	30	38	60	27	10	15
Gel (come)	14	110	29	34	71	31	0	10
Ye (eat)	38	49	29	22	63	33	8	41
De (say)	17	275	24	42	29	3	47	9
Uç (fly)	5	12	20	25	20	25	40	50
Yap (make)	5	142	20	4	20	23	0	27
Düş (fall)	6	24	17	0	33	50	50	21
Yat (recline)	9	9	11	33	11	67	11	0
Bit (finish intr.)	18	39	0	0	94	56	0	0
Al (take)	27	80	4	53	85	10	0	9
Git (go)	24	98	4	7	7	20	4	5

transcripts at age 2;1, Ali used some CVC in CVCV form, such as *düş-ü* for *düş* "fall" and *gel-i* for *gel* "come" (see also Çapan, 1988). This error, which occurred occasionally and usually in contexts that conveyed present tense, demonstrates how isolation of the verb root can be unsuccessful when grammatical morphemes and verb roots do not obey rules of vowel harmony.

Can a child who has yet to learn the inflectional morphemes of Turkish know that the roots of many Turkish words are CVC if roots are not frequently uttered in bare form? The evidence we gathered suggests that the answer is no. Although many Turkish words are simple CVCs and some appear in bare form (e.g., *bak* "look" is a verb Ali's mother used frequently), some are disyllabic, and yet the children in our study never produced a disyllabic root as a monosyllabic form(e.g., *kapa* "close" was never uttered as *kap*). They also did not combine verbs that end in consonant clusters CVCC with grammatical morphemes or utter them as CVCs (e.g., *kalk* "rise/get up" as *kal*) to a CVC form. They also did not combine grammatical morphemes with the verb root when it was in the VC form (e.g., *iç* "drink"). The errors children do not produce, along with instances in which they erroneously isolate verb roots when grammatical inflections do not obey vowel harmony, provide empirical evidence for the argument that Turkish-speaking children isolate the lexical root in their language by monitoring phonological regularities.

As depicted in Table 7.5, Elif's errors mostly consisted of omissions, but like Ali, as she added more inflections to her repertoire, her errors of substitution increased. About half of Elif's morphological errors involved verbs, and half involved nouns. Elif's speech errors also provided empirical evidence for use of vowel harmony in isolating lexical roots. As stated earlier, lexical roots generally do not change form under vowel harmony, but two verb roots, *de-* "say" and *ye-* "eat," change form when suffixed with certain inflections. For example, *de-* changes to *di-* when suffixed with the present and future tense inflections (e.g., *di-yor* (say-Pres) and *di-yecek* (say-Fut)) but not with others (e.g., *de-di* (say-Past), *de-mi-yor* (say-Neg-Pres)). At 2;1, Elif typically omitted verbal inflections, despite occasionally producing them. An inflected verb she produced demonstrates that she separated verbs into lexical and grammatical components rather than mimicking adult utterances without analysis. She produced the verb root *de-* (instead of *di–*) with the present tense inflection *–yor*, a form that does not occur in adult speech.

Although they were fewer in number, the types of omission and substitution errors Erel made were similar to Ali and Elif's errors (see Table 7.5). About half of Erel's morphological errors involved verbs, and half involved nouns. The types of omission errors Erel made were similar to Ali and Elif's errors. He did not use obligatory verbal suffixes on verbs (see 5).

(5) EX: niçin?
  Why?
 ER: *O koş-Ø o-ndan (correct form: O koş-ar o-ndan)
  He run- it-ABL He run-HAB it-ABL
  'He runs that's why (because of it/from it)

It should be noted that the children almost never made any duplication errors and any lexical category errors (i.e., use nominal inflections for verbs and use verbal inflections for nouns). In Elif's 511 nouns and verbs at age 2;1, she produced one duplication error applying the possessive case twice on the nominal stem *üst* "top" (i.e., she uttered **üst-ü-nü-nü* "top-poss- acc- acc"). Ali (655 noun and verb tokens) and Erel (473 verb and noun tokens) did not produce any duplication or lexical category errors.

Discussion

For both lexical roots and grammatical morphemes, we took two measures: one measure to determine overall percentage of inflected verbs and overall number of grammatical morphemes used, and a second measure to determine productivity with individual lexical roots and individual grammatical morphemes. Recall that despite of having two to three dozen verbs in their vocabulary, the children had few tokens for most of these verbs. Thus, although over 50% of the verbs were inflected, at Time 1 they were inflected with one type of grammatical morpheme. In general, Ali uttered verbs in past tense until age 2;6. At age 2;1, Elif had about two dozen verbs in her vocabulary, but they were mostly uttered in root form and when they were suffixed they were inflected in the past tense. Additionally, the analyses of children's morphological errors revealed that children uttered verbs (and nouns) in root form that were ungrammatical. The verb types in root form that we observed in Ali's speech and the verb types that he inflected in past tense did not correlate with the verb types that we observed in his mother's speech that were in root form and were inflected in past tense. Therefore, Ali could not be mimicking his mother in his usage of verbs in root form or verbs in the past tense. Furthermore, while adults use verb roots grammatically, children use them ungrammatically.

Despite the morphosyntactic differences between Turkish and English, children speaking ungrammatically produce verbs in root form. Although the three children's usage of multimorphemic words differed, their error patterns were similar. The stage during which children do not use morphology in Turkish was previously characterized as the pre-morphological stage without any explanation for the children's ability to isolate lexical roots (Aksu-Koç, 1988; Ketrez & Aksu-Koç, 1999).

GENERAL DISCUSSION

Lexical Roots

We hypothesized that if the process and means of language acquisition are the same across languages, then at a fundamental level, there will be similarities in the way children acquire Turkish and, for example, English. How are Turkish and English children similar in the way they acquire language? In order to map meaning onto each of the lexical and grammatical components of a word they try to isolate the lexical root first. In our study, similar to English-speaking children in the early stages of acquisition, the Turkish-speaking children produced verbs in root form in ungrammatical contexts.[5]

In English, verbs and nouns frequently appear in root form. In Turkish, lexical roots, for both verbs and nouns, appear with several grammatical inflections and a given lexical morpheme can have thousands of surface forms. Nevertheless, when they are in the imperative mode verbs occur in root form. One might argue that the presence of even a few lexical roots in the input might help the child. It is necessary, however, to notice that this only helps if the child has an (abstract) idea that lexical roots exist (Pinker, 1984). The question remains: how does the child know that a particular word is a bare form? Perhaps children possess a mechanism that detects phonological regularities of what constitutes a lexical root.

This could be as simple as innate (abstract) knowledge that the lexical root will be in the "center" of the word. There could be a more specific explanation. For example, if lexical roots are CVC, the child can use this to identify potential roots. If a child relies on phonological regularities, she will make predictable mistakes for stems that violate the general rule. We did not observe such errors in our data. Thus rather than relying on a simple algorithm to discover verb roots which would result in overgeneralizations and errors, children are more likely to use the complex phonological relationship between lexical and grammatical morphemes. We hypothesized that Turkish-speaking children will isolate lexical roots and use vowel harmony to decompose verbs into lexical and grammatical morphemes.

The speech errors children make are a source of empirical evidence for use of vowel harmony in differentiating lexical morphemes from grammatical morphemes. Some of the speech errors children make in the rare cases that vowel harmony does not apply provide some empirical evidence for its use. No data is available on the phonological properties of early child speech in Turkish or infants' ability to detect vowel harmony; however, researchers have shown that infants are capable of extracting regularities and phonotactic properties of their language, and it is likely that they use these properties to process speech (e.g., Jusczyk et al., 1993). Vowel harmony is a phonological property that is also found in Finnish and research has shown that in word segmentation with Finnish-adults are aided by vowel harmony (Suomi, McQueen, & Cutler, 1997).

Acquisition of Grammatical Morphemes

Following Slobin (1973) we hypothesized that some grammatical morphemes appear in early language sooner than others if there is a one-to-one correspondence between surface form and meaning. We further hypothesized that if two morphemes have similar semantic and morpho-syntactic properties, then their frequency may play a role in how early they are acquired. The most frequent grammatical morphemes in adult speech were the past tense inflection *-di* and the present tense inflection *-yor* for which there is a one-to-one correspondence between form and meaning. The present tense inflection which was also frequent in adult speech appeared soon after the past tense inflection. Other tense inflections, which appeared less frequently in adult speech than the past tense inflection, did not appear during early speech. Based on our data, despite being equally frequent as the past tense inflection, the optative mood marker, for which there was a two-to-one correspondence between meaning and form, did not appear in early child speech. Concurrent with previous research, the evidential mood marker for which the mapping is two-to-one and is not as frequent in adult speech as the optative mood marker, was also not observed in early child speech (Aksu-Koç, 1988; Ketrez & Aksu-Koç, 1999).

When it comes to mapping semantic meaning onto a grammatical morpheme the morphemes that are acquired the earliest are those that convey one semantic meaning. This suggests that the mechanism children possess functions on the assumption that each grammatical morpheme carries one piece of semantic information. This mechanism explains why the past and present tense inflections in Turkish are acquired earlier than the optative mood construction. It also explains why cross-linguistically Turkish-speaking children seem to acquire grammatical morphemes at an earlier age than children speaking Germanic and Romance languages (Pinker, 1981; Clahsen et al., 1994; Hamann, 1996). In languages with a richer the inflectional paradigms children stop using root verbs sooner than children who speak languages with impoverished inflectional paradigms (Phillips, 1995; Kim & Phillips, 1998; Pye, 2001).

Clearly, a few mechanisms aid the acquisition of lexical roots and grammatical morphemes. In addition to being sensitive to phonological regularity children also seem to possess a mechanism that allows a set of phonetic units corresponding to the grammatical morpheme rather than one phonetic unit (Aslin, Woodward, LaMendola, & Bever, 1996; Saffran, Aslin, & Newport, 1996; Selkirk, 1996; Suomi, McQueen, & Cutler, 1997). For example, in English the grammatical morpheme for plurality is expressed with three phonetic units [s], [z], [ez], and in Turkish grammatical morphemes change due to vowel harmony (e.g., for past tense *-di, -du, -dü*). We did not observe any incorrect use of phonetic units in past tense morphemes. If children possess a mechanism that operates on the assumption that one phonetic unit corresponds to each grammatical morpheme, then the present tense morpheme *-yor* should have been the first to be acquired because it, unlike other tense inflections, does not obey vowel harmony and does not change its phonological form. Thus, children seem to possess a mechanism that is flexible in the mapping between grammatical morpheme and phonological form. It is likely, for example, that the difficulty English-speaking children have with the morpheme *-s* is not the multiplicity of its phonetic forms but rather the multiplicity of semantic meanings mapped onto it (i.e., plural marker and possessive marker in nouns and subject-verb agreement marker in verbs).

Slobin (1973) hypothesized that grammatical morphemes that are at the end of words (i.e., suffixes) are easier to acquire then morphemes that are infixes because they are perceptually salient. This hypothesis might explain why, in our data, we observed tense and agreement markers but not negation or voice markers (e.g., causative, passive), which are infixes (see section (1.2) Morphosyntax of Turkish). However, if perceptual salience were the only mechanism at play, then we should have observed some subject agreement markers which appear at the rightmost edge of verbs, but we did not observe any agreement markers in the children's early transcripts. Additionally, as was stated in the Morphosyntax Turkish (section 1.2) the past tense marker *–di* has its own agreement paradigm and all the singular forms and first person plural consist of one syllable (e.g., *–dim, –din, –di, –dik*). Recall that first and second agreement markers are also frequent in adult speech. The lack of agreement markers in child speech suggests that children acquire grammatical morphemes according to an abstract ordering in which tense precedes agreement (as, for example, those outlined by Cinque, 1999).

The Turkish data makes it evident, that children do not simply mimic adult speech. Our findings suggest that in the acquisition of grammatical morphemes several mechanisms are at play, and these mechanisms do not seem possible unless children have an abstract knowledge of lexical and grammatical morphemes, of phonological differences between these two categories and of a universal ordering for grammatical morphemes which results in their gradual acquisition and production.

CONCLUSION

The two questions that we tried to answer in this paper are tied. Do children have grammatical and lexical categories when they begin the acquisition process? How do children acquire lexical and grammatical morphemes? Evidence from Turkish-speaking children's early language suggest that children start the acquisition process with an awareness of lexical and grammatical categories. Despite of the multiple surface forms of verb roots due to rich inflectional morphology, Ali and Elif preferentially omitted inflectional morphemes. Thus, the fact that two of the children in our study produce verb roots is consistent with acquisitional theories that claim that from the earliest stages of acquisition, children are equipped to learn lexical and grammatical morphemes. Children are likely to use vowel harmony in Turkish to isolate verb roots.

At the earliest stages, children acquire morphemes for which the mapping between form and meaning is one-to-one (i.e., semantic and syntactic information is not confounded). Turkish allowed us to test this hypothesis for grammatical morphemes. Concurrent with previous studies, we determined that of the three grammatical inflections that are frequent in adult speech the one that occurred earliest in child speech was the past tense morpheme for which the mapping between surface form and meaning was one-to-one. The present tense morpheme was acquired soon after. Another equally frequent morpheme, the optative mood marker, for which the mapping between surface form and meaning is one-to-two, was not observed in early stages.

ACKNOWLEDGMENTS

The research reported in this paper was supported by grants from the National Institute of Health (HD37818) to Stromswold and (F32DC007035) to Batmanian and the National Science Foundation (BCS-0002010) to Stromswold and Batmanian (formerly Batman-Ratyosyan). We are thankful to the family who participated in the longitudinal study. We also thank Ayhan Aksu-Koç, Shanley Allen, Mark Baker, Jennifer Ganger, Nihan Ketrez, Dan Slobin, Virginia Valian and Scot Zola, and the audience members at the 25th Annual Boston University Conference on Language Development for the help they provided.

NOTES

1. We will use the following linguistic abbreviations: * = ungrammatical, Abil = abilitative modal, Abl = ablative case, Acc = accusative case, Agr = agreement, Asp = aspect, Caus = causative, Cond = conditional, Dat = dative case, DerivMorph = Derivational Morphology, Epen = epenthetic vowel, Evid = evidential mood, Gen = genitive case, Hab = habitual tense, Inf= infinitive, Instr = instrumental case, Inter = interrogative, Imper = imperative, Loc = locative case, Mod = modal, N = noun, Nec = necessitative, Neg = negative, Np = noun phrase, Opt = optative mood, Past = past tense, Pers = person, Pl = Plural, Pres = present tense, Pass = passive, Recip = reciprocal, Refl = reflexive, Sg = singular, Tns = tense, Tran = transitive, Vp = verb phrase.
2. The vowel *–ü* in Turkish is pronounced like the German *–ü* or French *–u* whereas the vowel *–u* is pronounced like the cluster *–ou* in French.
3. Of the 76 verbs in transcripts, 48% of the verbs were either in CV or CVC form, 9% were in VC form, and 4 verbs ended in consonant clusters. The remaining 37% were verbs with disyllabic roots. Notably, however, 60% of the mother's 20 most frequent verbs had the CV or the CVC form.
4. Because of insufficient data, we were unable to conduct Chi-square tests to determine contexts in which a tense inflection was required and was used. However, Fisher exact tests revealed that use of past tense inflection and bare stem forms was not independent of proper context (i.e., past tense morphology in past tense context, and bare stems in non-past tense context), prior to age 2;6 when Ali was not productive with tense inflections or after ($\chi^2 = 18.17$ and $\chi^2 = 7.58$ both p's $< .05$).
5. We did not report them in this paper but the children also produced root nouns in ungrammatical contexts (see Batman-Ratyosyan & Stromsowld, 2001).

REFERENCES

Allen, S. E. M. (1996). *Aspects of argument structure acquisition in Inuktitut*. Philadelphia, PA: John Benjamins Publishing Co.

Aksu-Koç (1988). The acquisition of aspect and modality: The case of past reference in Turkish. Cambridge, England: Cambridge U Press.

Aksu-Koç, A. A., & Slobin, D. I. (1985). The acquisition of Turkish. In D. I. Slobin (Ed.), *The crosslinguistic study of language acquisition* (pp. 839–878). Hillsdale, NJ: Lawrence Erlbaum Associates.

Aslin, R. N., Woodward, J. Z., LaMendola, N. P., & Bever, T. G. (1996). Models of word segmentation in fluent maternal speech to infants. In J. L. Morgan & K. Demuth (Eds.). *Signal to syntax: Bootstrapping from speech to grammar in early acquisition* (pp. 117–134). Mahwah, NJ: Lawrence Erlbaum Associates.

Batman-Ratyosyan, N., & Stromswold, K. (2001). Early bare stems in an agglutinative language. In A. H.-J. Do, L. Dominguez, & A. Johansen (Eds.), *Proceedings of the 25th Annual Boston University Conference on Language Development*, 102–113. Somerville, MA: Cascadilla Press.

Bloom, L. (1970). *Language development: Form and function in emerging grammars.* Cambridge, MA: MIT Press.

Brent, M., & Siskind, M. S. (2001). The role of exposure to isolated words in early vocabulary development. Cognition, *81*, B33–B44.

Brown, P. (1996). Isolating the CVC root in Tzeltal Mayan: A study of children's first verbs. In E. Clark (Ed.), *Proceedings of the 28th Annual Child Language Research Forum* (pp. 41–52). Stanford, CA: Center for the Study of Language Information.

Brown, R. (1973). *A first language: The early stages.* Cambridge, MA: Harvard University Press.

Çapan, S. (1988). Acquisition of verb inflections by Turkish children: A case study. In S. Koç (Ed.), *Studies on Turkish Linguistics* (pp. 275–286). Ankara: Middle East Technical University Press.

Cinque, G. (1999). Adverbs and functional heads. Oxford, England: Oxford University Press.

Clahsen, H., Penke, M., & Parodi, T. (1994). Functional categories and clause structure in early German. *Language Acquisition 3*(4), 394–429.

Deprez, V., & Pierce, A. (1994). Crosslinguistic evidence for functional projections in early child grammar. In T. Hoekstra & B. Schwartz (Eds.), *Language acquisition studies in generative grammar* (pp. 57–84). Amsterdam: John Benjamins.

Ekmekçi, Ö. F. (1982). Language development of a Turkish child: A speech analysis in terms of length and complexity. *Journal of Human Sciences*, *1*(1), 103–112.

Erguvanlı, E. (1984). *The function of word order in Turkish grammar.* Berkeley, CA: University of California Press.

Hakkani-Tür D., Oflazer K., & Tür G. (2002). Statistical morphological disambiguation for agglutinative languages. *Journal of Computers and Humanities*, *36*(4).

Hamann, C. (1996). Null arguments in German child language. *Language Acquisition*, *4*(3), 155–208.

Hankamer, J. (1989). Lexical representation and process. In Marslen-Wilson, W. (Ed.), *Morphological parsing and the lexicon.* Cambridge, MA: MIT Press.

Jusczyk, P. W., Friederici, A., Wessels, J., Svenkerud, V., & Jusczyk, A. M. (1993). Infants' sensitivity to the sound patterns of native language words. *Journal of Memory and Language*, *32*, 402–420.

Ketrez, F., & Aksu-Koç, A. (1999). Once there was a verb. *Proceedings of the Eighth International Congress for the Study of Child Language.* San Sebastian, Basque Country. University of the Basque Country, Spain.

Kim, M., & Phillips, C. (1998). Complex verb constructions in child Korean: Overt inflections of covert functional structure. *Proceedings of the 22nd Annual Boston University Conference of Language Development.* Somerville, MA: Cascadilla Press.

Kornfilt, J. (1994). Some remarks on the interaction of case and word order in Turkish: Implications for acquisition. In B. Lust, M. Suñer, & J. Whitman (Eds.),

Syntactic theory and first language acquisition: Cross-linguistic perspectives, Vol. 1: *Heads, projections and learnability* (pp. 171–199). Hillsdale, NJ: Lawrence Erlbaum.

Kornfilt, J. (1997). *Turkish grammar*. London, England: Routledge.

Küntay, A. & Slobin, D. I. (1996). Listening to a Turkish mother: Some puzzles for acquisition. In D. I. Slobin, J. Gerhardt, A. Kyratzis, & J. Guo (Ed.), *Social Interaction, Social Context, and Language: Essays in Honor of Susan Ervin-Tripp* (pp. 265–286). Hillsdale, NJ: Lawrence Erlbaum Associates.

Lebeaux, D. (1988). *Language acquisition and the form of grammar*, (Unpublished doctoral dissertation), University of Massachusetts, Amherst.

MacWhinney, B., & Snow, C. (1985). The child language data exchange system. *Journal of Child Language, 12*, 271–296.

Pinker, S. (1981). On the acquisition of grammatical morphemes. *Journal of Child Language, 8*, 477–484.

Pinker, S. (1984). *Language learnability and language development*. Cambridge, MA: Harvard University Press.

Phillips, C. (1995). Syntax at age two: Cross-linguistic differences. In: C. Schtltze, J. B. Ganger & K. Broihier (Eds.), Papers on language processing and acquisition. *MIT Working Papers in Linguistics, 26*, 37–93.

Poeppel, D., & Wexler, K. (1993). The full competence hypothesis in clause structure in early German. *Language, 69*, 1–33.

Pye, C. (2001). The acquisition of finiteness in K'iche' Maya. In A. H.-J. Do, L. Dominguez, & A. Johansen (Eds.), *Proceedings of the 25th Annual Boston University Conference on Language Development* (pp. 645–656). Somerville, MA: Cascadilla Press.

Radford, A.(1990). *Syntactic theory and the acquisition of English syntax: The nature of early child grammars of English*. Cambridge, England: Basil Blackwell.

Saffran, J., Aslin, R. N., & Newport, E. (1996). Statistical learning by 8-month-old infants. *Science, 274*, 1926–1928.

Selkirk, E. (1996). The Prosodic structure of function words. In J. L. Morgan & K. Demuth (Eds.), *Signal to syntax: Bootstrapping from speech to grammar in early acquisition* (pp. 187–214). Mahwah, NJ: Lawrence Erlbaum Associates.

Slobin, D. I. (1973). Cognitive prerequisites for the development of grammar. In C. A. Ferguson & D. I. Slobin (Eds.), *Studies of child language development* (pp. 175–226). New York, NY: Holt, Rinehart and Winston, Inc.

Slobin, D. I., & Bever, T. (1982). Children use canonical sentence schemas: A cross-linguistic study of word order and inflections. *Cognition, 12*(3), 229–265.

Stromswold, K. (1999). The cognitive neuroscience of language acquisition. In M. Gazzaniga (Ed.), *The Cognitive Neurosciences*, 2nd ed. (pp. 909–932). Cambridge, MA: MIT Press.

Suomi, K., McQueen, J. M., & Cutler, A. (1997). Vowel harmony and speech segmentation in Finnish. *Journal of Memory and Language, 36*, 422–444.

Tomasello, M. (1992). *First verbs: A case study of early grammatical development*. Cambridge, England: Cambridge University Press.

Tomasello, M. (2000). Do young children have adult syntactic competence? *Cognition, 74*(3), 209–253.

Tomasello, M., Akhtar, N., Dodson, K., & Rekau, L. (1997). Differential productivity in young children's use of nouns and verbs. *Journal of Child Language, 24*(2), 373–387.

8

Scientific Theories and Fodorian Exceptionalism

ZENON W. PYLYSHYN

> *... if men had been born blind, philosophy would be more perfect, because it would lack many false assumptions that have been taken from the sense of sight (Galileo Galilei,* ca *1610).*[1]

Among the many things I have learned from Jerry Fodor is to take ideas seriously regardless of where they may lead, including when they sometimes lead to scientific heresies and barely credible mind-boggling claims, such as the somewhat flippant suggestion that all our concepts are innate (an idea that Fodor borrowed from Plato). But the really radical idea Fodor defended is this: If you really believe that *P* and if you really believe that *P entails Q*, then you should *at the very least consider the possibility* that Q. But Fodor went even further; he actually published papers defending Q, which often made him a pariah among the hordes of psychologists and psychophilosophers who did (and still do) believe that *not-Q*. I even found myself using a minor version of this gambit when I wrote a paper critical of contemporary theories of mental imagery which were (and are) believed by pretty much anyone who has written about mental imagery and which entail that we have actual 2-dimensional pictures in or near our visual cortex. "What the Mind's Eye Tells the Mind's Brain" was a paper that a major journal refused to publish but later admitted to making a mistake when the paper earned the *Science Citation Classic* award.[2]

Although I admit that I learned much of my philosophy from Jerry Fodor (often while defending him from my friends in artificial intelligence) but I learned the academic-social graces from others, so my knowledge is, as Fodor would put it, encapsulated.

I will start by saying something about the most basic acts in both science and philosophy—drawing distinctions. One of the first discoveries I made in my adventure in philosophy is that although words and ideas matter, what may matter even more is the distinctions we make—the right ones being those that carve the world into its natural kinds or (as Plato put it) "*Carve Nature at her Joints.*" Without a basic set of distinctions you can't begin navigating the quicksand of scientific or philosophical ideas. Without some understanding of the nature and the motivation for distinguishing concepts such as sense and reference, analog and digital, belief and knowledge, competence and performance, mass and weight, or energy and momentum, scientific progress would be impossible. I don't even mention distinctions that are important for everyday human thought and action, such as truth and falsity, right and wrong, seeing and imagining, thinking and saying, or even between thinking and imagining-saying. Without these, and many other basic distinctions, we could not begin the work of trying to understand cognition. For some purposes the distinctions just mentioned are needed to prevent our analyses from fall off the road to understanding. To these I have added a few new distinctions that have become central to my work and as well as some of my collaborate work with Fodor: *picking-out* versus *locating* or *indexing* versus *selecting-by-property*, and above all, the distinction between a *process* and a *processing architecture*.

Given how central distinctions-making is in science you might wonder why so little has been written about that process. The one exception I know is a little book by George Spencer-Brown called *Laws of Form* (Spencer-Brown, 1969) first published in 1969 and soon achieved a cult following. It introduced the logic of a conceptual operation based on *Cleaving* (the world or a representation of it) by making a *binary distinction*, as in the distinction between predicate and argument. Spencer-Brown's book begins with the assertion that "We take as given the idea of *distinction* and the idea of *indication*, and that we cannot make an indication without drawing a distinction." It's not clear whether "indication" is taken to mean "reference" or something else. I intend to leave that ambiguity stand because I will impose my own interpretation on what I take to be the underlying idea, which I have variously referred to as "picking out" or "indexing" or by using the technical neologism of "FINSTing" that I coined when I first needed that concept to explain certain phenomena of focal attention (Pylyshyn, 2001, 2007). It served to refer to the way certain things in the visual field are made accessible to thought, though not by virtue of any of their properties (e.g., they are not indexed as things that have certain locations, colors, shapes, motions, and so on) but by virtue of their *enduring individuality*, or what philosophers sometimes call their "numerical identity" which is also a cousin of the linguistic notion of *demonstrating* as in the use of the terms "this" or "that" or of using *deictic* gestures such as gazing or pointing. I believe that this may also be what Spencer-Brown alludes to in using the term "indication" and what I often call "indexing" (Pylyshyn, 2007). Since each of these existing terms carries some unwarranted assumptions, I often turn to the odd but neutral neologism FINST or FINSTing.[3]

There is a long standing philosophical discussion of a family of ideas related to FINSTs, which use a variety of terms for the idea, often talking of demonstrating, or of using demonstratives such as *this* or *that*, which pick out or refer to things without specifying or encoding any of their properties. This discussion includes an analysis of whether demonstrating requires concepts, such as ***sortals*** (See Stanford Encyclopedia of Philosophy at http://plato.stanford.edu/entries/sortals/) and even whether one needs to be conscious of the object demonstrated, as some have argued (Campbell, 1998). Such questions are relevant to our present discussion because if such demonstrating does require concepts then it will not serve the function that FINSTs were intended to serve, namely to provide the beginnings of a bridge between the distal physical world and the cognitive or conceptual world of mental representations, so that particular mental symbols can be associated with particular individual things in the world. The world-mind gap is one of the fundamental puzzles of cognitive science and one that Fodor and I have both obsessed about over the years. My introduction to it was not (at least not originally) through its philosophical role, but through a need to explain certain experimental phenomena, dealing with selective attention and in particular with experiments involving Multiple Object Tracking (MOT).

Although indexing or FINSTing is a very simple idea, it tempts the reader to make assumptions that are not warranted, given the role it intended to play and the body of empirical facts surrounding its use within the FINST theory. The theory begins with the idea that there must be a form of perceptual selection or access to in the world that is pre-conceptual—in other words the visual system must be able early to individuate proto-objects or what I have called *fings*[4] without regard to their type or their properties. Crucially this must even be so if these properties played a causal role in the individuation of the objects or even if these properties are among those that cause them to grab a FINST (Pylyshyn, 2007). I believe that some of my attempts to describe that simple idea may have caused some confusion, so rather than try to repeat these descriptions I begin with a concrete empirical demonstration of what I mean by the notions of FINST and FING. Consider the case of the simple motion illusion known as the Phi illusion (introduced by Wertheimer) which has been widely used to create apparent motion displays in marquees. In general, a succession of brief stills presented in close temporal-spatial proximity is seen as a continuously moving pattern, a phenomenon that is the basis for motion pictures and animated cartoons. The perceived smoothness and the reliability of the illusion depends almost entirely on spatiotemporal properties—motion tends to be seen as progressing from an object located at **L** at time **t** to the location of the nearest object at location $\mathbf{L} + \boldsymbol{\varepsilon}$ and time $\mathbf{t} + \boldsymbol{\alpha}$, where the motion is most reliably seen as smooth when $\boldsymbol{\varepsilon}$ and $\boldsymbol{\alpha}$ are small and correlated (this is sometimes called Kortes third "law" of apparent motion; Gepshtein & Kubovy, 2007). But in cases of multiple element motion the visual system must first determine which object in the first display is the same individual object as a particular individual object in the second display (to establish this identity is to solve what is known as the *correspondence problem*). What is surprising about the solution provided by the

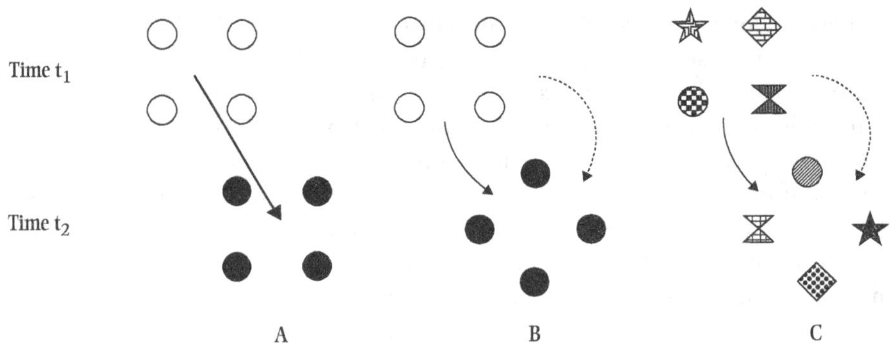

Figure 8.1 Illustrating solutions of the correspondence problem in apparent motion. *Panel A* illustrates the more typical case, where correspondence is established by linking "nearest neighbors." But *Panel B* shows how a particular constraint, known as the "rigidity constraint," based solely on local spatiotemporal properties, results in priority being given to the correspondence that would have occurred if the objects had been fixed to a rigid surface (Dawson & Pylyshyn, 1988). *Panel C* on the right shows that the correspondence is also solved without regard to properties of the objects themselves—the visual system ignores arbitrary shape, texture, and color changes in order to maintain the correspondence based on spatiotemporal properties—just as it does with fixed properties as in *Panel B*. Figure from Pylyshyn (2007). Used with permission.

visual system is that, except for some special cases, it does **not** depend at all on the properties of the static elements that are presented. The solution to the correspondence problem, and therefore the motion that is seen, is insensitive to correspondence of shape, color, or texture of the component elements, contrary to intuitive expectations. Figure 8.1 shows examples from group Phi motion. The caption explains what happens when pairs of the displays are actually show over time on a computer screen.

Another example of blindness to object properties that is characteristic of establishing same-individual is provided by the process I have already described as Multiple Object Tracking (MOT). This paradigm is described in several (Pylyshyn, 2003b; Pylyshyn, 2007). In studies of MOT an observer is shown a set of simple objects (typically 8 identical objects). Then some (usually half) of these objects become briefly distinct. Then all objects move unpredictably (although constrained by certain requirements, such as that they not move off the screen). This sequence is shown in Figure 8.2). After a certain amount of time (typically between 8 and 16 seconds) all objects stop moving and the observer is asked to indicate (by positioning and clicking a cursor) which of the identical objects had been flashed earlier (these are referred to as "targets"). Student volunteers are typically able to correctly identify all targets in this way under a variety of conditions. Since the objects are visually indistinguishable the only thing that allows them to be tracked is their continuing identity (their provenance), which is being tracked by the FINST mechanism which secures the continued identity of the objects.

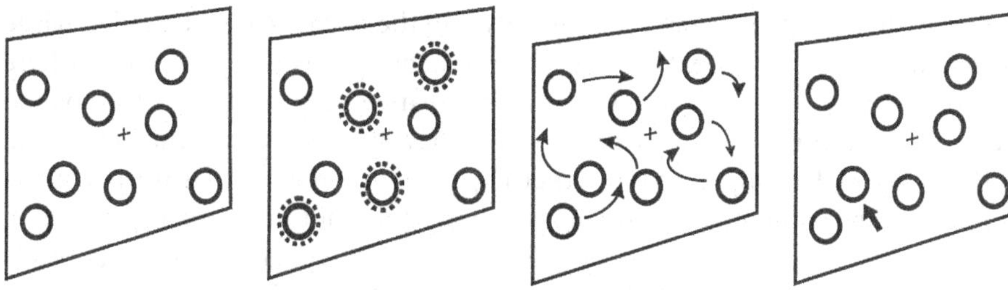

Figure 8.2 This sequence of displays shows what observers are shown. From left to right the first panel shows the initial display, in this case consisting of eight identical circles, located at random, non-overlapping locations on the monitor screen. The second panel shows that four of these circles are made briefly distinct—usually by flashing them on and off a few times. After this the circles begin to move smoothly but unpredictably on the screen, shown in the third panel. The motion was generated in various ways, typically by a stochastic process that also prevents the circles from intersecting or moving off the edge of the screen by simulating an inverse-square repulsion that keeps them apart and away from the edge. In other cases the motion is unrestricted except for maintaining an approximately uniform speed and smooth trajectory. After some period of time (about 10 seconds in our early experiments, but much longer times have been reported) the motion stops and the observer is asked to identify the "target' circles, either by moving the cursor and clicking on each "target" or sometimes by reporting the numeral that appeared inside the circle by judging whether a randomly chosen circle that is flashed was one of the designated targets. From Pylshyn (2007). Used with permission.

The basic experiment sketched has now been repeated hundreds of times in different independent laboratories and shows that observers are able easily to keep track of the target items under many different conditions. Even 5-year-old children and geriatric populations can do this task, although they tend to be limited to three rather than four targets (Trick et al., 2003; Trick et al., 2005). This experimental paradigm has been replicated very often (a list of peer reviewed reports published since 1999 and currently numbering over 150, not counting refereed conference presentations) has been maintained by Brian Scholl at Yale (a list is available at http://www.yale.edu/perception/Brian/refGuides/MOT.html).

Studies with many variants of this basic experimental paradigm have helped enrich our understanding of what goes on when we track objects over time and over changes of various kinds. For example it has been shown that even when the moving objects disappear briefly but completely during the period over which they are being tracked, tracking performance barely suffers providing that the disappearance occurs according to certain spatiotemporal principles—such as being occluded at their leading edge as oppose to shrinking continuously until they become invisible (Scholl & Pylyshyn, 1999). It has also been shown that if the tracked objects disappear and then reappear, tracking performance is degraded more as the distance of their reappearance from the place at which they disappeared increases, but not when the time of the reappearance is longer than predictable from their velocity and the distance they travelled while occluded. Other experiments showed a remarkable insensitivity to exactly where the invisible objects reappeared—including reappearing off their extrapolated trajectory—even reappearing behind where they disappeared (Franconeri et al., 2012; Keane & Pylyshyn, 2006)! Other experiments showed that if the objects that reappeared differed in their color or shape it did not interfere with tracking performance. Indeed, observers were often unable to report whether the objects had changed at all (B. J. Scholl et al., 1999; B. J. Scholl et al., 1999). Tracking appears to be robust and does not use information that could improve tracking performance. If the trajectory of a several independently-moving tracked object could be extrapolated when the objects disappear, it would help in establishing which of the reappearing objects was the lost target. The same goes for the objects' color, size or shape. Yet the visual system appears unable to use such information, for example by extrapolating the motion of several independently moving objects. This is true despite the fact that motion extrapolation of individual moving objects can be done with considerable *precision*. Even the impression that observers have that speeding up the motion makes tracking harder turns out to be an artifact of inter-object spacing because crowding increases as speed increases (Franconeri et al., 2010; Franconeri et al., 2008).

The conclusion of these and other studies is inescapable: tracking moving objects (even objects that change their properties while in motion) can be done without concepts and without noticing, encoding, and utilizing any of the visible properties of these objects other than their enduring individuality or their numerical identity. And a little thought shows that this is precisely how it must be if we are to bridge the physical-mental gap in perception.[5] Other reasons why this has to be the case are presented in a broader philosophical context in (Fodor & Pylyshyn, 2015).

Such findings support the idea that there is a stage in visual perception that is *structurally prohibited* from using relevant background knowledge in such essential tasks as individuating objects and maintaining their individuality while tracking them. Yet many people find this conclusion puzzling. Many of us have observed that if you are afraid in the dark you are more likely to see a swiftly moving black blur as an intruder rather than as a pet or shadow. In the linguistic domain if you hear a sentence like "They looked for hidden cameras, recorders and other -bugs in the room," you are more likely to hear "bugs" than "mugs" or to see the somewhat blurred word as "bugs." In fact if you see "bugs," even in error, that perceived word then prompts the recognition of a briefly flashed related word like "spy," at least within seconds of hearing the sentence (Swinney, 1979). Over the years there have been many experiments that confirmed these sorts of observations (the research movement known as the *New Look in Perception* amassed hundreds of such examples) that gave strong support to such commonsense ideas (Bruner, 1957). But the challenge here, as everywhere in science, is to make the right distinctions—ones that reveal the most general principles. In this case I have argued that we need to distinguish between *seeing* and *seeing as*—the latter of which involves both seeing and categorizing. *Seeing* is the core process that many of us have referred to as *early vision* or the encapsulated vision module (Pylyshyn, 1999). If it can be shown that a certain perceptual phenomenon depends on something you know or recall, or if in order understand the scene before you must appeal to your background knowledge, then this process is not modular. The litmus test of whether some process is modular is whether it is *Cognitively Penetrable*, a term introduced in (Pylyshyn, 1980a) to refer to whether some process can change its function in a rationally coherent way based on cognitive factors, such as knowledge or goals or beliefs.

The qualification "rationally" or perhaps "logically" is needed since otherwise any process might be seen as cognitively penetrable. We can, after all, change what see by moving our eyes or by turning on a light or by ingesting a psychotropic drug. But clearly that's not the type of alteration of function I have in mind when I refer to the cognitive penetration of perception. Nor would I include the kind of alteration that depended on whether I had ever seen a token of that type before—for example my ability to recognize a platypus might well depend on whether I had ever seen a platypus before. As I discussed at length in (Pylyshyn, 1999) the question of how to treat various forms of alteration of visual perception ought to be viewed in relation to the distinctions that the theory makes in order to focus on the essential causal properties of the phenomena. Of course, like most essential distinctions they are not easy to define. (Indeed very little of importance in empirical science, as opposed to mathematics, can be defined. For example, consider the concept "living," or even the simple concepts we all learned in school from definitions provided by teachers or textbooks—like "circle." Definitions such as the latter do not provide ways of understanding or recognizing the defined concept because they apply only to idealized worlds where lines are infinitesimally thin and loci of moving "pencils," such as straight lines, are more precise that any instruments would allow us actually to draw or recognize. That's why Chomsky introduced the distinction between competence and performance (Chomsky, 1965), which influenced my thinking about explanation in cognitive science (Pylyshyn, 1973a).

COGNITIVE ARCHITECTURE: PENETRABILITY AND MODULARITY

Explanation: The Role of Constants, Variables, Parameters, and Explanatory Power

Scientific theories are built on a vocabulary that is proprietary to a particular science and may differ from one area of science to another. A science that purports to be universal, like physics, may also claim that the explanatory vocabulary is fixed, although new terms (perhaps not new concepts) may be inter-defined. In most other sciences, particularly the so-called special sciences, such as psychology, biology or geology, the concepts may be specific to the domain of that science, where the boundaries of the domain are subject to empirical study, The idea is that as the science matures it discovers the relevant Natural Kinds or categories or vocabulary of concepts over which its most general laws will be expressed. It cannot be overemphasized that this vocabulary must respond to the empirical facts, not to the hopes and intentions of the scientists or their sponsors (e.g., granting agencies). To accept a different term for *mass* and for *weight* is to propose a basic distinction of nature; without that distinction Galilean physics and later Newtonian physics would not have been possible.

So what are the basic distinctions in cognitive science? For present purposes I will focus on the distinction between cognitive process and cognitive architecture. Because there are similar distinctions, sometimes with the same names, in other disciplines such as computer science, there is always the danger of assuming that all aspects of the distinctions carry over from these familiar contexts as well. For example the process-architecture distinction is similar to the distinction between hardware and software, or even between virtual hardware that is often simulated on a conventional computer and the software that runs on it. But none of these analogies capture the essential aspect we need in the computational view of mind—described in (Pylyshyn, 1978, 1980a, 1980b, 1984), so I will try to circumscribe my reliance on such analogies.

More on the Physical-Mental Gap: Some Consequences of the Present View

The claim that things in the visible world can establish connections to mental tokens nonconceptually (causally, reflexively, or "mechanically") by capturing a FINST, is a claim that has some rather surprising consequences. Selection of things through a FINST link, is not selection of something as a member of some category or as falling under some concept, even though this may be the usual meaning of "select." For example when a FINST picks out, say, the rabbit before me and links it to a mental symbol, it does not select it under the concept "rabbit" (i.e., as a token of the type "rabbit") or under any other concept (including, for example, "the white furry thing located by that tree"). Similarly a FINST does not distinguish between selecting a whole intact rabbit, a rabbit's properties (such as its fur, shape, movement, color), or a set of undetached rabbit parts (to use the example introduced by Quine, 1969), because the selection is not based on any category at all. But if I do not select it under one of those concepts then there is a real sense in which I don't know what it is that has been selected (i.e., what type of thing it is or where it is) even though I know *which* individual thing it is. If the FINST is to serve as the first link in the chain between the world and our concepts, then we can't initially "know" anything about the cause of first step in the chain (the "quasi referent"). This is the price we must pay if we are to understand how our conceptualizations can be grounded in causal contact with the world. If we knew what we were selecting, then what we select would fall under a concept. In that case the selection would constitute a test (or a judgment) that applied the concept, rather than what a FINST assignment must be, which is more like an interrupt in which the object causally imposes itself on our perceptual system (which is why I talk of FINSTs being captured or grabbed by things in the perceived world).

This then raises an interesting question: what does this early stage of the vision system deliver to the mind? Clearly it does not deliver beliefs of the form "there is a truck coming from my left"; it must deliver something that can be expressed in a much more primitive code or "language." Exactly how primitive this output must be has been the subject of a great deal of philosophical analysis, including the work of Quine and Strawson (Strawson, 1959). I suggest that it is even more primitive than what can be expressed in what Strawson called a *feature-placing language* that (Clark, 2004) adopted in his theory of sentience—i.e., the claim that our initial representation has the form "Feature *f* at location **λ**." According to the view I have proposed, what it delivers may be expressed (roughly) in terms of a demonstrative such as "this" (although the evidence suggests that there may be four or five such distinct demonstratives: this1, this2, this3, this4). In other words it delivers a reference to a selected sensory individual (call it x) to which the argument of a predicate can then be bound, so that properties may be subsequently predicated of x—presumably starting with such predicates as Object(x) or Location(x; **λ**) where **λ** is bound to some location.

Of course there must be some empirical constraints on what can in fact be selected in this way. For example, what is selected must have the causal power to capture a FINST index. Moreover, there is evidence that not just any well-defined cluster of physical features can be selected and tracked—some, like certain arbitrary parts of things, may not lend themselves to being selected and tracked by FINSTs (e.g., in Scholl, Pylyshyn, & Feldman 2001, we showed that the endpoints of lines cannot be selected and tracked), and others may be selected but because of the way they move cannot easily be tracked (e.g., vanMarle & Scholl 2003 showed that objects that appear to liquefy and "pour" from one place to another or that stretch and slink in wormlike fashion can't be tracked). The exact nature of the physical constraints on primitive selection and tracking are empirical questions. As scientists we may carry out research to determine the sorts of properties that tend to grab FINSTs. Since these are primitive causal connections, we may not be able easily to specify what these connections are connections to in the general case; they could be connections instantiated by any possible series of links in some causal chain that starts at the object (FING) and ends with the appropriate stimulation of the retina. Whatever we may eventually discover to be possible properties that cause the assignment of a FINST index, the index itself does not deliver that property or category as part of its function: It just delivers a link to the primitive selection, to the relevant FING, and our tracking experiments suggest that it can do this for up to four or five individuals. According to the view presented here, it is this selection that enables a reference to the selected FING. Moreover, if the FINST was captured by virtue of the object having property P, what the FINST connects to need not be the property P, but the bearer of P (the FING that has property P). A FINST refers to something that has properties with the causal power to capture it, even though it need not refer to those particular properties (e.g., it might refer to the object that has a unique brightness without referring to its luminance at all).

Notice that unlike certain accounts of the fixing of reference (such as the Kripke, 1980 analysis of the fixation of proper names), where there is an appeal to an initial "dubbing" event, the grabbing of a FINST does not involve any intentional act. Since, according to the present story, "grabbing" a FINST is entirely a causal process, the question arises, *which link in the causal chain determines what is selected*? Or which of the possible links in the chain is the one that is being demonstrated or FINSTed. Not one that was *intended* by someone! In vision, for example, the chain includes the light leaving some light source(s), being reflected from some surface(s), passing through the cornea of the eye and stimulating the rods and cones of our retina. Why not say that the light source or some element of texture of the reflecting surface, or the specks of dust in the air through which the light passed, is what the FINST refers to? I claimed that when properties are encoded, they are encoded as properties of particular *Fings*—the FINSTs allows them to be represented as P(x) where P is the property-type and x is the object (FING) indexed by the FINST. So it matters which causal link is associated with what the FINST refers to, since that is what the property P is predicated of or is associated with. Insofar as selection is a causal process, one might take the position that asking what is causally selected is no different from asking which link in the chain is the cause of the firing of the relevant rods and cones—all links are equally part of the causal story. But that isn't true of *referring*. There has to be something that is referred to. As soon as we have a predicate that specifies a property, some particular unique thing is represented as having the property in question. So what determines the particular link in the causal chain that has the predicated property? There are several views on this question, which I will not discuss here. It is one of the big philosophical questions about how reference is naturalized, to which Fodor has contributed a great deal (see the discussion in Fodor, 2009; Fodor & Pylyshyn, 2015).

Whatever a FING is, it clearly does not meet the requirements for *individuating* or for *being an individual*, as understood by Strawson, Quine, and most other philosophers who wrote on the topic. Being selected and FINSTed, are constructs we introduced in FINST theory. FINSTs do not come with conditions of individuation required by these philosophers. That's because FINGs are not true individuals in the general sense; they are what the visual system gives us in order to bootstrap us into relevant causal contact with the world. This is similar to the situation that faces us in many other logically indeterminate functions such as the visual perception of 3-D shapes from 2-D images, as well as perception of (apparent) motion from a sequence of stills. In both cases vision appears to reflect the natural constraints of our kind of world. Such computations appear to solve what seems like conceptual or inferential tasks, yet they do so using wired-in /concept-free and inference-free methods (the term "natural constraint" was introduced by David Marr and discussed in Marr, 1982; Pylyshyn, 2003b, pp. 52–58).

An even more basic and far-reaching examples of how that the automatic FINSTing function has wide consequences in cognitive development is the case of the child's acquisition of the names of things. In order to learn what mother is referring to when she says "rabbit" the child-learner must meet several prerequisites. First, she needs to know that the noise /ræbɪt/ is being uttered with the intention of naming or referring. Second the child needs to have individuated possible candidates for the referent. The latter process has been studied with some surprising findings by Lila R. Gleitman and her students and colleagues (Gleitman & Papafragou, 2005; Gleitman et al., 2005). The "possible candidates" problem may be solved by the FINST mechanism which causes one to four object (granted complex object rather than the simple ones we studied in MOT) to automatically grab an index that then becomes available to connect to a name.[6] Notice that this simple process assumes a primitive individuating and "grabbing" mechanism that is presumably not only non-conceptual but also innate. Such a mechanism (or something like it) is essential for solving what we might call Quine's Gavagai problem—that is, how does the child (or any learner) know which of the myriads of possible things and properties in the world is the referent when learning names by ostention? Much is known about the conditions under which a child solves this problem (and many of them studied by Lila R. Gleitman and her students and colleagues (Gleitman et al., 2005).

Finally: Modularity of Visual Perception

We now have the ideas on which the concept of visual modularity is based. Arguments against the modularity of vision have typically focused on empirical demonstrations suggesting that what we see something *as* is influenced by what we expect, want, believe and so on. These demonstrations are often persuasive. But they are also trade on the vagueness of everyday terms like "see." Is we are to have a science (and philosophy) of seeing, we need—as always—to circumscribe the large range of ways in which "see" is used. In particular we need to distinguish what might be called "seeing" from "seeing as." Clearly what we see something as (or see it as being a token of a certain type or as belonging to a certain category) must be cognitively penetrable since we can only see a particular thing as an instance of some category C if we some knowledge or memory of what C's typically *look like*. Even the strangest objects can be seen as having some shape or other—a shape that can, at least in some approximate way, be conceptualized—i.e., can be formally described in terms of some primitive features and relations (here has been a lot of work in Computer Vision on shape grammars, deformable surface grids, skeletal or axial-based representations and so on. Of course ordinary words rarely allow a unique description. But that's a fact about the vocabulary of language, not about the need for concepts to encode a shape and other object properties sufficient for recognizing the object. Studies in similarity judgments and identification error in recall of shapes strongly suggest that these shapes are build out of conceptual components (Biederman, 1987).

So we arrive at the idea that the first stage of visual processing consists of a nonconceptual and cognitively impenetrable (i.e., encapsulated or modular) process that provides little more than a set of FINST indexes linking things in the visual world with internal symbols. Just having these links allows the next stage in perception to progress because it allows a top-down process to begin the task of recognizing the indexed objects or object parts. As local properties are detected, this mechanism allows the properties to be ascribed to particular FINGs (via their Object Files) which then makes it possible (at least in principal) to determine whether properties of several indexed FINGS belong to the same object cluster or to different objects. If they belong to the same object it gives us a way to solve the conjunction problem (Treisman, 1988) that had baffled those who believe that detecting the presence of a conjunction of properties on the same object requires special high-level mechanisms. In the current view an indexed object has an object file where its properties are stored as they are encoded and if there are several properties in that file, then the associated object is deemed to have a conjunction of those properties (this idea is sketched in Chapter 4 of Fodor & Pylyshyn, 2015).

The reader might have noticed that there is little room in a scheme such as the one sketched in Figure 8.3, for the role of knowledge, expectations and other cognitive states (which Jerome Bruner characterized as "Value and needs as organizing factors in perception") to play a direct role.[7] Which brings us back to another similarly-influential proposal of a modular structure in early vision (Fodor, 1983; Pylyshyn, 1999). In recent times the debate has often centered on experiments allegedly showing how certain cognitive states (especially expectations of what might be seen when the stimulus signal is impoverished or under certain imagined conditions). The problem with these studies is that they fail to distinguish between perceptual effects and effects attributable to pre-perceptual attentional selection and post-perceptual inference as to what was see. Without such distinctions the claim of cognitive penetration of vision is empty (an excellent analyses of alleged top-down effects in vision is provided by Firestone & Scholl, 2016). Much depends on what we take to be the basic phenomena of vision. While Fodor and I have assumed that there is a core subprocess within vision that we and others have called *early vision*. It is up to empirical science to explore its boundaries and discover whether the resulting core process is cognitively penetrable. One might also take the converse tactic of leaving the term "vision" to be whatever generates an illuminating natural taxonomy. Such a natural boundary divides the mind into natural kinds whose boundaries are a matter of empirical discovery—which I suggested might correspond to what we have called the early vision module. This tactic assumes (what surely nobody would deny) that *some* part of vision is encapsulated (even the most extreme anti-modularity fanatics would not deny that for example, what happens on the retina or the auditory ossicles in the middle ear or perhaps even in the visuomotor pathways in the brain (Milner & Goodale, 2007), may be insensitive to the organisms attention; attenuation of signals in particular pathways had been observed some time ago (Hernandez-Péon et al., 1956).

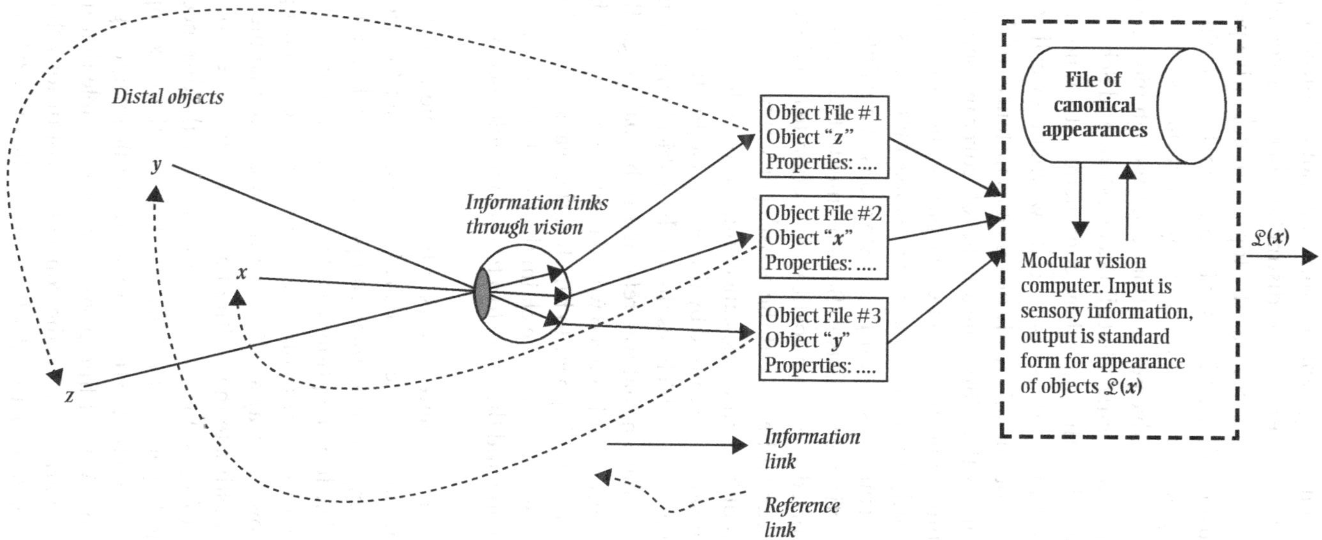

Figure 8.3 The FINST model is shown here as a non-conceptual modular system, augmented to compute canonical equivalence classes of sensory-based visual patterns, beginning with indexed elements (objects or object parts) and some of their shape properties stored in individual object files and looked up in a table of canonical shape form. Based on Fodor & Pylyshyn (2015, p. 125).

But (Firestone & Scholl, 2016) offer a bold and intriguing suggestion; that *all* of what we call vision constitutes such a natural taxonomic kind, once we have cleared away confounding issues such as my proposal that pre-perceptual focal attention and post-perceptual judgment and inference not be considered as part of *vision* for the purpose of assessing the distinction between cognition and perception, and therefor of what counts as the top-down cognitive penetration of vision. To these exclusions they add a list of 6 methodological "Pitfalls" that produce effects that may appear like effects of cognition on vision, but should not be counted as cases of cognitive penetration (even though most recent anti-modularity claims fall prey to them). What allows Firestone and Scholl to defend their bold thesis persuasively is that they show convincingly that, notwithstanding the large number of studies that purport to show what they call "top-down effects," these studies cannot unequivocally be taken as instances of that effect.

Many of the recent studies that Firestone and Scholl cover in their broad survey are found to be victims of one or another of the *pitfalls* the authors list. In fact some of them are so patently unsustainable as to be embarrassing. As an illustrative example, many experiments suggest that distortions of a painted scene (or face) may be due to the artist's astigmatism. But a moment's thought reveals that it cannot be that a distorted retinal or cortical projection led to distortions in rendering the scene on canvas because if the artist's percept was distorted then so would his perception of his paintings, cancelling out the effect. For example the El Greco illusion was said to be caused by the severe astigmatism from which the painter El Greco suffered and which resulted in his subjects appearing to him to be extremely thin—an illusion that was reflected in his paintings. But if that was the case we would expect that El Greco would render the people in his paintings *wider* so that they too would appear to him like the real subjects did. In this (and many similar cases) we can reject the claim that the effect on the painting is due to a cognitive distortion and therefore to a top-down influence of a cognitive state on vision.

Several other pitfalls are reviewed by Firestone and Scholl in their excellent review (rather than cite all these published papers, I refer to reader to the Firestone and Scholl paper, as well as their comprehensive bibliography on this topic (at http://www.yale.edu/perception/TopDownPapers/). For example, one set of experiments they discuss asks observers to wear a heavy backpack and estimate how high or how far away a hill is that they were to imagine climbing. Or it asked observers to estimate the width of a gap that must be crossed while wearing weights on their feet, or for a swimmer to estimate the distance of a goal in water while wearing flippers on their feet, and so on. These examples typically call for estimating some properties of an action while imagining some relevant contextual situation. What I find particularly remarkable that in all such cases there are the close parallels with the conceptual confusions that arise in discussions of mental imagery, many of which I identified in (Pylyshyn, 2002; Pylyshyn, 2003a; Pylyshyn, 2003b). There seem to be a small number of what may be universal confusions which infect our theorizing in cognitive science, maybe as a result of taking the content of our naïve conscious experience as providing a privileged access to the form of mental representations. That certainly appears to be the case, both in the study the factors affecting how we perceive the world as well as in our theorizing about the nature of mental imagery. Is that why Vision Science and Cognitive Science are so hard and so full of "pitfalls"? Maybe. In the meantime I recommend the epigraph quoting Galileo at the top of this chapter.

So Where Do We Stand?

A system that embodies some of these schematic ideas not only offers a promising approach to characterizing the processes that might go on in the visual module, but also provides some ideas for dealing with such hard problems as the *conjunction problem* and the *world-mind gap*. And it also explains a lot of otherwise-unexplained (and often implausible) empirical results purporting to show that vision is indistinguishable from thought. What more could you ask for? As current crop of climate deniers remind us, *it's only a theory*.

AUTHOR'S NOTE

"Exceptionalism" as in "American Exceptionalism" is a phrase rarely heard nowadays. It was a popular term during the Cold War and was used both as praise and as condemnation of the United States. In that sense it served as a Rorschach test. According to the *Atlantic Magazine* (McCoy, 2012) the phrase was introduced by either Joseph Stalin or Alexis de Tocqueville, or both, and was more recently adopted by Newt Gingrich and Rick Santorum as a jingoistic term of praise. I use it here in the same ambiguous sense in referring to my exceptional friend Jerry Fodor.

NOTES

1. I thank my friend Peter Slezak for drawing my attention to this revealing quote from (Galileo Galilei, 1610/1983). Galileo astutely noticed what I have spent years trying to say, though not as well as Galileo did: namely that our intuitions about how vision works, and even more egregiously about what a mental image is and how it functions in mental life, are deeply flawed because they take our experience of these acts as being a direct access to how the information is represented and processed. This is also true of the language faculty where introspection of language understanding and speaking are very misleading, since, like seeing and thinking, the process is largely unconscious.
2. The Institute for Scientific Information invited me to write a short essay on why I merited the award. I said truthfully (Pylyshyn, 1982), that the paper won that title because it had been cited a gazillion times, mostly in the context "(Pylyshyn, 1973b) claims that P but nobody in their right mind agrees." This appears to be a theme for both Fodor and me. In our latest, we write at the outset that, "Most of this book is a defense and elaboration of a galaxy of related theses that, as far as we can tell, no one but us believes."
3. This quaint term is an acronym for "Finger of INSTantiation" because its original theoretical role was to be a mechanism for binding (i.e., *instantiating*) objects

in the visual field to mental symbols—such as the symbolic arguments of mental predicates or functions (see, e.g., Pylyshyn, 2007).

4. I introduced this innocuous term (Pylyshyn, 2007) in order to reassure the reader that it is not a technical term of art but rather a place-holder for a theoretical concept within the FINST theory (a term meaning roughly "whatever is FINSTed") that may or may not turn out to be a natural kind.
5. Little will be said here about this gap (or perhaps chasm). It remains one of the Big Problems in Cognitive Science and the Philosophy of Mind except that it will require not only a theory of early vision, but also an account of how primitive mechanisms such as FINSTs, can provide access not only to individual FINGs, but also to combinations of such primitive objects and properties so as to begin to explain how a child can learn that the thing she is looking at constitutes not only a particular token, but also a type so that she can learn that the word "dog" refers to things of that type rather than to that unique individual (or even to a collection of properties as Quine (1977) argued). Such an account will need to appeal to and/or explain how similarity is established, how "stimulus generalization" occurs and how complex concepts can be assigned to things in the world.
6. Although (Gleitman et al., 2005) report that the child does not entertain many candidate FINGs prior to fixing on the referent of the word, which suggests that there may be a more complex non-conceptual process between seeing an object and binding it to a word (see Note 6 and Figure 8.3).
7. Jerome Bruner's influential 1947 paper, "Value and Need as Organizing Factors in Perception" (cited in Bruner, 1957), initiated a revival of an analysis of perception that derives from linguist and anthropologist Benjamin Lee Whorf and Edward Sapir and that still resonates today.

REFERENCES

Biederman, I. (1987). Recognition-by-components: A theory of human image interpretation. *Psychological Review, 94*, 115–148.

Bruner, J. S. (1957). On perceptual readiness. *Psychological Review, 64*, 123–152.

Campbell, J. C. (1998). Sense and consciousness. *New Essays on the Philosophy of Michael Dummett, 55*, 195–211.

Chomsky, N. (1965). *Aspects of the theory of syntax*. Cambridge, MA: MIT Press.

Clark, A. (2004). Feature placing and proto-objects. *Philosophical Psychology, 17*(4), 443–469.

Dawson, M., & Pylyshyn, Z. W. (1988). Natural constraints in apparent motion. In Z. W. Pylyshyn (Ed.), *Computational processes in human vision: An interdisciplinary perspective* (pp. 99–120). Stamford, CT: Ablex Publishing.

Firestone, C., & Scholl, B. J. (2016). Cognition does not affect perception: Evaluating the evidence for "top-down" effects. *Behavioral and Brain Sciences* (In Press).

Fodor, J. A. (1983). *The modularity of mind: An essay on faculty psychology*. Cambridge, MA: MIT Press, a Bradford Book.

Fodor, J. A. (2009). *The language of thought II*. LOT2: The Language of Thought Revisited. Oxford University Press, 2008. Oxford, England.

Fodor, J. A., & Pylyshyn, Z. W. (2015). *Minds without meanings: An essay on the content of concepts*. Cambridge, MA: MIT Press.

Franconeri, S., Jonathan, S. J., & Scimeca, J. M. (2010). Tracking multiple objects is limited only by object spacing, not by speed, time, or capacity. *Psychological Science, 21*, 920–925.

Franconeri, S., Lin, J., Pylyshyn, Z., Fisher, B., & Enns, J. (2008). Evidence against a speed limit in multiple-object tracking. *Psychonomic Bulletin & Review, 15*(4), 802–808.

Franconeri, S. L., Pylyshyn, Z. W., & Scholl, B. J. (2012). A simple proximity heuristic allows tracking of multiple objects through occlusion. *Attention, Perception and Psychophysics, 72*(4), 691–702.

Galileo Galilei. (1610/1983). The starry messenger. In S. Drake (Ed.), *Telescopes, Tides & Tactics*. Chicago, IL: University of Chicago Press.

Gepshtein, S., & Kubovy, M. (2007). The lawful perception of apparent motion. *Journal of Vision, 7*(8). doi:10.1167/7.8.9

Gleitman, L., & Papafragou, A. (2005). Language and Thought. In K. J. Holyoak & R. G. Morrison (Eds.), *The Cambridge handbook of thinking and reasoning* (pp. 633–661). New York, NY: Cambridge University Press.

Gleitman, L. R., Cassidy, K., Nappa, R., Papafragou, A., & Trueswell, J. C. (2005). Hard Words. *Language Learning and Development, 1*(1), 23–64.

Hernandez-Péon, R., Scherrer, R. H., & Jouvet, M. (1956). Modification of electrical activity in the cochlear nucleus during "attention" in unanesthetized cats. *Science, 123*, 331–332.

Keane, B. P., & Pylyshyn, Z. W. (2006). Is motion extrapolation employed in multiple object tracking? Tracking as a low-level, non-predictive function. *Cognitive Psychology, 52*(4), 346–368.

Kripke, S. (1980). *Naming and necessity*. Cambridge, MA: Harvard University Press.

Marr, D. (1982). *Vision: A computational investigation into the human representation and processing of visual information*. San Francisco, CA: W.H. Freeman.

McCoy, T. (2012). How Joseph Stalin invented "American Exceptionalism." *The Atlantic*, March 15, 2012.

Milner, A. D., & Goodale, M. A. (2007). *The visual brain in action* (2nd ed.). New York, NY: Oxford University Press.

Pylyshyn, Z. W. (1973a). The role of competence theories in cognitive psychology. *Journal of Psycholinguistics Research, 2*, 21–50.

Pylyshyn, Z. W. (1973b). What the mind's eye tells the mind's brain: A critique of mental imagery. *Psychological Bulletin, 80*, 1–24.

Pylyshyn, Z. W. (1978). Computational models and empirical constraints. *Behavioral and Brain Sciences, 1*, 93–99.

Pylyshyn, Z. W. (1980a). Cognition and computation: Issues in the foundation of cognitive science. *Behavioral and Brain Sciences, 3*(1), 111–132.

Pylyshyn, Z. W. (1980b). Cognitive representation and the process-architecture distinction. *Behavioral and Brain Sciences, 3*(1), 154–169.

Pylyshyn, Z. W. (1982). This week's citation classic: Pylyshyn, Z.W. What the mind's eye tells the mind's brain: A critique of mental imagery. *Social Science Citation Index, 42*, 164.

Pylyshyn, Z. W. (1984). *Computation and cognition: Toward a foundation for cognitive science*. Cambridge, MA: MIT Press (Also available through CogNet).

Pylyshyn, Z. W. (1999). Is vision continuous with cognition? The case for cognitive impenetrability of visual perception. *Behavioral and Brain Sciences, 22*(3), 341–423.

Pylyshyn, Z. W. (2002). Mental imagery: In search of a theory. *Behavioral and Brain Sciences, 25*(2), 157–237.

Pylyshyn, Z. W. (2003a). Return of the Mental Image: Are there really pictures in the brain? *Trends in Cognitive Sciences, 7*(3), 113–118. Retrieved from http://ruccs.rutgers.edu/faculty/pylyshyn/tics_imagery.pdf

Pylyshyn, Z. W. (2003b). *Seeing and visualizing: It's not what you think*. Cambridge, MA: MIT Press/Bradford Books.

Pylyshyn, Z. W. (2007). *Things and places: How the mind connects with the world* (Jean Nicod Lecture Series). Cambridge, MA: MIT Press.

Quine, W. V. O. (1969). *Ontological relativity and other essays*. New York, NY: Columbia University Press.

Quine, W. V. O. (1977). Natural kinds. In S. P. Schwarts (Ed.), *Naming, necessity, and natural kinds*. Ithaca, NY: Cornell University Press.

Scholl, B. J., & Pylyshyn, Z. W. (1999). Tracking multiple items through occlusion: Clues to visual objecthood. *Cognitive Psychology, 38*(2), 259–290.

Scholl, B. J., Pylyshyn, Z. W., & Feldman, J. (2001). What is a visual object Evidence from target merging in multiple object tracking. *Cognition, 80*, 159–177.

Scholl, B. J., Pylyshyn, Z. W., & Franconeri, S. L. (1999). *The relationship between property-encoding and object-based attention: Evidence from multiple-object tracking*. (Unpublished manuscript.)

Scholl, B. J., Pylyshyn, Z. W., & Franconeri, S. L. (1999). When are featural and spatiotemporal properties encoded as a result of attentional allocation? *Investigative Ophthalmology & Visual Science, 40*(4), 4195.

Spencer-Brown, G. (1969/1977). *Laws of form*. London, England: George Allen and Unwin (Reprinted in 1977 by Julian Press).

Strawson, P. F. (1959). *Individuals: An essay in descriptive metaphysics*. London, England: Methuen.

Swinney, D. A. (1979). Lexical access during sentence comprehension: (Re)consideration of context effects. *Journal of Verbal Learning & Verbal Behavior, 18*(6), 645–659.

Treisman, A. (1988). Features and objects: The fourteenth Bartlett memorial lecture. *The Quarterly Journal of Experimental Psychology, 40A*(2), 201–237.

Trick, L. M., Audet, D., & Dales, L. (2003). Age differences in enumerating things that move: Implications for the development of multiple-object tracking. *Memory & Cognition, 31*(8), 1229–1237.

Trick, L. M., Perl, T., & Sethi, N. (2005). Age-related differences in multiple-object tracking. *Journals of Gerontology: Series B: Psychological Sciences & Social Sciences, 2*, 102.

vanMarle, K., & Scholl, B. J., (2003). Attentive Tracking of objects versus substances. *Psychological Science, 14*(5), 498–504.

PART II

Concepts and the Language of Thought

9

Fodor and the Innateness of All (Basic) Concepts

MASSIMO PIATTELLI-PALMARINI

An Italian colleague, some years ago, defined me as "un Fodoriano di ferro" (an iron-clad Fodorian). I do not know whether I am so clad, but I have no compunction in admitting I am, indeed, a Fodorian. Jerry's work and innumerable conversations I have had with him over many years, not to mention the privilege of writing a book with him (Fodor & Piattelli-Palmarini, 2010, 2011), have constantly and profitably inspired my own thinking.

I first met Jerry when I organized the Chomsky-Piaget Royaumont debate (Piattelli-Palmarini, 1980)[1] and was there and then favorably impressed by the cogency of his critique of traditional learning theories and unfavorably impressed by the fact that not one of the numerous other participants (with the exception of Noam Chomsky and Jacques Mehler) seemed to have understood what he was saying and how important it was. Ever since, I have had many occasions to present Jerry's argument and to teach it, almost invariably witnessing, at first, a mixture of misunderstanding and disbelief. As Jerry was (and is) the first to admit, there is something paradoxical to his strongly innatist conclusion, but it's also inevitable, in the absence of an equally mighty counterargument. None has been offered so far, so I plead for acceptance. The present regrettable return of neo-empiricism, in the shape of Bayesian learning models (Tenenbaum & Griffiths, 2001; Xu & Tenenbaum, 2007),[2] statistical regularities (Chater, Reali, & Christiansen, 2009), collective exchanges and the extraction of generic cognitive patterns (Chater & Christiansen 2010; Tomasello, 2000, 2003, 2006), recurrent networks (Elman, 1991), and similar notions, makes it relevant to re-propose, clarify, and update his argument. This is what I will be trying to do in this chapter, as a fitting (I hope) homage to Jerry.

FODOR'S ARGUMENT, IN ESSENCE

In order for induction (conceptual learning) to succeed, the "learner" must possess already the concept that she is (allegedly) learning. This applies to all primitive concepts, that is, basically, to all single "words" in the mental lexicon. Any non-composite concept that one can acquire has a full translation into a pre-existing concept of mentalese (Fodor, 1975, 2008). This is deeply paradoxical and counterintuitive, yet this is the way it is.

Prototypical cases: "Brown cow" contains brown and cow. It's a composite concept, built by compositionality. Two words, in fact. But it is not "more powerful" in any interesting sense. Certainly not in the sense of a buildup of progressively more powerful concepts as suggested by Piaget and his school.

It's crucial to acknowledge that this containment relation, unlike in the case of brown cow, does not apply to basic concepts. In particular, just to take the most typical examples: DOG does *not* contain ANIMAL, RED does *not* contain COLORED, KILL does *not* contain CAUSE-TO-DIE (Fodor, 1970).[3]

As Fodor has amply developed in his subsequent writings, what is being ruled out is the classic notion that the meaning of a concept consists in a set of simple primitives derived from sense data (à la Locke), or a network of mandatory inferential links to other concepts, or a web of beliefs.

Basically, the repertoire of primitive concepts is *very* large (in the order of tens of thousands), is *very* heterogeneous, because *all* they have in common is that they are basic concepts (Piattelli-Palmarini, 2003). Nothing else. They are "atomic" (their meaning is not dependent on the meaning of any other basic concept—no networks). They typically are of "middle" generality, neither as specific as "Turkish angora cat," nor as general as "living being" (Rosch et al., 1976). Basic concepts are not just "potentially" present (whatever that means) in the process of learning, they are *actually* there and ready to be tested.

ON LEARNING

Unless it's just a metaphor (as it often is) covering all instances of somehow managing to do later something cognitive that you did not manage to do earlier, bona fide learning, in the classical sense, requires many trials, some kind of differential feedback (it's OK, it's not OK) and a continuous ascending function of increasing success (the famous curves of the behaviorists).[4] Monitoring progress in learning requires some precise hypothesis on what is being learned (one needs split experiments, and solid counterfactuals). If what is at stake (as it should be) is an intensional relation, then there must be a relation based on content (meaning)[5] between input and output. Bare extensional relations are not enough.

The only psychological process of learning that makes some sense, and that has been well studied, is hypothesis testing: the confirmation or dis-confirmation of hypotheses.

As a paradigm case, let's take the kind of experiments carried out, long ago, by Jerome Bruner and his school (Bruner, 1973; Bruner, Goodnow, & Austin, 1956, see Figure 9.1). The experimentalist uncovers and ostensibly sorts cards, one after the other, into two stacks, one is the pile of the exemplars, the other of the non-exemplars. Exemplars of what? Some property (concept, predicate) X that the subject has to "learn" on the basis of the evidence presented and partitioned into those two mutually exclusive sub-sets (is an instance, is not an instance; it's OK, it's not OK; satisfies, does not satisfy; etc.).

In those cards there were four relevant properties, each presenting limited variability: number (1, 2, 3), color (white, gray, black), shape (circle, square, cross), and border (thin, medium, thick). These experiments revealed that there is a spontaneous, quite predictable, hierarchy of hypotheses that subjects try out, one after the other. First come the atomic properties: crosses, squares, circles. Then binary conjunctive properties: three circles, white rectangles, and so on. Then ternary conjunctions: two white squares, three gray circles. The borders are usually disregarded and only come later. Disjunctions (crosses *or* black figures, pairs of figures *or* white circles) come last, if ever.[6]

These data confirmed that there is no learning without a priori constraints on the kind of hypotheses that are actually tested and on their order of precedence. This is a point famously stressed by Carl G. Hempel (Hempel, 1966), Nelson Goodman (Goodman, 1983), Hilary Putnam (Putnam, 1975) and many more ever since.[7] Curiously, there has been scant consensus on the *origin* of these constraints on learning. Fodor and Chomsky have attributed them to innate factors

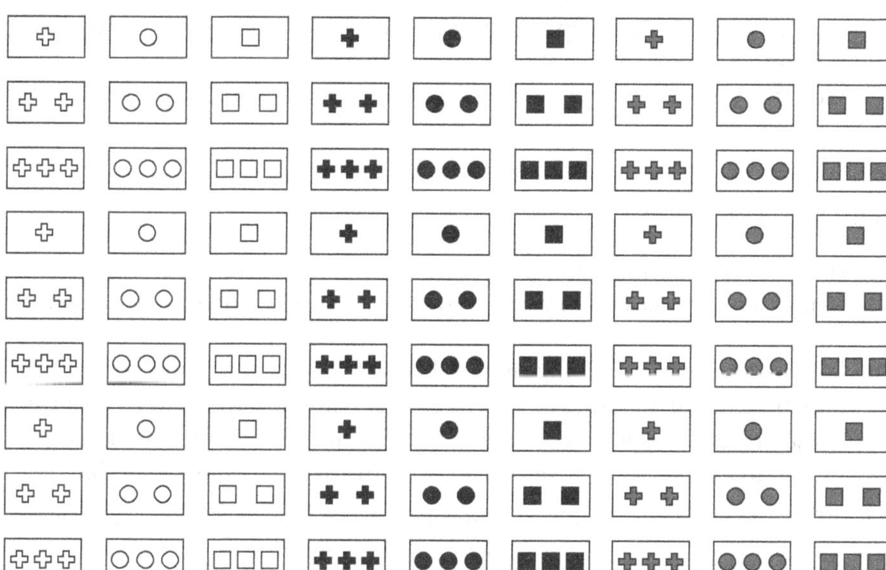

Figure 9.1 Materials from Bruner, Goodnow, & Austin (1956). See text for discussion.

(an explanation that I share), Goodman has invoked an "entrenchment" due to a record of collective past success, while Quine attributed it to Darwinian natural selection. Putnam brings several factors into the picture, some of them social, some cognitive, and keeping innate factors to a minimum.

Let's notice, however, as Fodor rightly does, that all these hypotheses—whatever they are and in whatever order they are tested—in order to be tested, must be "there" already, not just potentially present (as Piaget claimed), but actually there.

Many authors, notably including Piaget and Seymour Papert (who participated in the Royaumont debate, and on whose theses we will come back in a moment) insist that the process of hypothesis testing and the process of hypothesis formation are strictly associated. Piaget was not thrifty in postulating a variety of processes to explain the origin of concepts. These bear names such as thematization, reversibilization, reflective abstraction, abstracting reflection, and more. In the following years, the connectionists and other enemies of innatism have proposed other variants of such processes, for instance "representational redescription" (Elman et al., 1996; Karmiloff & Karmiloff-Smith, 2001; Karmiloff-Smith, 1992).

Fodor is adamant in wanting to keep the source of hypotheses separate from any process of hypothesis confirmation/dis-confirmation, finding himself, for once, in agreement with Hilary Putnam. Their agreement, however, collapses when the nature of the source comes into play. For Fodor the source is innate, while for Putnam meanings are "not in the head." In the published Royaumont exchange with Putnam (see the section Fodor's Suggestions (in the Royaumont Debate)), Jerry accuses him of being "thunderously silent" (*sic*) on the source of concepts and hypotheses. Not only does Fodor stress that the concepts and the hypotheses that are actually tested have to be innate, but adds that the test must be based on their content. This is a crucial point in his argument.

FODOR'S ARGUMENT

All basic concepts are innate.

This is based on three converging, but distinct, lines of evidence and reasoning:

(1) No induction (no learning) is possible without severe a priori constraints on the kinds of hypotheses that the learner is going to try out (this is hardly objectionable, at least ever since the Hempel-Goodman's paradox and Putnam's paradox, and confirmed experimentally in many experiments).

(2) The failure of Locke's project. Concepts cannot be grounded on a restricted set of sensory impressions. More on this anon.

(3) Richer (more powerful) concepts cannot be developed out of poorer ones by means of learning (in any of the models of learning that have been proposed so far).

THE THREE MOST POPULAR LEARNING PARADIGMS

Paradigm 1

The learner already has a repertoire of relevant concepts (predicates, hypotheses), $X_1, X_2, \ldots X_n$

He/she/(it if an automaton) tries them out in some order of decreasing a priori plausibility and selects the best guess compatible with all the evidence seen so far.

Inductive logic will tell you (not an easy task)[8] which hypotheses will be tried out first, second, third, and so forth, and what constitutes "sufficient" confirmation.

But inductive logic is totally silent on the *origins* of the repertoire.

This suggests the innatist explanation for the origin of concepts. We have some understanding of how it works. In fact, it's the only paradigm which we begin to understand.

Paradigm 2

The learner has a repertoire of vaguely relevant, but weaker concepts (properties, predicates, hypotheses), $x_1, x_2, \ldots x_n$ (Piaget, Papert, and connectionism; see the section Papert and Connectionism).

He/she/it must find the means to develop (acquire, generate, compute) a "more powerful" concept X.

Thesis: The methods for testing concepts do also tell you how the more powerful concept is generated (see Piaget's theory). Call this: Feed-back, variational re-computation, abstraction, representational re-description, whatever.

Fodor shows that no such schema could possibly work.

Why? I'll try to expand the following issue a bit.

Sub-Option 1

The learner generates X by sheer luck, and X fits the available evidence by sheer luck. Otherwise he/she dies, along with all the descendants (the neo-Darwinian approach; if it's a computer program, then it's a piece of junk soon to be forgotten).

In fact, most of the time, X is wrong. Only rarely do such guesses work (extreme neo-Darwinian approach).

Then, no learning has taken place, just the biological reproductive fixation of lucky blind guesses.

Special case: The target concept X and the wildly guessed concept Y happen to be co-extensive. They mean quite different things (they are semantically distinct). We may have to conclude that Y (not X) has been "fixed" by natural selection. Deciding what has been fixed may not be easy, requiring split experiments and a host of counterfactuals (Gallistel, 1990; Gallistel & King, 2011). The "lucky" component of this fixation may (just may) apply to genuine triggers, as described by the ethologists. But from ethology, we know that a trigger and what it "triggers" need not have any structural relation. Pecking on a red spot on the mother's gorge brings about the regurgitation of food. We will go back to the issue of triggers in a moment. Triggers are not semantically related to the ensuing result. No learning has occurred. In fact this is a paradigm of lucky, senseless wild guesses, hardly a scheme for genuine learning. Let's then consider a different tack.

Sub-Option 2

The learner, somehow, "tracks" the content of X, and why it is adequate with respect to (true of) the available evidence. The process of progressive convergence is, somehow, guided. The content of X, and some sensitivity of the process to the truth/falsehood of X, supply the required "guidance" (tracking).[9]

Nothing else could supply it. It so turns out that what selection is selection *for* can only be a correct detection of truth-values, individuating which real or possible extensions make the concept true or false. In that case, indeed, there is learning (inductive fixation/rejection) of the *meaning* of X. But X cannot be fixed/rejected unless it is *actually* available to the learner and it is exploited in the process by the learner. We are back to the previous paradigm. Nature endows the learner with a sensitivity for truth and meaning, these are innate predispositions. In the innatist theoretical frame also mental *contents* and *rules* are innate, not just sensitivities and predispositions. This is a point deemed to be unacceptable by the anti-innatists, but any approach that appears to avoid this innatist conclusion is doomed to fail.

Let's continue: the learner "works on" the previously available, weaker (primitive) concepts by means of combinations, re-descriptions, thematizations, whatever (or Quinean bootstrapping à la Susan Carey; Carey, 2001, 2009) and thereby generates a genuinely new and more powerful concept Y.

One possibility: Y is literally a composite concept, composed out of the x's (brown cow) and what it means is that way of composing them. We have the syntax and the compositional semantics of the composition, no less, no more. This is perfectly OK, but then the new concept is not "more powerful" in any interesting way. Moreover, not all concepts can be composite, one has to compose something into them in order to get them, therefore some concepts must be primitive. On these we must concentrate our approach.

This is where the failure of Locke's program becomes important. If all primitive concepts could be derived from sense-impressions and if the "rule of composition" were just association, then, we would have a bona fide empiricist psychology (apt to be simulated in a connection machine). But, as amply shown, this is not a possibility.[10] The acquisition of even very simple concepts requires things like a theory of mind, the understanding of relevant aspects of a situation, understanding the syntax of the sentences that contain them, as shown by a rich literature on the acquisition of the lexicon (see Paul Bloom, 2000, 2001; Lila R. Gleitman [Medina, Snedeker, Trueswell, & Gleitman, 2011; Gleitman & Landau, 2013]; Legate & Yang, 2013).

So, we must offer a very different story:

ANOTHER SUGGESTION DOOMED TO FAIL

One is acquiring genuinely "more powerful" primitive concepts by means of *definitions* that contain and organize more basic ones. It's not "just" the syntax and the compositional semantics, but the articulated net of obligatory, possible, and impossible inferences that the definition specifies. But Jerry has shown, quite persuasively, that this is not a viable possibility. Genuinely more powerful concepts cannot be exhaustively "defined" in terms of less powerful ones (this is what goes under the name of the "plus-X" problem). KILL cannot be CAUSE-TO-BECOME-NOT-ALIVE, because many situations that make the second true are not also true of the first (a calumny propagated on Monday that causes the suicide of the affected person on Saturday, and similar cases). So, is it must be KILL = CAUSE-TO-BECOME-NOT-ALIVE + X? What can X be? Well, as Jerry shows, X cannot be other than KILL itself. So no gain. Truth conditions on formulas containing more powerful concepts cannot be characterized with formulas containing genuinely less powerful concepts. Evidence, suitably labeled (what Jerry calls Mode of Presentation (MOP)), can "activate" them, but not "engender" them, for all the above reasons. Lila R. Gleitman and collaborators have carried out, over the years, an impressive series of experiments showing that word meanings are very frequently acquired upon one single exposure, under clear conditions that correspond to MOPs (in Fodor's terminology).

In other words: the manipulation of primitive concepts can (in fact, it typically does) produce "brown cow" from "brown" and "cow," and the syntax of the composition, but no repetition of "This cow is brown," and "This cow is brown," and "This cow is brown" . . . can generate "All cows are brown," unless you have the universal quantifier ("every," "all") already in your conceptual repertoire. You must have a record of past observations of As and Bs involving some general uniform way of representing "All __ are __." Otherwise you cannot do that, no matter how many As and Bs you observe.

We are now ready for Jerry's final line of the argument.

Learning a concept is learning its unique semantic properties. At some stage you must entertain the following formula (F) in mentalese:

(F) For every x, P is true of x, if and only if Q(x).

Q is supposed to be a "new" concept of mentalese. The one you have (allegedly) "learned," while P is some concept you had already. As a necessary, but not sufficient condition, P must be coextensive with Q, if (F) is correct. But this is plainly not enough: P must be coextensive with Q *in virtue of the intensional properties of P, in virtue of the content of P.* Otherwise (F) is not a correct semantic formula. So Q is *synonymous* with P, but P you had already, ex hypothesis, so you also had Q. End of the story.

To repeat: So Q is synonymous with P. So you had Q already in your "language of thought," because you had P. So Q is not "learned." Iterate this for every primitive concept, keeping in mind the failure of Locke's program.

Conclusion: All primitive concepts are innate. And (due to the failure of Locke's program) they are not all mere constructs from sensory impressions.

It is a shocking conclusion, but it is also unavoidable.

FODOR'S SUGGESTIONS (IN THE ROYAUMONT DEBATE)

So, where do new concepts come from?

Three possibilities:

(1) God whispers them to you on Tuesdays.
(2) You acquire them by falling on your head.
(3) They are innate.

In order to lay out an indubitable, certified example of a more powerful logic and contrast it with a less powerful one, Jerry mentioned sentential logic versus first-order logic.[11] The second, but not the first, countenances quantifiers. The punch line was that no matter how many instances of "brown cow" (and thereby of confirmations of the predicate "brown cow") the learner encounters, the hypothesis "all cows are brown" (or "most cows are brown, or "few cows are brown" and so on) will never become accessible via learning, unless the learner does possess quantifiers already and he/she/it is ready to apply them to the available evidence.

Hilary Putnam (Putnam, 1960) in his critique of Carnap's (notoriously weak) theory of induction, has a fine example pointing in the same direction as Jerry's. Imagine an induction automaton A that has all the number predicates and all the relative frequencies predicates. It is presented with a series of balls of different colors. It may correctly produce and test and finally, with a good coefficient of confirmation, converge on the hypothesis: "one out of five balls is red." However, it will never produce and test the more powerful (and true) hypothesis "every fifth ball is red." In order to do this, we need a more powerful automaton B, that possesses all the predicates that A possesses, plus quantifiers and predicates of ordering. The power to make the transition between A's hypothesis and B's hypothesis has to be built into the machine from the start, it cannot be the piecemeal result of learning from successive inputs of colored balls.

The Piagetian notion of "constructing" (*sic*) a piecemeal ascending succession of genuinely more and more powerful logics by means of abstraction, generalization, and new assemblages of concepts, via hypotheses testing and learning, is untenable.

So is the idea of a transfer from some other (non-semantic) source (pragmatics, usage, general intelligence, constraints satisfaction, social exchanges, etc.). Pace several authors who claim this to be the case, Jerry rightly qualifies it as a miraculous solution.

MEANING VERSUS SORTING

A theory to which Jerry has come back in successive work, persuasively demolishing it, is that possessing the concept C should be assimilated, or reduced, to some pragmatic capacity to sort things-in-the-world into the Cs and the-not-Cs. If that were the case, then such sorting should be done on the C-things "as such," not on any collateral property P, by happenstance extensionally co-instantiated with C. What must be involved is an intensional sorting. Take the (supposedly) extensionally equivalent predicates CAT versus THE-ANIMAL-TO-WHICH-AUNTIE-IS-ALLERGIC. These correspond to different sorting criteria, even if cats indeed are the animals to which auntie is allergic (see Jerry's book "Psychosemantics" [Fodor, 1987]) and LOT Revisited (Fodor, 2008). But sorting cats "as such" is something only God can do. No one can recognize just any old cat (or very small ones, or very big ones, or dead ones, or yet-unborn ones, etc.) under any circumstance (in a very dark night, or under polarized light, etc.). Sorting things into Cs and not-Cs is enormously context-dependent. Appealing to "standard (or normalcy) conditions" will not solve the problem, as the notorious unsolvable problems with verificationism have amply shown. Moreover, sorting is also C-dependent. Standard conditions for cat-sorting are not the same as for fish-sorting, oboe-sorting etc. But, even admitting that there are predicate-independent standard sorting conditions for sorting Cs, they cannot be part of the *content* of C. Normalcy conditions do not compose. For example, take NIGHT-FLYING BLUEBIRDS. Got it? Sure. But normalcy conditions here are patently conflictual. Maybe you recognize them by way of a unique song they sing, or a unique smell they produce, but that cannot be constructed by composing the normalcy conditions for things that fly at night and those of birds in general and those of bluebirds in particular. Even if that song (or smell or whatever) is criterial and unique, it's so in virtue of the property of being night-flying bluebirds. No way out. So, even if the sorting criterion did apply to primitive concepts (but it doesn't), it would not apply to composite ones.

Why does it not apply to primitive concepts? Because, as we saw, also sorting things into primitive concepts requires normalcy conditions, and there are no concept-independent general criteria of normalcy conditions. These conditions would have to compose, because concept possession and identification are systematic and productive, but no epistemic criteria do compose. Jerry, therefore, urges us to conclude that possession conditions for concepts aren't epistemic, they are (as he would put it) metaphysical. Confusing issues of metaphysics with issues of epistemology in the domain of semantics is a capital sin, against which Jerry has thundered relentlessly for a good part of his career. Having concept C "just" *is* being able to *think* about Cs *as such*. Sorting, inferences, perceptual accessibility, ease of representation, relevant beliefs, and so forth, are all secondary to this. Abilities to think about Cs do compose, as they should. Minds are for thinking and concepts are for thinking with. Can some kind of dispositions do the job? No, because, in Jerry's own words: "Mere dispositions don't make anything happen." What causes a fragile glass to break is not its being fragile. It is its being dropped.

ABOUT TRIGGERS

A very interesting cautionary proviso was made by Jerry in the debate. After concluding that, for all the reasons he had offered, we have to admit that all primitive concepts are innate, he adds:

> Unless there is "*some notion of learning that is so incredibly different from the one we have imagined that we don't even know what it would be like, as things now stand.*"[12]

A different notion of learning, usually now replaced by the term acquisition, has indeed been offered in the domain of language. It's the very idea of principles and parameters, where learning is assimilated to the fixation by the child of a specific value for a restricted set of binary parameters (Roeper & Williams, 1987; Gibson & Wexler, 1994; Breitbarth & Van Riemsdijk, 2004). One of its first proponents, Luigi Rizzi, has recently succinctly explained the very idea:

> "*The fundamental idea is that Universal Grammar is a system of principles, expressing universal properties, and parameters, binary choice points expressing possible variation. The grammar of a particular language is UG with parameters fixed in a certain way. The acquisition of syntax is fundamentally an operation of parameter setting: the child fixes the parameters of UG on the basis of his/her early linguistic experience. This approach introduced a powerful conceptual and formal tool to study language invariance and variation, as the system was particularly well suited to carefully identify what varies and what remains constant across languages. And in fact, comparative syntax using this tool boomed as of the early 1980's, generating theoretically conscious descriptions of dozens of different languages.*" (Rizzi, 2013)

This is certainly a "different" notion of learning, which I could characterize as revolutionary, if the term had not been tarnished by too frequent and too sloppy use. Some years ago, in fact, Noam Chomsky has said that the notion of principles-and-parameters is the most significant and most productive innovation that the field of generative grammar has introduced into the study of language. It was initially introduced as quintessentially applicable to syntax, with some plausible specific candidates. Later on, and for some researchers still today, parameters are rather circumscribed to the functional morpho-lexicon (Borer & Wexler, 1987; Boeckx & Leivada, 2013; Boeckx, 2006), narrow syntax being genuinely universal and not parameterized. But this is not the place to delve into the interesting complex issue of the nature of parameters. See the special issue of Linguistic Analysis "Parameters: What are They? Where are They?" Edited by Simin Karimi and Massimo Piattelli-Palmarini (2018, forthcoming). A vast specialized literature is available. What counts here, in the present context, is a conceptual consequence of that model, well condensed by Wexler and Gibson with the notion of "triggers" (Gibson & Wexler, 1994).

This notion comes from ethology and Jerry has used it as a prompt retort in the Royaumont debate. Piaget and his close collaborator Bärbel Inhelder in previous writings and during the debate had suggested that language is literally constructed (*sic*) upon primitives present in motor control.[13] The French biologist Jacques Monod was quick to question this hypothesis (and was absolutely right in doing so) by saying that, in that case, quadruplegics should never develop language, but they do. Inhelder replied that very little movement suffices, "even just moving your eyes" (*sic*). Jerry's quick retort was that, even admitting that it's so, then eye movement is a trigger, exactly in the sense given to the term by ethologists, and not a motor primitive suitable for construction.

It is none other than Jerry, in fact, who has wisely stated that cognitive science begins with the poverty of the stimulus (for a recent multi-disciplinary characterization of this notion see (Piattelli-Palmarini & Berwick, 2013). He and I surely are among those who are grateful to the generative enterprise for having dissociated the child's acquisition of her mother language from any model invoking inductive processes and trial-and-error. The effects on the growth of the child's mind of receiving relevant linguistic data bear a close analogy to the effect of triggers. One significant quote from Chomsky (among many more in his work) stresses this point:

> "*Language is not really taught, for the most part. Rather, it is learned, by mere exposure to the data. No one has been taught the principle of structure dependence of rules (. . .), or language-specific properties of such rules (. . .). Nor is there any reason to suppose that people are taught the meaning of words. (. . .) The study of how a system is learned cannot be identified with the study of how it is taught; nor can we assume that what is learned has been taught. To consider an analogy that is perhaps not too remote, consider what happens when I turn on the ignition in my automobile. A change of state takes place. (. . .) A careful study of the interaction between me and the car that led to the attainment of this new state would not be very illuminating. Similarly, certain interactions between me and my child result in his learning (hence knowing) English.*" (Chomsky, 1975)

In subsequent years, in his 1998 book on concepts, modestly ensconced in a footnote, Jerry says:

> "*As far as I can tell, linguists just take it for granted that the data that set a parameter in the course of language learning should generally bear some natural*

> *unarbitrary relation to the value of the parameter that they set. It's hearing sentences without subjects that sets the null subject parameter what could be more reasonable? But, on second thought, the notion of triggering as such, unlike the notion of hypothesis testing as such, requires no particular relation between the state that's acquired and the experience that occasions its acquisition. In principle any trigger could set any parameter. So, prima facie, it is an embarrassment for the triggering theory if the grammar that the child acquires is reasonable in the light of his data. It may be that here too the polemical resources of the hypothesis-testing model have been less that fully appreciated.*" (italics as in the original; Fodor, 1998a, p. 128, n 8, "Linguistic footnote")

This is a very interesting remark, and in a personal communication (September 2013) Jerry says he still thinks it's correct and that he does not have anything much to add to it.

Being a fan of principles-and-parameters and of something like triggers (more or less a` la Gibson and Wexler), I have been puzzled by this remark.

Chomsky's reaction, if I understand it correctly,[14] is basically that evidence is always pre-structured somehow by the innate endowment and, even if marginally, by a previous history of exposure. Evidence thus pre-filtered always has some structure, and it may impact the speaker's internal state (I-Language) producing an internal change of state (see Yang's 2002 book for a detailed picture). No appeal to induction is needed. Moreover, there is no such thing as a "relation" between the speaker X and X's I-language, such that X examines linguistic data, accesses some internal representation of his/her I-Language, and tries out hypotheses "about" it. There is, simply, an internal state of the speaker's mind-brain, and this state may vary marginally over time, under the impact of relevant evidence. No "relation," no induction, but just the speaker's being in a certain internal state.

A much simplified vignette of parametric acquisition upon single exposure to a relevant sentence, the one Fodor is alluding to, would be, for the child acquiring Italian: "andiamo e poi giochiamo" ((we) go and then (we) play). No subject is manifestly expressed, therefore the local language is a pro-drop language. The child fixes the pro-drop parameter to the value +. Real life instances are more complex than this simplified case (Berwick & Niyogi, 1996; Janet Dean Fodor, 1998b; Kam & Janet Dean Fodor, 2013; Roeper & Williams, 1987), but it may convey the essentials. No classic hypothesis testing, no induction, no trial and error. But it's not an arbitrary pairing either, as it would be for a bona fide trigger. Fodor's perplexity bears upon the relation between the form of this sentence and the ensuing specific parameter value fixation by the child.

My own take on this issue is that the child innately has in her mind, in the domain of language, several formulae like the following (borrowing the expression "doing OK with" from Putnam).

(L) Given linguistic input L, I will be doing OK with parameter value X.

Several conditions apply to the application of (L) (see Gibson & Wexler, 1994; Thornton and Crain, 2013), but parametric acquisition comports something like (L).[15] Single stimulus learning is an idealization, a useful one, but still an idealization. Some frequency threshold of incoming linguistic stimuli my well have to be attained for parametric fixation to succeed (Yang, 2002, 2011a, 2011b; Legate & Yang, 2013). The nature of the relation between L and X is pre-wired, not the result of induction, but L is not an arbitrary trigger either. In fact, some distinction ought to be introduced between releasing mechanisms and bona fide triggers. In immunology we have many examples of releasing mechanisms that bear a functional relation to the final output. When the organism makes contact with pathogens, this encounter normally switches on automatic reactions of recognition, binding, and rejection, approximately in this order. The net result is the maintenance of a healthy state. Many spontaneous reflexes are appropriate responses to the releasing stimulus: pupil contraction and closing one's eyes in the presence of a blinding flash of light, head and body retraction in the presence of a dangerously incoming object, vomiting when unhealthy substances are swallowed, and so on. Granted, these stereotyped reactions can also be activated by irrelevant stimuli. For example, the intravenous injection of minute quantities of harmless egg yolk induces a powerful immune reaction. Such systems can be "fooled" and a certain degree of arbitrariness of the releasing input does exist, but it's far from being total arbitrariness. In the case of parametric acquisition, therefore, we do not have to choose exclusively between induction and arbitrary triggers. The structure of suitable linguistic inputs and the ensuing fixation of an appropriate parametric value bear complex non-arbitrary relations that researchers have been investigating with some success (Yang, 2011a, 2011b).

It is indeed "a notion of learning that is so incredibly different from the one we have imagined," just as Fodor has wisely suggested. The innate components of the process are manifold, complex, and language specific. There is no "construction" from any kind of non-linguistic primitives.[16]

OBJECTIONS AND COUNTERS

There have been two interesting objections to Fodor's innatist conclusion that I think are worth summarizing here, along with Fodor's retort. The first was made there and then, at Royaumont, by Seymour Papert, the second in a commentary written by Hilary Putnam after the debate and published in the proceedings. Let's start with Papert.

SEYMOUR PAPERT ON CONNECTEDNESS

Papert introduced in the debate, in his critique of Fodor's innateness, a device called a perceptron, in hindsight an early, elementary artificial intelligence device, similar to present-day connection machines. Basically, this device has an artificial retina connected to a parallel computer. There are several interconnected local mechanisms, none of which covers the whole retina. None of these has any "global knowledge." The device computes weighted sums of the local "decisions" reached by each sub-machine. The result is global decisions, not localized in any sub-part. Papert insists that there is a "learning function" sensitive to positive and negative feedback supplied from the outside, until the device converges onto a correct global decision, such that new relevant instances can be presented and recognized correctly. What can it learn? The answer is, according to Papert, far, far from obvious. It can easily learn to discriminate, say, between "triangular" and "square," by looking at local angles in the retinal image. What about the property or predicate "connected"? Can it learn to decide whether the image is made up of one single connected piece, or several connected pieces? The answer (far from obvious) is that it can.

Imagine an investigator (a Fodorian) who, therefore, concludes that "connectedness" is innate (prewired in the machine), but the wiring diagram cannot reveal anything that corresponds to "connectedness." Big surprise! Therefore, Papert concludes, one has to be very careful in concluding that a predicate is innate.

EULER'S THEOREM

A centerpiece of what Papert presented is a theorem due to Leonhard Euler, proving that, if and only if the algebraic sum total of all the curvatures along the borders of a blob is 2π, then the blob is connected, whatever its shape. Otherwise it's not (Figure 9.2).

If it is n2π, then we have *n* distinct connected objects.

No piece of the machinery "possesses" (is sensitive to) the concept of connectedness, nor does it "contain" such concept. Angles do not even have to be measured continuously along the borders of each blob, provided all are, in the end, measured and given the appropriate algebraic sign (say, plus if the rotation is clockwise, minus if counter-clockwise).

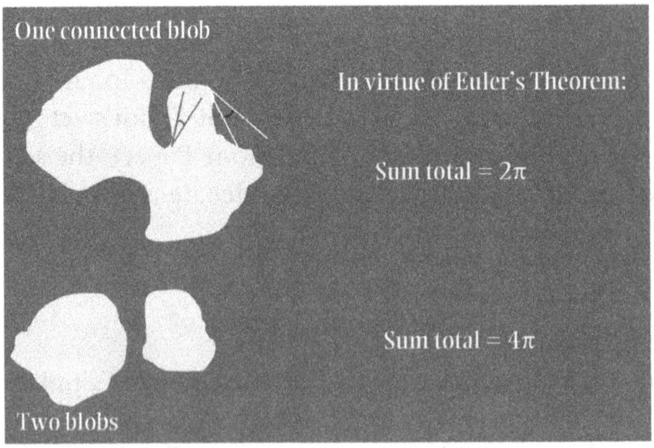

Figure 9.2 Application of Euler's Theorem. See text for discussion.

The pieces of the perceptron just detect angles and measure them. And the machine as a whole then sums up all the angles. It does not have to "know" (keep track of) whether two non-successive observations of angles are concerned or not with the same blob. As a result (not of innateness, but of the process itself), the machine is sensitive to connectedness, in exactly the right way, thanks to Euler's theorem. Connectedness is neither innate (prewired) nor learned. It is the inevitable consequence of the dynamics of the process, the development of the process, and the deep property discovered by Euler. He insisted that terms like "concept," "notion," "predicate" are generic and misleading. We need better ones.

Papert's Piagetian "lesson" is that, similarly, the cognitive capacities of the adult may well be neither innate, nor learned. They have a developmental history, as shown by Piaget. They emerge from other, different, components in a process of construction. Whatever is innate will not resemble in the least what you find in the adult's mind. The real search will have to track precursors, intermediate entities, and constructions. The point is: The perceptron, indeed, has the concept "connected," but it's precisely and exhaustively defined on the basis of other predicates the machine is sensitive to (local angles of curvature and their algebraic sum). So contra Fodor's thesis of the innateness of all concepts, this global concept is genuinely constructed from strictly local ones. If you had searched the "genome" of this machine to find where "connected" was encoded, the answer would have been: Nowhere! Yet the machine has it.

FODOR'S COUNTER

The machine has the concept "connected," since it necessarily (not by sheer luck) applies the concept correctly to all and only the connected blobs. You would not have noticed that it had this concept, and why, if you were not as clever as Euler.

But it does have the concept "connected," exactly for the reasons explained by Papert, based on Euler's theorem.

If all one has to decide are mere extensional criteria (the behavior of the device, what it prints out), then one will never know whether the device is "answering": "Yes, the figure is connected," or "Yes, I am printing 'yes,'" or "Yes, I am printing 'Yes, it's connected,'" or innumerable other possible printouts. It's just like in the extensional behaviorist approach to learning. Suppose the mouse manages in the end to learn to make the right turn in a maze. Has it learned to turn right, or to turn (say) North, or to move his left legs faster, or to go away from the light, and so on? The learning curve cannot make any distinction among these possibilities. One needs to plan split experiments (turn the maze 90 degrees, or turn it away from the light, or flood it with water, etc.) and develop relevant counterfactuals.[17] If Papert's device is remotely similar to a human mind, then judging about connectedness, it must have an internal representation of something like

(C) If, and only if, total sum = 2π, then the figure is connected.

Otherwise, we succumb to Fodor's essential indeterminacy.

Being able to determine which predicates a device has involves it having a repertoire of internal representations and several computational options. That is, a whole system of predicates, and the quantifiers (as Putnam's case of the two automata shows). It may not be easy to determine which innate predicates a child has, but ease of discovery cannot count as a criterion for the existence of innate concepts. A cognitive system does not have only the concepts that it's easy for us, cognitivists, to ascertain that it has.

So, in Papert's case, it may take a mathematical genius like Euler to actually discover that the device, somehow, has the predicate "connected" and determine the way this relates to total curvature, but since (C) is the only criterion one can envision, then the device must have (C) as an internal state.

PUTNAM'S CRITIQUE OF FODOR

It is true that learning must be based on dispositions to learn (or "prejudices") that are not themselves the result of learning, "on pain of infinite regress." Everyone, notably including the empiricists, granted that. The causal explanation for these dispositions is, quite plausibly, some functional organization of our brain. But one has to be careful here: if any device that can give correct answers on property P is said to possess some P-related predicate, then thermometers "possess" the predicate "70 degrees," and speedometers possess the predicate "60 miles per hour." This is patently absurd. The dividing line is with systems that can learn about properties, and that can master a language. This needs, however, great caution. Let's imagine two devices: the "inductron" and the "Carnaptron." The inductron is capable of making only an extremely limited set of inductions (say, just 1), monitoring the attainment of a certain threshold of confirming instances, over all instances.

The more sophisticated Carnaptron accepts or rejects certain "observation sentences" in simple first order language, under appropriate circumstances that it can detect. It might well be monitoring some Bayesian degree of confirmation and print out probabilities that the sentences are true. It uses an inductively defined computation program, whose definition is over the set of sentences it can accept or reject.

It is minimally appropriate to describe the Carnaptron as "having" a language. One can well generalize this to a hypothetico-deductive device that carries out eliminative inductions, given a simplicity ordering of hypotheses in a certain language. Now, the inductron is a dismally weak machine, totally unable to account for the mastery of natural language, it does not "have predicates." What is in need of careful analysis is whether a hypothetico-deductive device has predicates.

Let's repeat Fodor's formula:

(A) For every x, P is true of x iff Q(x)

Formula (A) is in "machine language" or "brain language" (or in the "Language of Thought" [LOT], according to Fodor). So is Q, by hypothesis. What about P?

Fodor purports to have shown that P must have a full translation into LOT. So P is synonymous with some predicate that LOT possesses already.

Let's see how this can be false.

Imagine a programmable digital computer, a hypothetico-deductive machine like the one we just saw. Its machine language has "add," "subtract," "go to step N," "store x in register R," etc. But no quantifiers. Generalization (A) cannot even be stated in that language. Formula (A) is, therefore, contra Fodor, not in "machine language." Maybe, it's in some formalized Inductive Logic Language (ILL), according to some program of eliminative induction. Suppose, then, that Fodor really means ILL, not LOT. Well, his argument does not hold even in this weaker case, because Fodor's P must be equivalent to some subroutine in machine language (something a compiler can understand and process). Call it S. Even granting that the brain-mind learns P by making an induction, the conclusion will not be formula (A), but formula (B).

(B) I will be doing OK with P, if S is employed.

This (but not [A]) can be stated in ILL, provided ILL contains machine language and has the concept "doing OK with." But this does not require ILL to have synonyms for "cow," "face," "kills," and so on. Fodor's argument has failed.

The punchline is that notions such as "rules of use" or "doing OK with" have been (tacitly) unduly extrapolated, making Fodor's argument seem cogent. But it's not. "Doing OK with P" is not problematic: The device does not really have to "understand" it the way we do. It's just a signal to the device to add S to its repertoire of subroutines. Nothing more. The hypothetico-deductive device, or a collection of such devices, is the best model we have (contra associationist models), but not for the reasons offered by Chomsky and Fodor. Better reasons can be offered. And indeed they must.

PUTNAM'S OWN CRITIQUE OF ASSOCIATIONISM

Associationist models can accommodate a very rich repertoire of categorizations and inductions. This is not the problem. But they are all first-level inductions.

These models cannot accommodate higher-order inductions (cross-inductions).

They cannot accommodate inductions on types of inductions (such as: "Conclusions about internal structures from just the observation of manifest properties are usually unreliable"). In order to do this, observations and first-order inductions must be represented in some uniform way. We need classes-of-classes, classes-of-classes-of-classes, and so on, not just things and properties, but inductions as such have to be represented internally in some uniform way, and they have to be quantified over.

The model ceases to be an associationist one. We have to model it as having a (complex) language. This is a persuasive (though not "knockdown") argument that Chomsky and Fodor should have used.

But associationism is not as hopeless as Chomsky and Fodor claim it is.

It allows one to carry out any amount of independent first-level inductions, on any amount of categories and classes, and some can be pretty complicated. One, then, needs a great variety of dedicated subunits, indeed the modules proposed by Chomsky and Fodor. Maybe a strong modular hypothesis is compatible with associationism, after all.

Putnam's punch-line is: Unless one accommodates a lot of cross-inductions and high-level inductions, uniform representations, quantification, and multi-purpose strategies, one has no reasons to exclude an associationist (empiricist) model for language and language learning. Chomsky's and Fodor's rejection of learning altogether, and of anything resembling general intelligence, weakens their anti-empiricist position instead of consolidating it.

FODOR STRIKES BACK

Putnam's suggestion, basically is: What if learning a concept is not learning its content, but something else? Say, its rules of use. Well, the same kind of argument applies. Then "rules of use" are not the result of learning either. You tell me what is necessary and sufficient to learn a concept. Call it X. Fodor's argument will show you that X (whatever X is) has to be innately available to the "learner."

The inference from perceptrons to thermometers and speedometers can be blocked easily. As Putnam points out, perceptrons are supposed to be learning devices, and it is the case that every device capable of learning must have predicates, because these devices must test hypotheses, and there can be no hypotheses without predicates. We have many reasons to say that the inductron has no predicates (as Putnam also concedes). One is that we could not say which predicates it has, even imagining it had some. "Yes there are tree white blobs," "Yes, I am printing 'There are three white blobs,'" "Yes, I am now typing 'Yes,'" and innumerable more, are all coextensive with the machine's rudimentary "behavior."

Only a machine that has access to a sufficiently rich set of computational options can be said to have predicates. Predicates come in systems.

Putnam offers a slippery slope argument: since we cannot set a precise threshold for deciding when a device is sufficiently complex to say it has predicates, then we can only resort to Wittgenstein's criterion of a full set of rules of use and inductive definitions (a full language). Indeed, we have no precise threshold, but we cannot tell precisely, either, when an acorn turns into an oak. This does not mean that nothing is an oak. All we want is to be able to stop the inference short of thermometers and speedometers, even if, for the sake of the argument, we would admit that the inductron has predicates. No real problem here.

The assumptions behind Fodor's argument for predicate innateness are extremely simple and unquestioned by anyone. To repeat: Learning a predicate is learning its meaning/content. Language learning involves (inter alia) the projection and confirmation/refutation of hypotheses.

Putnam proposes his formula (B):

(B) I will be doing OK with P, iff S employed.

S, indeed, (Fodor agrees) must be specifiable in machine language (in LOT), but Putnam is "thunderously silent" on the origins of S. If learning P really is identifiable with learning S, then the device truly concludes (B). Now, the question is: What can S be? Answer: Some procedure for sorting things into those that satisfy P and those that don't, by reference to whether they exhibit, or not, some property Q. Just add (as one must) that exhibiting Q determines satisfaction of P as a consequence of the meaning/content of P, and you are back to Fodor's formula (A). Fodor's (A) and Putnam's (B) are in fact equivalent. The difference is that Putnam (after Wittgenstein, and with all procedural semanticists) suggests that meaning is "rules of use," and this leads to some operational definition of P. Even granting that, for the sake of the argument, the rules of use (or operational definitions) have to be innate. The argument goes through regardless. You tell me what you think is learned when P is learned, call it X and I show you that X (whatever that is) must be assumed to be innately available to the learner. Period.

In "The Meaning of Meaning" Putnam had taken a different stand. Learning P is to be connected to two things: Some prototypical exemplar and a certain extension. The second component is "not in the mind," not under the governance of the individual speaker. It's determined socially. Only the progress of science may ultimately determine what is necessary and sufficient for a thing to be a P. Fodor remarks that, even in this theory, P is not learned. There is no internal subroutine S, but a complicated collective causal story to be told.[18] In this causal-collective story something may well satisfy S and still not be P (Putnam's famous examples of grug, molybdenum, twin-water etc.). In this story, in fact, P is neither innate nor learned.

It's quite possible, in this story, that nobody "really" ever understood the meaning of P, and nobody will for another (say) two centuries. There is no such thing as essential conditions for being a P (there is no S and no Q). "Meaning ain't in the head." A totally different story.

On cross-induction, Putnam is right: Cross-induction forces us to go well beyond associations, and to impute mental representations and mental computations to the organism, but this does not entail that the organism *only* needs them when it makes cross-inductions. It needs them long before that (for instance for beliefs about the past, the future, false beliefs, counterfactuals etc.). Mental ontology must be separated from epistemology. Suppose we are forced to admit the existence of molecules only when we consider phenomena of solubility. We would not conclude that only soluble materials are composed of molecules. One thing is to admit that there is a mental medium of computation (the Representational-Computational Theory of Mind), another to suppose that a lot of specific and structured contents are innately present in this medium. Fodor and Chomsky endorse both these hypotheses, while Putnam (and other defenders of "General Intelligence") accept the first, but not the second. Looking at other species, we see a lot of specialized behaviors and specialized innate dispositions (not much of "general intelligence"). It's reasonable to infer that our species is not so different, that our mind-brain is heavily modular.

PUTNAM'S REJOINDER

Quite a lot is known about general learning strategies (Bayesian probability metrics, inductive logic, extrapolation of functions, etc.). So the notion of multipurpose learning strategies is no more vague than Chomsky's "language faculty," or "universal grammar."

Putnam denies that a grammar is a description of properties of his brain, but he does not deny that a grammar is represented in his brain. Putnam says: "*The geography of the White Mountains is represented in my brain, but the geography of the White Mountains is not a description of my brain.*"

There is a referential component to meaning, and it is rooted in causal interactions of whole communities of speakers to stuff out there.[19] Concepts are not in the head. There is also a use component. And Fodor's argument fails even if we limit ourselves to just that component. A subroutine is the description of the employment of a concept, not the concept (the predicate) itself. Even if use were all there is to meaning, then Fodor's argument would show that "*Mentalese contains devices for representing the employment of all predicates, not that mentalese already contains all predicates.*"

CONCLUSION (MINE)

There still is widespread and often fierce resistance to the notion that there can be innate mental *contents*. It's considered OK that there are innate dispositions and innate cognitive processes (devices for representing the employment of mental states, to put it in Putnam's terms), but the innateness of all basic concepts appears to be in a different league of plausibility. Also unacceptable to many is the hypothesis that several formulae like my (L) are innately available to the child, allowing non-inductive parametric language acquisition. A general consideration that supports this hypothesis bears upon a mini-max solution to the problem of language acquisition (Vercelli & Piattelli Palmarini, 2009). Two biological solutions to the problem can be envisioned, in the abstract: (1) Make every linguistic trait innate and be ready to accommodate a quite heavy genetic load. (2) Make every linguistic trait learned and be ready to accommodate a lot of neuronal plasticity and a long and tortuous path of inductive attempts. We think that it could be shown, quantitatively, that the best compromise between these two possible solutions is parametric acquisition. A restricted set of innate pre-wired dispositions to apply formulae like (L), implying the detection of the relevance of input L toward the parametric value X, and rapid convergence upon parametric value X, as shown by Charles Yang. This should assuage Fodor's perplexity about triggers.

As to the learning of concepts, Fodor's argument is perfectly cogent and is supported by the awesome rate of acquisition of words by the infant and the modalities of acquisition. Recent work by Lila R. Gleitman and co-authors (Medina, Gleitman et al., 2011; Gleitman & Landau, 2013) reports accumulating evidence that child and adult word learning share essential commonalities. Moreover, learners form a single conjecture, retaining neither the context that engendered it nor alternative meaning hypotheses. This rules out, in my opinion at least, a Bayesian process of convergence upon the meaning of basic concepts, unless there is a vast repertoire of strong a priori probability assignments, requiring no multiplicity of exposures and no carrying over of alternative candidates. This repertoire would have to be itself innate. Recent experiments tell us that even in the recognition of possible words, the ability to compute nontrivial statistical relationships "*becomes fully effective relatively late in development, when infants have already acquired a considerable amount of linguistic knowledge. Thus, mechanisms for structure extraction that do not rely on extensive sampling of the input are likely to have a much larger role in language acquisition than general-purpose statistical abilities*" (Marchetto & Bonatti, 2013).

Other suggestions (motor schemata, pragmatic inferences, rules of use, patterns of interpersonal exchanges, general cognitive processes, and so on) bring us back to Fodor's reply to Putnam: You tell me what you think is learned when a concept is learned, call it X, and I show you that X (whatever that is) must be assumed to be innately available to the learner. In a nutshell, it seems to me clear and unquestionable that learning word meanings is a process of activation, not of construction by means of progressive guesses and trial-end-error. Obviously, one cannot activate something that is not there already. Therefore . . .

AUTHOR'S NOTE

I am indebted to Robert Berwick, Luca Bonatti, Stephen Crain, Lila R. Gleitman, and Dan Osherson for their constructive criticisms and suggestions that led to various revisions and improvements.

NOTES

1. This book has been translated into 11 languages, and I am told it is still adopted as a textbook in several courses in several places. But it is presently unavailable, except on the used books market, because Harvard University Press has decided not to reprint it. So be it. I will summarize or transcribe several of the relevant passages here.
2. One caveat must be entered here: some of these authors admit an innate load on the acquisition of concepts, their task being rather to explain the acquisition of

beliefs and the learning of general categories. Fodor has written extensively on beliefs, but I cannot include a discussion of this issue in the present chapter.

3. I will conform to Fodor's notation, writing the concepts in capitals.
4. Recent work by Lila R. Gleitman and co-authors (and several prior theorists including, over the years, Irving Rock, Gordon Bower, Charles Randy Gallistel, Roddy Roediger) shows that these "gradual learning graphs" are just a misuse of statistics because, in fact, each individual is essentially a one trial learner (i.e., has the right immediate epiphany when the right situation comes along) and the 'gradual learning' curves are generated only by an illegitimate use of cumulative statistics on data pooled across subjects/items (Gallistel, 1990, Gallistel & King, 2011; Medina, Gleitman, et al. 2011; Gleitman & Landau, 2013).
5. As I am writing (October 2013), Jerry Fodor and Zenon Pylyshyn have finished a new monograph, "Minds without Meanings: An Essay on the Content of Concepts," which has as its goal to expunge the very notion of meaning and replace it with a causal connection between the speaker, the lexical-syntactic form of a concept in mentalese, and the truth-makers of the concept. They write: "*reference supervenes on a causal chain from percepts to the tokening of a Mentalese symbol by the perceiver.*" I cannot go into this new development in the present context. I will continue to use meaning here, leaving it open whether it can be replaced by the ingredients of the new Fodorian approach.
6. Early perceptrons did also show that the learning of exclusive "or" (XOR) is impossible, for deep reasons bearing on the separability of predicates. These issues have been well examined in the meantime (Harnish, 2002).
7. The necessity of prior constraints on hypotheses is a separate thesis from the impossibility of a construction of what is to be learned from preexisting mental primitives. Abstractly, one can imagine that the constraints operate on such constructions or, at the opposite, that constructions from simpler primitives may apply without constraints (making the process of learning indefinitely slow, but this is yet another kind of consideration).
8. For a formal treatment of the complexity of learning theory and inductive logic see (Jain, Osherson et al. 1999) and the rich bibliography therein.
9. A subtle development of this notion and its importance is due to the late Robert Nozick, especially in (Nozick, 1981).
10. For a development of his critique of an empiricist approach to semantics, see Fodor (2003).
11. This turned out to be strategically unobjectionable but tactically misleading, because several minutes of discussion were wasted by some participants with remarks on the history of the discipline of logic, missing the point that Jerry was making. This segment did not survive into the published version.
12. It strikes me how similar this consideration about learning is to one made much more recently, in a different context (one we have extensively dealt with in our book on Darwin) by a highly qualified evolutionary biologist: Leonard Kruglyak, Professor of Ecology and Evolutionary Biology at Princeton University, about the genotype-phenotype relation for complex diseases, but the same can be said, I think, for complex traits more generally. He says (*Nature, 456* (6) November 2008, p. 21:"*It's a possibility that there's something we just don't fundamentally understand, that it's so different from what we're thinking about that we're not thinking about it yet*".

13. One recent revamping of hypotheses linking language and motor control is based on the discovery of mirror neurons, an approach that explicitly goes against Generative Grammar. For an early statement, see (Rizzolatti and Arbib, 1998, 1999); for a more detailed evolutionary reconstruction see (Arbib, 2005, 2012) for counters to this approach to language, see (Tettamanti and Moro, 2012; Lotto et al. 2009; Piattelli-Palmarini and Bever 2002). For doubts that mirror neurons exist at all in humans, see (Lingnau, Caramazza et al. 2009). For a counter to the modularity and specificity of language based the EEG of motor control and lexical inputs see (Pulvermueller et al. 2005). A rather different recent approach to syntax and motor control is to be found in the reference in footnote 16.
14. Personal communication.
15. Of course, the trigger must be a linguistic input. Italian children do not fix the null subject parameter by eating spaghetti. Interesting issues relate to the set-subset problem. Simplifying drastically, if a child started with a parameter choice that identifies a "larger" language, while the local language is a "smaller" one, then no input would cause her to revise this choice. If, on the contrary, the child starts with a value that identifies a "smaller" language, then a sentence that discloses the local language to be "larger" would cause her to revise that choice. Pro-drop languages are "larger", because they do also admit sentences with explicit subjects, while non-pro-drop languages are "smaller" because they do not admit sentences without overt subjects. The vignette presented here is, therefore, to be taken with caution. It's only meant to convey the basic intuition. (I am grateful to Stephen Crain for stressing this point in personal communication). For a recent re-analysis of the very notion of parameter, see (Thornton and Crain 2013).
16. For a brave recent attempt to connect syntax and motor schemata see (Roy et al., 2013)
17. The radical under-determination of what actually has been learned by sheer quantitative data on progressive behaviors eventually made behaviorism implode. See (Gallistel, 1990, 2002)
18. This issue, the nature of inter-personal transmission over time of causal links between a formula in mentalese and its extension, is developed in detail in the new manuscript by Fodor and Pylyshyn (see footnote 4)
19. For insightful developments of this collectivist approach to semantics see the work of Tyler Burge (Burge, 1979, 1996)

REFERENCES

Arbib, M. A. (2005). From monkey-like action recognition to human language: An evolutionary framework for neurolinguistics. *Behavioral and Brain Sciences, 28*, 105–167.

Arbib, M. A. (2012). *How the bran got language: The mirror system hypothesis*. Oxford, England and New York, NY: Oxford University Press.

Berwick, R., & Niyogi, P. (1996). Learning from triggers. *Linguistic Inquiry, 27*, 605–622.

Bloom, P. (2000). *How Children Learn the Meanings of Words*. Cambridge, MA: Bradford Books/The MIT Press.

Bloom, P. (2001). Précis of "How Children Learn the Meanings of Words". *Behavioral and Brain Sciences, 24*, 1095–1103.

Boeckx, C. (2006). *Linguistic minimalism: Origins, Concepts, methods and aims.* Oxford, NY: Oxford University Press.

Boeckx, C., & Leivada, E. (2013). Entangled parametric hierarchies: Problems for an overspecified universal grammar. *PLoS One*, *8*(9), e72357.

Borer, H., & Wexler, K. (1987). The maturation of syntax. In T. Roeper & E. Williams (Eds.), *Parameter setting* (pp. 123–172). Dordrecht, Holland: D. Reidel.

Breitbarth, A., & Van Riemsdijk, H. (Eds.). (2004). *Triggers*. Berlin, Germany: Mouton de Gruyter.

Bruner, J. S. (1973). *Going beyond the information given*. New York. NY: Norton.

Bruner, J. S., Goodnow, J. J., & Austin, G. A. (1956). *A study of thinking.* New York, NY: Wiley.

Burge, T. (1979). Individualism and the mental. In P. A. French, T. E. Uehling, & H. K. Wettstein (Eds.), *Midwest studies in philosophy* (Vol. 4, Studies in Metaphysics), pp. 73–122. Minneapolis, Minnesota: University of Minnesota Press.

Burge, T. (1996). Individualism and psychology. In H. Geirsson & M. Losonsky (Eds.), *Readings in language and mind*. Cambridge, England: Blackwell publishers.

Carey, S. (2001). On the very possibility of discontinuities in conceptual development. In E. Dupoux (Ed.), *Languge, brain, and cognitive development: Essays in honour of Jacques Mehler* (pp. 303–324). Cambridge, England: A Bradford Book/The MIT Press.

Carey, S. (2009). *The Origin of Concepts*. Oxford, England and New York, NY: Oxford University Press.

Chater, N., Reali, F., & Christiansen, M. H. (2009). Restrictions on biological adaptation in language evolution. *Proceedings of the National Academy of Sciences USA*, *106*, 1015–1020.

Chater, N., & Christiansen, M. H. (2010). Language acquisition meets language evolution. *Cognitive Science*, *34*, 1131–1157.

Chomsky, N. (1975). *Reflections on language*. New York, NY: Pantheon Books.

Elman, J. L. (1991). Distributed representations, simple recurrent networks, and grammatical structure. *Machine Learning*, *7*(2), 195–225.

Elman, J. L., Bates, E. A., Johnson, M. H., Karmiloff-Smith, A., Parisi, D., & Plunkett, K. (Eds.). (1996). *Rethinking innateness: A connectionist perspective on development.* Cambridge, MA: A Bradford Book/The MIT Press.

Fodor, J., & Piattelli-Palmarini, M. (2010). *What Darwin got wrong.* New York, NY: Farrar, Straus and Giroux.

Fodor, J., & Piattelli-Palmarini, M. (2011). *What Darwin got wrong (Paperback, with an update, and a reply to our critics)*. New York, NY: Picador Macmillan.

Fodor, J. A. (1970). Three reasons for not deriving "kill" from "cause to die." *Linguistic Inquiry*, *1*(4), 429–438.

Fodor, J. A. (1975). *The language of thought*. New York, NY: Thomas Y. Crowell.

Fodor, J. A. (1987). *Psychosemantics*. Cambridge, MA: Bradford Books/MIT Press.

Fodor, J. A. (1998a). *Concepts: Where cognitive science went wrong.* New York, NY and Oxford, United Kingdom: Oxford University Press.

Fodor, J. A. (2003). *Hume variations*. Oxford, England: Clarendon Press/Oxford University Press.

Fodor, J. A. (2008). *LOT 2: The langauge of thought revisited.* Oxford, England and New York, NY: Oxford University Press.

Fodor, J. D. (1998b). Unambiguous triggers. *Linguistic Inquiry, 29*(1), 1–36.

Gallistel, C. R. (1990). *The organization of learning*. Cambridge, MA: MIT Press.

Gallistel, C. R. (2002). Frequency, contingency and the information processing theory of conditioning. In P. Sedlmeier & T. Betsch (Eds.), *Frequency Processing and Cognition* (pp. 153–171). Oxford, England: Oxford University Press.

Gallistel, C. R., & King, A. P. (2011). *Memory and the computational brain: Why cognitive science will transform neuroscience*. Hoboken, NJ: Wiley-Blackwell.

Gibson, E., & Wexler, K. (1994). Triggers. *Linguistic Inquiry, 25*(3), 407–454.

Gleitman, L., & Landau, B. (2013). Every child an isolate: Nature's experiments in language learning. In M. Piattelli-Palmarini & R. C. Berwick (Eds.), *Rich languages from poor inputs* (pp. 91–106). Oxford, England: Oxford University Press.

Goodman, N. (1983). *Fact, fiction and forecast*. Cambridge, MA and London, England: Harvard University Press.

Harnish, R. M. (2002). *Minds, brains, computers: An historical introduction to the foundations of cognitive science*. Malden, MA: Blackwell Publishers.

Hempel, C. G. (1966). *Philosophy of natural science*. Englewod Cliffs, NJ: Prentice Hall.

Jain, S., Osherson, D. N., Royer, J. S., & Sharma, A. (1999). *Systems that learn, 2nd Edition: An introduction to learning theory (learning, development and conceptual change)*. Cambridge, MA: Bradford Books/ MIT Press.

Kam, X.-N. C., & Fodor, J. D. (2013). Children's acquisition of syntax: Simple models are too simple. In M. Piattelli-Palmarini & R. C. Berwick (Eds.), *Rich languages from poor inputs* (pp. 43–60). Oxford, England and New York, NY: Oxford University Press.

Karmiloff, K., & Karmiloff-Smith, A. (2001). *Pathways to language: From fetus to adolescent*. Cambridge, MA: Harvard University Press.

Karmiloff-Smith, A. (1992). *Beyond modularity: A developmental perspective on cognitive science*. Cambridge, MA: MIT Press.

Legate, J. A., & Yang, C. (2013). Assessing child and adult grammar. In M. Piattelli-Palmarini & R. C. Berwick (Eds.), *Rich Languages from Poor Inputs* (pp. 168–182). Oxford UK: Oxford University Press.

Lingnau, A., Gesierich, B., & Caramazza, A. (2009). Asymmetric fMRI adaptation reveals no evidence for mirror neurons in humans. *Proceedings of the National Academy of Sciences, 106*(24), 9925–9930.

Lotto, A. J., Hickok, G. S., & Holt, L. L. (2009). Reflections on mirror neurons and speech perception. *Trends in Cognitive Sciences, 13*(3), 110–114

Marchetto, E., & Bonatti, L. L. (2013). Words and possible words in early language acquisition. *Cognitive Psychology 67*, 130–150.

Medina, T. N., Snedeker, J., Trueswell, J. C., & Gleitman, L. R. (2011). How words can and cannot be learned by observation. *Proceedings of the National Academy of Sciences USA, 108*(22), 9014–9019.

Nozick, R. (1981). *Philosophical explanations*. Cambridge, MA: Belknap Press.

Piattelli-Palmarini, M. (2003). To put it simply (basic concepts). *Nature, 426*(11), 607.

Piattelli-Palmarini, M. (Ed.). (1980). *Language and learning: The debate between Jean Piaget and Noam Chomsky*. Cambridge, MA: Harvard University Press.

Piattelli-Palmarini, M., & Berwick, R. C. (Eds.). (2013). *Rich languages from poor inputs*. Oxford, England: Oxford University Press.

Piattelli Palmarini, M., & Bever, T. G. (2002). The fractionation of miracles: Peer commentary of the article by Michael Arbib: "From monkey-like action recognition to human language: An evolutionary framework for neurolinguistics." *Behavioral and Brain Sciences*. Electronic supplements, from http://www.bbsonline.org/Preprints/Arbib05012002/Supplemental/Piattelli-Palmarini.html

Pulvermueller, F., Hauk, O., Nikulin, V. V., & Ilmoniemi, R. J. (2005). Functional links between motor and language systems. *European Journal of Neuroscience, 21*, 793–797.

Putnam, H. (1960). Minds and machines. *Collected Papers*, Vol. 2: *Mind, language, and reality*, (pp. 362–384). Cambridge, England and New York, NY: Cambridge University Press.

Putnam, H. (1975). Probability and confirmation. *Philosophical Papers*, Vol. 1: *Mathematics, matter, and method*, (pp. 293–304). Cambridge, England: Cambridge University Press.

Rizzi, L. (2013). Editorial: Introduction: Core computational principles in natural language syntax. *Lingua, 130*, 1–13.

Rizzolatti, G., & Arbib, M. A. (1998). Language within our grasp [Viewpoint]. *Trends in NeuroSciences, 21*(5), 188–194.

Rizzolatti, G., & Arbib, M. A. (1999). From grasping to speech: Imitation might provide a missing link: Reply. *Trends in Neurosciences, 22*(4), 152.

Roeper, T., & Williams, E., Eds. (1987). *Parameter setting*. Dordrecht, Holland: D. Reidel.

Rosch, E., Mervis, C. B., Gray, W. D., Johnson, D. M., & Boyes-Braem, P. (1976). Basic objects in natural categories. *Cognitive Psychology, 8*, 382–439.

Roy, A. C., Curie, A., Nazir, T., Paulignan, Y., Portes, V. d., Fourneret, P., & Deprez, V. (2013). Syntax at hand: Common syntactic structures for actions and language. *PLoS One, 8*(8), e72677.

Tenenbaum, J. B., & Griffiths, T. L. (2001). Generalization, similarity, and Bayesian inference. *Behavioral and Brain Sciences, 24*, 629–640.

Tettamanti, M., & Moro, A. (2012). Can syntax appear in a mirror (system)? *Cortex, 48*(7), 923–935.

Thornton, R., & Crain, S. (2013). Parameters: The pluses and the minuses. In M. Den Dikken (Ed.), *The Cambridge Handbook of Generative Syntax* (pp. 927–970). Cambridge, England: Cambridge University Press.

Tomasello, M. (2000). Do young children have adult syntactic competence? *Cognition, 74*, 209–253.

Tomasello, M. (2003). *Constructing a language: A usage-based theory of language acquisition*. Cambridge, MA: Harvard University Press.

Tomasello, M. (2006). Acquiring linguistic constructions. In W. Damon, R. Lerner, D. Kuhn, & R. Siegler (Eds.), *Handbook of child psychology. 2. Cognition, perception, and language* (pp. 255–298). New York, NY: Wiley.

Vercelli, D., & Piattelli Palmarini, M. (2009). Language in an epigenetic framework (Chapter 7). In M. Piattelli-Palmarini, J. Uriagereka, & P. Salaburu (Eds.), *Of minds and language: A dialogue with Noam Chomsky in the Basque Country* (pp. 97–107). Oxford, England: Oxford University Press.

Xu, F., & Tenenbaum, J. (2007). Word learning as Bayesian inference. *Psychological Review, 114*(2), 245–272.

Yang, C. (2011a). Learnability. In N. Seel (Ed.), *Encyclopedia of learning sciences* (pp. Chapter 837). New York, NY: Springer.

Yang, C. (2011b). Learnability. In T. Roeper & J. de Villiers (Eds.), *Handbook of language acquisition* (pp. 119–154). Boston, MA: Kluwer.

Yang, C. D. (2002). *Knowledge and learning in natural language*. New York, NY and Oxford, England: Oxford University Press.

10

The Immediacy of Conceptual Processing

MARY C. POTTER

When we are looking around or reading or listening to speech, how quickly do we understand what we are looking at or reading or hearing? And how long do we remember what we understood? Here I describe my experimental work on the immediacy of conceptual understanding in viewing pictures and reading words, sentences, or paragraphs. I propose that such understanding relies on a conceptual short-term memory that gives momentary access to rich associated knowledge, allowing cogent material to become structured and the rest to be quickly forgotten.

I started with a specific question: What do we remember from each eye fixation as we look around? Each fixation lasts only about 300 ms, followed by a 30 ms saccade to the next glimpse, with each glimpse presumably masking the previous one. What are we picking up from each of these glimpses that arrive in a continuous sequence in normal vision? It was difficult to address this question because people typically make multiple fixations on a given scene: how can one determine what is seen and understood in just one of the fixations? To solve this problem, Ellen Levy and I (Potter & Levy, 1969) presented participants with a series of unrelated pictures, each replacing the previous one in the center of the screen, a method later named rapid serial visual presentation (RSVP) by Ken Forster (1970; cf. Potter, 1984).

The pictures were of a wide variety of scenes and objects, new to the participants. We showed sequences of 16 pictures, at durations between 100 ms and 2 seconds per picture, simulating successive fixations without requiring eye movements. After each sequence we tested viewers' ability to recognize the pictures they had seen, mixing the old pictures with new ones. Viewers could not remember most pictures presented for 300 ms or less, even though they reported that the RSVP pictures had all seemed to be clear and meaningful. If the rate was slowed to one picture per second, participants could remember them almost perfectly. Thus, single presentations of 300 ms (a typical fixation duration) were often forgotten, even though the pictures were easy to remember if seen for the equivalent of three fixations (Potter & Levy, 1969).

We hypothesized that viewers momentarily understand the briefly presented pictures and then immediately forget most of them. If that was so, viewers should be able to detect a target picture specified in advance by a name or title, such as "boat" or "couple smiling," if they could respond before they forgot the picture. New participants were given a name for the target picture (or were shown the actual picture) just before they viewed each rapid sequence, responding immediately if they saw the specified target. As we hypothesized, targets were easy to detect at 300 ms per picture, equivalent to a single fixation. They were even fairly easy to detect at 100 ms/picture (Potter, 1975, 1976). Moreover, seeing the target name in advance was almost as good as showing the picture in advance, suggesting a remarkably abstract and instant understanding of glimpsed objects and scenes.

Although a brief glimpse is often sufficient for immediate understanding, it is insufficient for memory beyond the moment, unless the viewer gives that scene more extended attention. The visual system seems built for rapid understanding in order to guide immediate action, but memories of these glimpses are only consolidated if there is more time to think about the picture (half a second or more) or if a particular picture is picked out as important.

I tested the idea that viewers have an immediate abstract understanding of what they see in a series of experiments comparing pictured objects and their written names. If visual understanding is conceptual as well as perceptual, then one theory about encoding of words and images is wrong. According to dual coding theory (Paivio, 1971), words are part of a verbal/associative system that is separate from an imagery system responsible for perception. In that model the two systems are linked by associations between images of concrete objects and their names. The model explained an old finding: written words can be named aloud about 250 ms faster than pictured objects can be named. In Paivio's model, this asymmetry reflects the time to activate the associative link from the imagery system to the verbal system.

A central claim of the theory (Paivio, 1971) is that abstract ideas are represented only in the verbal system. My research suggested that this is wrong. Instead, words and perceptual images are each associated with abstract concepts: We have an abstract idea of a hat just as we have an abstract idea of democracy. Rather than having two codes, we have at least three codes, including an abstract code that is the language of thought (Fodor, 1975). In speaking, we use a natural language, but in thinking we rely on the conceptual code that underlies both language and perception.

In a crucial experiment Barbara Faulconer and I (Potter & Faulconer, 1975) contrasted Paivio's dual coding model (1971) with the conceptual alternative.

Participants viewed line drawings of single objects or their written names: in two conditions. In.one condition they named the word or pictured object; in the other condition they made a yes-no decision whether the item (e.g., a hat or *hat*) was a member of a specified superordinate category (e.g., "clothing"). Whereas the response time to name was 250 ms longer for the picture than for the word (just as Paivio had found), there was no difference in making the yes-no category response. In fact, the category judgment was actually 50 ms faster for the picture. This outcome was consistent with my hypothesis that abstract category information is represented in a conceptual code, not a verbal code.

An even stronger test of the hypothesis that both verbal and perceptual information are linked to abstract concepts was a study on sentence comprehension (Potter, Valian, & Faulconer, 1977). We showed that the meaning of a sentence can be probed as readily by a pictured object as by the object's written name. Participants listened to a sentence such as "The dogsled raced across the hard-packed snow" and participants then decided whether the picture (or word) that followed (e.g., bear) was relevant to the sentence. Even in the linguistic context provided by a sentence, a pictured object conveyed meaning just as rapidly as a written word. That confirmed our hypothesis that the meaning of a sentence is conceptual, not specifically tied to language. An unpublished study with Bernard Elliot used the same logic, but this time starting with pictures. Participants viewed a color photograph of a scene (such as bird on a branch), followed by a relevant or irrelevant word or line drawing (e.g., nest vs. scissors). Once more, the verbal and pictorial probes were responded to equally fast: the concept conveyed by the photograph was matched to the concept of the probe word or drawing. Both sentences and scenes are coded abstractly, as well as into modality-specific codes.

Such work is relevant to cognitive development, as I once proposed:

> "The perceptual process leading to the recognition of real objects is not different in principle . . . from symbol recognition. Each requires a process of inference, of going from sensory experience to the appropriate concept. Activation of the concept of a given object or event does not depend on any particular sense experience but abstracts away from perception just as it abstracts away from words and drawings in the experiments described earlier . . ..In a sense, then, a particular perception is a 'mundane symbol' for an object. The abstract notion we have of the object is brought to mind by a glimpse or a touch, just as it can be brought to mind by words and other symbols. So the symbolization process is foreshadowed by the development of perception" (Potter, 1979, pp. 62–63).

This notion of perceptual experiences as mundane symbols, later elaborated by DeLoache (1987, 2004), put the emphasis on abstract concepts rather than uninterpreted perceptual experience or language as the foundation of cognition—words and perceptual experiences evoke concepts. Our later finding that pictures could replace nouns in RSVP sentences without impairing comprehension

(Potter, Kroll, Yachzel, Carpenter, & Sherman, 1986; Potter & Kroll, 1987) supported the idea that language comprehension involves the building of conceptual representations, word by word.

An extension of thinking about the abstract codes that underlie both verbal and pictorial perceptions was to consider the representation of a second language. I proposed that the same conceptual code is used in both of a speaker's two languages. To test this hypothesis we compared naming pictures and words in the participant's first language (L1) and second language (L2; Potter, So, Von Eckardt, & Feldman, 1984). We hypothesized that L2 words are attached directly to their concepts as they are learned (the concept shared by objects and their names). In that case, generating an L2 word should be just as fast (or perhaps faster) when cued by the picture as by the L1 name. Other models had proposed that L2 words are learned by associating known words in L1 with new words in L2; were that the case, L2 words should be produced faster when a learner is cued by the L1 word than when cued by a pictured object. But that was not the case: words in L2 were cued as rapidly by a picture as by their L1 name, as we predicted. That was true not only for experienced bilinguals, but also for relative novices with only two years of high school study. Later work by Kroll and others (e.g., Kroll & Stewart, 1994) showed that students do use an association between L1 and L2 as a crutch in the early stages of learning a second language, perhaps during the period that they are focusing on the orthography and pronunciation of the new word rather than its meaning. They drop the L1 crutch as the new words become more familiar.

The hypothesis that readers and listeners immediately activate a conceptual representation of context words led to experiments in sentence comprehension. In understanding a sentence, listeners should quickly compute meanings of phrases and clauses. One way of testing this is to look at how listeners understand an adjective-noun sequence. Is the noun phrase understood by retrieving the meanings of the adjective and noun separately and then combining them (as many theories of language processing assume), or is it understood by retrieving the meaning of the phrase more interactively? Participants listened to a spoken sentence, and we probed a noun in the sentence by showing a picture of the noun or of an unrelated object directly after the noun had been spoken (Potter & Faulconer, 1979). For example, in the sentence "It was getting late when the man saw the [burning] house ahead of him" (the adjective was omitted on half the trials), the positive probe was a picture of a house Crucially, however, the picture was either of a canonical house or of a burning house. The instruction was to say yes if the object in the picture was named by the noun, ignoring any match or mismatch to the rest of the sentence.

We reasoned that it should be easy to ignore the adjective if indeed people process the adjective and noun separately before combining them. We predicted, however, that listeners would immediately combine the meaning of "burning" with the meaning of "house" and thus respond faster to a picture of a burning house than a canonical house. When there was no adjective, we predicted the reverse pattern. Consistent with this prediction, response times to sentences with the adjective were faster to the modified, conceptually matching pictures than to the canonical pictures; the reverse was true when there was no adjective. We concluded that the interpretation of a constituent of a sentence, such as a noun phrase, is arrived at by rapid conceptual combination, rather than by a two-step process.

In a study using spoken two-clause sentences (Von Eckardt & Potter, 1985) we asked whether there is a shift in abstraction at a clause boundary from a surface lexical level in the current clause to a more conceptual level in the completed clause, as several theorists had proposed (cf. Fodor, Bever, & Garrett, 1974). Participants listened to the sentence and were given a word probe or a picture probe at the end; they responded yes if the probe was in the sentence. Response to a probe of the first clause was slower than to the second, more recent, clause, consistent with the original finding that had led to the clause shift hypothesis. Crucially, however, this clause effect did not interact with whether the probe was a picture (conceptual). In each clause, responses to picture and word probes were equally fast. Thus, in this experiment, as in numerous other more recent studies, abstract comprehension was shown to be incremental, word by word to the extent possible, not postponed until the end of the clause.

Language understanding requires an added layer of conceptual processing to inform syntactic processing, to disambiguate the meaning of words, and to anticipate the meaning of an unfolding sentence. I looked at this process by presenting short RSVP paragraphs that were hard to understand without a key word or phrase such as "laundry" to explain the context (this idea originated with Dooling & Lachman, 1971). The following is one of the four paragraphs we used.

> "It seemed like hours since I had called. Finally it arrived, richly colored but a little thin. I wondered if it was too hot to touch. The smell was so strong that I couldn't help but try. I pulled at one section, but it was difficult to remove, so I tried another. Elastic fibers developed, attaching it to the rest. I exerted more force. As I pulled, however, droplets of hot oil splashed off one side, burning my hand. I dropped it. Perhaps I could last a couple more minutes. I was hungry but the pizza was very hot" (from Potter, Kroll, & Harris, 1980).

The last sentence containing the key word "pizza" was presented either at the beginning of the paragraph, in the middle, or at the end as here, or the word "pizza" was replaced by "it." We showed that immediate recall of the paragraph was markedly improved from the point at which the key information was introduced. The key information was beneficial even when participants were reading in RSVP at about 10 words per second and consequently were remembering relatively little. What they did remember was determined by what they could understand, and that depended on having the conceptual key. The probability of reporting the key topic (e.g., pizza) was above 80% for RSVP readers at 4, 8, or even 12 words per second, whether it appeared at the beginning, middle, or end of the paragraph. In contrast, for readers given the same total amount of time to read a normally printed paragraph, the topic was always recalled when it appeared in the first sentence, but began to be missed when in the middle and was often missed when at the end, particularly when time was short. (The topic was almost never guessed when it was omitted.) The results show that readers can understand words and process the meaning of sentences and their relation to general knowledge, even when reading at 12 words per second—further evidence for rapid understanding.

More evidence for the immediacy of conceptual processing was shown in another experiment (Potter, Moryadas, Abrams, & Noel, 1993) in which participants recalled RSVP sentences such as "The child fed the dack at the pond" that included a word that could be misspelled by changing just one vowel to turn it into another word or a nonword (e.g., *duck, deck, dack*). Participants were told that they might encounter misspellings or inappropriate words and were instructed to write what they had seen. When the sentence meaning was consistent with just one of those words (as in this example), the other word or the nonword tended to be misperceived as the context-appropriate word. In that study we argued for a two-stage modular-interactive model of word processing, beginning with a modular lexical processor the output of which is a weighted set of candidates that then interacts with the conceptual context to select the most likely candidate, all within a fraction of a second.

Resolution of lexical ambiguity is another process that requires very rapid use of conceptual context, as in understanding "Joe walked up the grade to get to class" versus "Joe chalked up the grade to lack of studying." Swinney (1979) hypothesized that multiple meanings of ambiguous words are briefly and usually unconsciously activated before the right one is selected. I simulated this process in an experiment (Potter, Stiefbold, & Moryadas, 1998) in which, instead of encountering an ambiguous word, the participant was given a choice between two different words, one above the other, at one point in an RSVP sentence ("Maggie carried the kitten in a basket/pencil to her house"). The words of the context sentence were presented for 133 ms and the two critical words for only 83 ms. Participants were instructed to write down the sentence with the word that fit. Remarkably, they were able to do so on most of the trials, showing again the immediacy with which conceptual representations become available in sentence processing.

A question that I tried to answer in another study (Potter & Lombardi, 1990) was why even long sentences are easy to remember verbatim, whereas recall of unrelated words is limited to about seven. Everyone agrees that this is because sentences are more meaningful or more redundant, but just how does this fit with known limits of working memory? We proposed that the sentence is retained as a conceptual representation built as one reads or hears it, and is recalled by regenerating a sentence to express that representation. One reason that such regeneration often ends up verbatim is that we tend to reuse recently activated words when expressing a thought. To test this hypothesis, we asked participants to read a sentence such as "The knight rode around the palace looking for a place to enter." Before they recalled the sentence, the participants read five more words and responded yes or no to a probe word (was the probe in the list?). On half these lists we included a lure word, in this example the word "castle," that could rather naturally replace a word in the sentence. Consistent with our hypothesis, the lure word "castle" did replace "palace" on more than 25% of recalls. (In a control condition without the lure word there were few intrusions of "castle.") In two later studies (Lombardi & Potter, 1992; Potter & Lombardi, 1998) we found a similar tendency to reuse recent syntactic structures, an effect that Bock (1986) termed syntactic priming. Along with the memory advantage of familiar conceptual information in sentences, these two mechanisms—reuse of words and reuse of syntax—can account for the apparently verbatim immediate recall of sentences that are actually being regenerated from conceptual memory.

Collectively, these studies suggest that conceptual knowledge plays an immediate role in perception and thought, coming into play faster than standard models of long-term memory retrieval suggest. I proposed a new form of memory termed *conceptual short- term memory* (CSTM) to account for the speed and appropriateness with which our prior knowledge shapes current perception and thought (Potter, 1993, 1999, 2009, 2012). When we identify a new stimulus, not only its concept but also other associated information in long-term memory is immediately activated, allowing new conceptual structures to be formed. You glimpse the disappearing tip of a black tail and think, *That is the cat they told me about.* Activated information that does not become structured is quickly forgotten and may never become conscious. The need for such a rapid but transient process is evident in many of the experimental tasks I have described, such as the duck-deck-dack task in which the sentence context influences word perception when reading at 10 words per second and the two-word selection task that simulates ambiguity resolution. In daily life, this skill is in view in crossword puzzlers given a combination of cues that may instantly pull out the needed word, leaving uncompleted thoughts to evaporate.

The role of CSTM is equally evident in another domain discussed earlier, our ability to understand in an instant a picture we have never seen before. In recent work I returned to this question. With the dramatic improvement in computers since our earlier work one can present full color pictures much more rapidly than was possible earlier. In one study we addressed the question of whether people can actually understand pictures presented at about 10 per second (107 ms per picture), as our earlier work had suggested. At such rates, do people see only features such as colors and shapes or object parts such as beaks and feathers, without binding them to specific objects, as proposed by Evans and Triesman (2005)? To show that people really can detect objects at that rate, we presented two different exemplars of a specified category (such as bird) and required viewers to report the names of each exemplar (e.g., swan, eagle). Viewers were able to report both objects on most trials, even when one directly followed the other (Potter, Wyble, Pandav, & Olejarczyk, 2010).

To try to assess the temporal limits of our ability to perceive scenes, we looked at even higher rates of presentation of sequences of 6 or 12 pictures. We found that people can understand pictures presented at incredibly high rates. As in our earlier studies, the task was to detect the presence (or not) of a named picture. Although performance dropped steadily as the duration decreased from 80 to 53 to 27 to 13 ms per picture, the measure *d'* (which corrects for guessing) was above chance even at 13 ms (Potter, Wyble, Hagmann, & McCourt, 2014). Crucially, the participants had only a name for the target (such as "couple smiling"), so their yes-no decision had to be based on concepts, not on precise visual features. Moreover, performance was lower but still above chance when the name was given immediately *after* the six-picture sequence, rather than before. Thus, it is possible for even a 13 ms glimpse to get to a conceptual level before it is truncated or overwritten by the following stimulus, and to persist long enough to be queried immediately after the sequence has ended. Our rich acquired knowledge over a lifetime lets us understand a huge range of pictured scenes, and the knowledge is apparently structured so that access can occur in a first feedforward pass up the hierarchical visual system.

To conclude, one of the mysteries of cognition is how we manage to think and understand as well as we do, given the claimed limitation of working memory to four or fewer items. Our memory capacity is considerably higher when proactive interference is minimized, however (Endress & Potter, 2014). Moreover, our immediate ability to abstract meaning, shown in the present review of evidence for CSTM, suggests that the "items" in everyday working memory arrive enriched by rapid understanding.

In short, I've come to have respect for the major role of well-structured, long-term knowledge in giving rapid access to the concepts that do the work of thought.

REFERENCES

Bock, J. K. (1986). Syntactic persistence in language production. *Cognitive Psychology, 18*, 355–387.

DeLoache, J. S. (1987). Rapid change in the symbolic functioning of very young children. *Science, 238*, 1556–1557.

DeLoache, J. S. (2004). Becoming symbol-minded. *Trends in Cognitive Sciences, 8*, 66–70. doi:10.1016/j.tics.2003.12.004

Dooling, D. J., & Lachman, R. (1971). Effects of comprehension on retention of prose. *Journal of Experimental Psychology, 88*, 216–222.

Endress, A. D., & Potter, M. C. (2014). Large capacity temporary visual memory. *Journal of Experimental Psychology: General. 143*(1), 548–565. doi: 10.1037/a0033934

Evans, K. K., & Treisman, A. (2005). Perception of objects in natural scenes: Is it really attention free? *Journal of Experimental Psychology: Human Perception and Performance, 31*, 1476–1492.

Fodor, Jerry A. (1975). *The language of thought*. Crowell Press.

Fodor, J., Bever, T., & Garrett, M. (1974). *The psychology of language: An introduction to psycholinguistics and generative grammar*. New York, NY: McGraw Hill.

Forster, K. I. (1970). Visual perception of rapidly presented word sequences of varying complexity. *Perception & Psychophysics, 8*, 215–221.

Kroll, J. F., & Stewart, E. (1994). Category interference in translation and picture naming: Evidence for asymmetric connections between bilingual memory representations. *Journal of Memory and Language, 33*, 149–174.

Lombardi, L., & Potter, M.C. (1992). The regeneration of syntax in short term memory. *Journal of Memory and Language, 31*, 713–733.

Paivio, A (1971). *Imagery and verbal processes.* New York, NY: Holt, Rinehart, and Winston.

Potter, M. C. (1975). Meaning in visual search. *Science, 187*, 965–966.

Potter, M. C. (1976). Short-term conceptual memory for pictures. *Journal of Experimental Psychology: Human Learning and Memory, 2*, 509–522.

Potter, M. C. (1979). Mundane symbolism: The relations among objects, names, and ideas. In N. R. Smith & M. B. Franklin (Eds.), *Symbolic functioning in childhood* (pp. 41–65). Hillsdale, NJ: Erlbaum.

Potter, M. C. (1984). Rapid serial visual presentation (RSVP): A method for studying language processing. In D. Kieras & M. Just (Eds.), *New methods in reading comprehension research* (pp. 91–118). Hillsdale, NJ: Erlbaum.

Potter, M. C. (1993). Very short-term conceptual memory. *Memory & Cognition, 21*, 156–161.

Potter, M. C. (2009). Conceptual short term memory. *Scholarpedia, 5*(2), 3334.

Potter, M. C. (2012). Conceptual short term memory in perception and thought. *Frontiers in Psychology, 3*, 113. doi: 10.3389/fpsyg.2012.00113

Potter, M. C., & Faulconer, B. A. (1975). Time to understand pictures and words. *Nature, 253*, 437–438.

Potter, M. C., & Faulconer, B. A. (1979). Understanding noun phrases. *Journal of Verbal Learning and Verbal Behavior, 18*, 509–521.

Potter, M. C., & Kroll, J. F. (1987). Conceptual representation of pictures and words: Reply to Clark. *Journal of Experimental Psychology: General, 116*, 310–311.

Potter, M. C., Kroll, J. F., & Harris, C. (1980). Comprehension and memory in rapid sequential reading. In R. Nickerson (Ed.), *Attention and Performance VIII* (pp. 395–418). Hillsdale, NJ: Erlbaum.

Potter, M. C., Kroll, J. F., Yachzel, B., Carpenter, E., & Sherman, J. (1986). Pictures in sentences: Understanding without words. *Journal of Experimental Psychology: General, 115*, 281–294.

Potter, M. C., & Levy, E. I. (1969). Recognition memory for a rapid sequence of pictures. *Journal of Experimental Psychology, 81*, 10–15.

Potter, M. C., & Lombardi, L. (1990). Regeneration in the short-term recall of sentences. *Journal of Memory and Language, 29*, 633–654.

Potter, M. C., & Lombardi, L. (1998). Syntactic priming in immediate recall of sentences. *Journal of Memory and Language, 38*, 265–282.

Potter, M. C., Moryadas, A., Abrams, I., & Noel, A. (1993). Word perception and misperception in context. *Journal of Experimental Psychology: Learning, Memory, and Cognition, 19*, 3–22.

Potter, M. C., So, K.-F, Von Eckardt, B., & Feldman, L. (1984). Lexical and conceptual representation in beginning and proficient bilinguals. *Journal of Verbal Learning and Verbal Behavior, 23*, 23–38.

Potter, M. C., Stiefbold, D., & Moryadas, A. (1998). Word selection in reading sentences: Preceding versus following contexts. *Journal of Experimental Psychology: Learning, Memory, and Cognition, 24*, 68–100.

Potter, M. C., Valian, V. V., & Faulconer, B. A. (1977). Representation of a sentence and its pragmatic implications: Verbal, imagistic, or abstract? *Journal of Verbal Learning and Verbal Behavior, 16*, 1–12.

Potter, M. C., Wyble, B., Hagmann, C. E., & McCourt, E. S. (2014). Detecting meaning in RSVP at 13 ms per picture. *Attention, Perception, & Performance, 76*(2), 270–279. DOI 10.3758/s13414-013-0605-z

Potter, M. C., Wyble, B., Pandav, R., & Olejarczyk, J. (2010). Picture detection in RSVP: Features or identity? *Journal of Experimental Psychology: Human Perception and Performance, 36*, 1486–1494.

Swinney, D. A. (1979). Lexical access during sentence comprehension: (Re)consideration of context effects. *Journal of Verbal Learning and Verbal Behavior, 18*, 645–659.

Von Eckardt, B., & Potter, M. C. (1985). Clauses and the semantic representation of words. *Memory & Cognition, 13*, 371–376.

11

On Language and Thought

A Question of Formats

DAVID J. LOBINA AND JOSÉ E. GARCÍA-ALBEA

The question of how language and thought relate can be approached in various ways, from developmental studies of preverbal abilities and comparative analyses of animal cognition to the experimental probing of the overall effects acquiring a given natural language may have on cognition. These approaches all have both virtues and vices, but Jerry Fodor certainly made good use of some of these methods in his 1975 book, *The Language of Thought*. In the first half of that book, Fodor shows, not only why there has to be a private language, but also how there could be one to begin with and what it must be like to boot. More interestingly for our purposes, in the second half of that now-classic publication Fodor turned his attention to the linguistic and psychological evidence one can muster to work out what sort of structure the internal code exhibits, a type of analysis that today doesn't receive as much attention as it perhaps deserves (the approach was also employed in some parts of the earlier Fodor, Bever & Garrett 1974). Among the features Fodor (1975) ascribes to the *language of thought* (LoT), we find a rich enough expressive base so that the acquisition of any natural language can be accommodated; a layered organization of both mental representations and mental operations; and, in regards to mental representations, the possibility (perhaps now abandoned) that single items in language may often be associated to complex representations in thought.

We offer a similar study to what Fodor undertakes in that second half, but in certain respects a much more modest one. Our aim is simply to discuss whether any of the formats—that is, types of representations—natural language exhibits can be regarded as being constitutive of the representations thought presumably subsumes, and we will conclude that this doesn't seem to be the case (we will mention a caveat at the end of the paper, though). In particular, we shall focus on two distinct but interrelated issues: (a) the role of inner speech in thought and thinking, and (b) whether the employment of inner speech, syntactic structures, or semantic representations can account for the flexibility involved in spatial cognition (or more specifically, the ability to combine information from different modalities into one thought, a central feature of cognitive flexibility).

To this end, the next section offers working definitions of both language and thought and explains how we will relate these two terms, something that is surprisingly seldom done in studies on language and thought. In turn, the third section provides the bulk of our study, as in this part of the essay we will probe the role linguistic representations may play in the psychological processes under scrutiny here. Finally, the last section summarizes and briefly discusses a complementary issue (the aforementioned caveat).

SETTING THE STAGE

By natural language we will here understand, following the line defended by the generative grammar enterprise in the last 60 or so years (Chomsky, 2005 offers a pertinent outline), the faculty of language (FoL), an architectural state of the mind composed of the following components: a combinatorial operation (recently termed *merge*); a set of lexical items (bundles of syntactic, phonological, and perhaps semantic, features); and two interfaces that connect the faculty with other systems of the mind, namely the sensorimotor (SM; roughly, the sound/sign systems) and the conceptual-intentional (C/I; even more roughly, the thought/meaning systems).

According to this architecture, the language faculty generates a number of *prima facie* very different types of representations (TRs):

TR1: one for the SM systems, termed a PHONetic representation (a flat and linear object ready for externalization, be in speech or in any other modality)

TR2: one for the C/I systems, this one called SEMantic and the central concern of many linguists (a hierarchical object encompassing the containment relations inherent in a linguistic object; despite its name, SEMs are *syntactic* structures)

TR3: an actual semantic representation put together, we assume, at the C/I interface and meant to establish (or compute) the interpretation and meaning of a sentence (and perhaps also for the benefit of other systems of the mind, a possibility that will engage us a great deal when we discuss the semantics of language and its relationship to conceptual representations)

TR4: a phonological representation mediating between *merge* and the SM interface before the linearization of a SEM into a PHON (the linearization process perhaps operating over this phonological representation)

In simpler terms, TR1 is a phonetic representation, TR2 is syntactic, TR3 is semantic, and TR4 is phonological. Representations 1, 2, and 4 have been more thoroughly studied than TR3—this is at least the case for the specific semantic representations we will discuss (mostly due to Paul Pietroski)—but the general point we shall make in what follows applies to all four. To wit, these four domains manipulate atomic elements that, combined with principles and constraints specific to each domain, yield computations and representations that appear to be extraneous to thought (since the relevant properties would appear to be specific to language). This is perhaps most obvious in the case of phonological representations, given its primitives (phonemes, etc.) and structural tiers (segmental, syllabic, metrical), and as such we won't have much to say about them here. The other three representations (TRs 1, 2, and 3), however, have all been argued, as we shall see, to play some sort of role in thought and thinking (sometimes a very central role), and therefore we need to make our case that these too are extraneous to thought—and three times over, naturally. Be that as it may, what we should make clear at this point is that SEMs are the central representations within the design of the language faculty, for the other representations (PHONs, the phonological, and the semantic) are supposed to be derived from SEMs, either right after the syntactic engine has finished operating or in tandem with it (perhaps even strictly *pari passu*).

Regarding PHONs and SEMs, perhaps the better understood of the four, we note that these representations are compiled from different sets of lexical features (morpho-phonemic features in one case, syntactic in the other) and obey different types of computational principles (inter alia, linearity for PHONs, locality for SEMs), thereby yielding rather different derivations, the result of the different constraints the SM and C/I systems impose upon the underlying generative procedure. What are these constraints precisely? The very nature of the physical channel employed for communication accounts for why the SM systems require a flat and linear representation, while the C/I systems demand a structured object that more or less matches the corresponding thought representation. The reason for the latter is that syntactic structure goes some way into establishing the meaning and interpretation of a sentence (but not the whole way; we would need to consider some other features in order to compute semantic values appropriately and fully, such as contextual information, speakers' intentions, etc.).

In describing the language faculty in such manner, one is naturally pointing to the computational-representational paradigm so dominant in studies of the mind, but we will not be preoccupied with questions of the truth and accuracy of these mental representations (the concern of most philosophers), nor will we have anything to say about the complexity or memory cost of the computations involved in actual behavior (the purview of most cognitive psychologists). Instead, we aim to determine whether any of the representations language provides are suitable for the fixation of belief, which is what we will take thinking to be primarily (with problem solving, planning, etc. constituting particular instances of belief fixation). We should stress that we will not focus on the *process* of thinking per se, even if various psychological data will have to be discussed in what follows. Rather, we will concentrate on what thought representations must be like so that belief fixation is at all possible, which is a slightly different issue. So what is thought like, then?

Thought will here be closely associated to two features of the LoT that have engaged Fodor a great deal: (a) what philosophers usually call content, or a proposition (Peacocke, 1986), the kind of objects over which propositional attitudes range, and the objects of beliefs themselves; and (b) the representational vehicle, or format, of a thought representation, that is, the structural properties of a thought representation. To have a thought, according to this view, is to entertain a proposition along with the relevant representational vehicle, while to think is to combine propositions and representational vehicles in various ways, from embeddings and combinations of various kinds (via connectives, for instance) to the premises-and-conclusion organization typical of reasoned thought. To that end, a thought representation must be fully explicit and complete, bear truth values, and exhibit a constituent structure (viz., a predicate with its arguments). Consequently, we take the question of "what thought representations must be like" to be primarily about the format properties of such representations, and not so much a question about its content properties.[1]

The internal structure of propositions will be one of the key issues here; propositional constituents, in this sense, must allow for the flexibility and creativity involved in belief fixation, as evidenced in the very common phenomenon in which different types of perceptual inputs (modalities) can be combined with each other and with many other beliefs during the construction of thought, something that needs to be explained. The usual way to do this is to take thought constituents to be what some philosophers and psychologists call concepts, the mental particulars that underlie propositions (mental *particulars* in the sense that the concept CAT is a different *particular* from the concept DOG; we don't mean the phrase *mental particular* to refer to *tokens*, cf. Fodor 1998). Concepts are abstract, therefore amodal, stable and thus reusable, and must be embedded in a conceptual repertoire of not insignificant structure, a conglomerate of properties that would allow for the combination of mental representations into ever more complex representations. As Fodor has argued since at least his 1975 book, conceptual structure is a bit like natural language: thought has a syntactic structure and a compositional semantics; this, in a nutshell, is the LoT story.

So thought necessitates explicit, complete, and structured thought representations, but what does that actually entail? Consider Gareth Evans's Generality Constraint. Put simply, this constraint states that thoughts must be structured (G. Evans, 1982, p. 100), not in terms of their internal elements, a suggestion Evans resists, but in terms of "their being a complex of the exercise of several distinct conceptual *abilities*" (p. 101; his *emphasis*). By this turn of phrase, Evans is drawing attention to the apparent fact that if one can entertain a thought in which a given property, call it *F*, can be ascribed to one individual, *a*, this is the result of two abilities: understanding *F* and understanding *a*. Consequently, if one also understands property *G* and individual *b*—that is, one understands sentences *Fa* and *Gb*—then there are "no conceptual barriers" (p. 101) to entertain sentences *Fb* and *Ga*. This, in short, is an ability "to think of an object in a series of indefinitely many thoughts" (p. 104); a fortiori, if one can entertain the thought that *a* has the property *F*, it will also be possible to entertain many other thoughts along that line, namely, that *a* also has the property *G*, and the property *H*, and the property *I* and so on and so forth.

The Generality Constraint is related to, but is not quite the same as, what Fodor has come to call the systematicity of thought (Fodor, 1987), the claim that our ability to entertain some thoughts is intrinsically connected to our ability to entertain similar thoughts (Fodor & Pylyshyn, 1988). According to Fodor & McLaughlin (1990), this property is a reflection of constituent structure, given that the stated similarity amongst thoughts is a matter of the form these thoughts have, and not of their actual content.[2] Thus, if one can entertain the thought that $P \rightarrow (Q \rightarrow R)$, to draw from McLaughlin (2009), one should *ipso facto* be able to entertain the thought that $(P \rightarrow Q) \rightarrow R$; thoughts, McLaughlin tells us, come in clusters (p. 253). These law-like psychological generalizations, to keep to McLaughlin's way of describing all things systematic and cognitive, are brought about thanks to a compositional syntax and semantics, the compositionality of thought thereby also producing a possibly infinite number of such thoughts (the productivity of thought).

Conceptual representations, then, must be such that mental processes of the type usually postulated in cognitive science can be supported at all, and that imposes pretty stringent (structural) requirements on the so-manipulated representations. Indeed, it must be the case that thought processes manipulate structured objects that combine and give rise to new objects in ways that respect their structural properties and interrelations, much as is the case in the derivations of formal logic; the chains of day-to-day inferences we typically carry out would not be licensed otherwise (cf. Davies 1986).

From the structure of propositions, then, to the structure of cognitive abilities, but what does *all this* yield exactly? The preceding description provides the desiderata that a medium of thought must meet; namely, a representational system of thought must be able to: (a) appropriately represent the contents of thoughts; (b) accurately distinguish the contents of different thoughts; (c) faithfully represent the propositional attitudes; and (d) play a causal role in mental processes. These four requirements, which are effectively conditions of adequacy on the structure of representational vehicles, provide the relevant yardstick against which linguistic representations can be measured, even if the third desideratum won't preoccupy us that much. Indeed, it is these other desiderata—content representation and differentiation, and its role in belief fixation—that are most important, the actual point under study here being whether linguistic representations are able to represent content accurately, distinguish similarly-structured contents appropriately, and subsume the relevant mental processes. Natural language will have to meet these desiderata if it is to account for the cognitive abilities that we will analyze in this paper.

A QUESTION OF FORMAT

Three relevant linguistic representations have been identified: PHONs, SEMs (syntactic structures), and the semantic representations put together, we assume, at the C/I interface. The first such linguistic format we would like to discuss is that of PHONs, the morpho-phonemic chains underlying the externalization of language—and, of course, the very strings that hearers receive during language comprehension. We now turn to this, while the other two representations (TRs 2 and 3) will be discussed in the following section.

PHONs and Inner Speech

PHONs are flat and linear internal representations, produced one element after another without pause during externalization (on account, naturally, of the limitations of the physical channel in which we communicate), and when fully externalized are hard to disentangle with no assistance (one must be in possession of the grammar of the language being heard, otherwise the input is perceived as a continuous sound).

The problem involved in parsing the input appropriately also applies to the act of reading, the written record another modality in which linguistic material can be produced. Indeed, the sequence in (1) would require some time to parse, hindering the speed of normal reading, while (2) presents an analysis into constituents of sorts (in this case, into words), others being of course possible (say, into morphemes). Examples such as (2) are roughly what scholars usually have in mind when talking of *sentences*, and we will keep to that usage here.

(1) theauthorofstephenherodecidedtowritefinneganswakeinparis.
(2) the-author-of-stephen-hero-decided-to-write-finnegans-wake-in-paris.

Even though PHONs are meant for externalization, they are not exclusively external phenomena; on the contrary, one of the internal objects the FoL outputs is this very flat and linear representation, the mental representation the linguistic production system would operate over. Sentences, thus, have a mental reality as much as syntactic structures (SEMs) do. Further, the generation of PHONs precedes their externalization, the latter a phenomenon that is undertaken by the sensorimotor systems proper, while linearization (the creation of a PHON) would be the result of operations that are certainly internal to the faculty of language—possibly the terrain of the phonological component and some other operations.[3] Having located PHON representations within our mental architecture, the question now is the following: what sort of role could PHONs play in cognition?

Much seems to be in store against the possibility that PHONs are directly involved in thought, however. Firstly, PHONs are, by definition, not hierarchical objects and as such they wouldn't qualify as thought representations. Moreover, and as mentioned in the previous section, PHONs are composed of the wrong sort of primitives for the purpose of thought. The morpho-phonemes in (2), each one of them a word of the English language, wouldn't do the job, for concepts are meant to be abstract mental particulars, and therefore something other than the lexical items (or complexes of lexical features) of a particular language (incidentally, stripping PHONs of this lexical overdress wouldn't take one far, for PHONs would still remain flat and linear representations).

The actual act of externalizing language, moreover, a case of overt linguistic behavior if anything is, couldn't possibly constitute thought itself; thought *must* involve bouts of conceptualization that precede actual behavior—planning before action, to cite Miller, Galanter, and Pribram (1960). That is, there has to be a mental activity of some sort prior to speaking (one's mind), for what we utter at any one time is usually not as rich as the content we seem to be entertaining at that precise moment, and that must be explained somehow. Indeed, speakers oftentimes produce strings that can be quite confusing for hearers, especially if the appropriate context is absent, as would be the case if (2) were uttered out of the blue, given that at least two different interpretations can be assigned to that sentence (that is, a hearer could construct two rather different structures while processing it, as we shall see).

No such confusion can exist for speakers, and not only because the underlying thought must be clear enough to them—after all, there isn't such a thing as a structurally ambiguous thought. In addition, the linguistic object generated prior to linearization (and prior to externalization) is in fact a hierarchical structure of more or less the right kind. That is, the language faculty *does* generate an object that represents the message to be conveyed correctly (a structurally unambiguous SEM), but this object is linearized and externalized in such a way as to potentially confuse a hearer. In the case of (2), the corresponding SEMs would be two syntactic objects respecting the hierarchical relations laid out, simplifying significantly, in (3) and (4), where the bracketing and underlining are supposed to mark the right dependencies (we will only discuss SEMs in the next section, though, for now we just wish to draw the relevant contrast between PHONs and SEMs).[4]

(3) The author of *Stephen Hero* [<u>decided</u> [to write *Finnegans Wake*] [<u>in Paris</u>]].
(4) The author of *Stephen Hero* [<u>decided</u> [<u>to write</u> *Finnegans Wake* <u>in Paris</u>]].

Now, while we wouldn't want to claim it is anyone's view that the actual act of speaking constitutes, *ipso facto*, an act of thought (or thinking), some scholars *have* suggested that inner speech, or interior monologue, is implicated in thought somehow. It was a belief of behaviorists such as Watson and Skinner, as discussed by Carruthers (2012), that inner speech *is* the only type of thinking there is, as stimulus-response pairings are the only psychological/mental reality to be accepted, according to this now-almost-defunct perspective, and one's own (inner) speech *is* a response to a given stimulus (or rather, it *can* be; much speech can be initiated without any stimuli). The behaviorist view on inner speech and thought is now perhaps too strong to accept, but it would not go amiss to show why exactly it is an unwarranted take on things. Furthermore, some philosophers and cognitive scientists, while not going as far as the behaviorists, do assign inner speech a substantial role in thought (or cognition), and some of the pessimism to be levelled at Skinner, Watson, and company, applies to more modest proposals too.

A number of reasons suggest that the role of inner speech in thought is very uncertain, and possibly not very substantial. Firstly, it doesn't at all follow from the apparent fact that one's train of thoughts can be laid out in interior monologue (or in writing for that matter) that inner speech is the actual vehicle in which those thoughts were had. This is effectively the line Slezak (2002) and Lurz (2007) both take while discussing related issues, the point being, to paraphrase the latter, that the internal representation of thought is almost certainly of a slightly different nature to its explicit, linguistic representation. Further, and more importantly, inner speech is a type of linguistic production, but to oneself, and as such it is under the same constraints as normal speech; that is, inner speech is the ultimate result of a process that starts with the linearization of a SEM structure, resulting in a PHON representation that is then produced. Consequently, the same sort of reasons that legislate against PHONs being the vehicles of thought would apply to inner speech too (wrong primitives, wrong structure, etc.).

That being the case, this conclusion doesn't necessarily mean that inner speech plays no role in thought. Carruthers (1996, 1998), for instance, assigns it the *prima facie* significant role of allowing us to entertain propositional thought in consciousness, thereby yielding a more sophisticated type of thinking (Carruthers, 1996, p. 50). According to this view, while inner speech may not constitute the actual vehicles of thought, it can nonetheless be employed to "imagine" sentences, this envisioning granting access to the structure of the sentences so produced (that is, access to the underlying SEM representation generated prior to linearization)—and, in turn, access to the proposition the sentence expresses. As Carruthers (1998) has it, inner speech would constitute a conduit to the structure of thought, rather than being its vehicle.

This is not all that reasonable, however, for internally producing a sentence (what Carruthers terms *imagining* it) doesn't actually grant one conscious access to its underlying structure. To be sure, speakers know exactly what they mean when they speak, overtly or covertly, but this knowledge, as has been pointed out in the literature before, is tacit and implicit, and never accessible to consciousness (see Davies, 2015 for an up-to-date discussion of implicit types of knowledge, including the linguistic kind). That is, a typical speaker doesn't know the facts in virtue of which a sentence means what it means, let alone the theory of how those facts determine meaning.

In fact, the sort of knowledge that accounts for what a sentence means (or doesn't mean) is rather intricate in nature, and its elucidation requires focused and intensive research, something that introspection alone cannot yield. Ordinary speakers are hardly aware of such simple facts as the hierarchical prominence of a specifier vis-à-vis the head of a phrase and its complements, and the question is clearly moot regarding the more intricate details of syntactic objects (movement, control/raising pairs, etc.).[5] Since, however, it is these very details that one would have to access in order to be aware of a sentence's underlying SEM representation (and, in turn, of the structure of the proposition so expressed), it is that syntactic information that Carruthers would have us access in inner speech.[6]

We suspect that Carruthers is over-intellectualizing what ordinarily happens during inner speech. To be sure, it is probably a common occurrence for academics of a certain training and disposition to employ inner speech (or writing) to organize their thoughts, to give form to/rehearse an argument, to identify what proposition a premise or a conclusion expresses, etc., but that would *prima facie* not appear to be a feature of normal, ordinary life.

Even if inner speech may not be constitutive of thought, nor grant us access to the structure of our thoughts, we can certainly still envision some sort of role for inner speech to play in cognition, a belief that seems to be widespread in the literature. Plausibly enough, Fitch, Hauser and Chomsky (2005) describe inner speech as a facilitating cognitive tool, and claim its effects are visible in problem solving, the enhancement of social intelligence by the rehearsing of thought, memory aids, focusing attention, so forth—what they call the private uses of language (p. 186). We don't disagree with this more moderate position, but the allusion to problem solving, or reasoning more generally, merits some discussion, as that ties in with a consensus of sorts within rationality studies, with an obvious connection to inner speech and belief fixation.

According to this consensus, human rationality is subsumed by two kinds of systems (J. S. B. T. Evans & Frankish, 2009). One of these is fast, mandatory, intuitive in nature, and apparently inaccessible to consciousness. The second system is slow and deliberative, accessible to consciousness, and according to Carruthers and colleagues (but perhaps not widely shared by other dual theorists), largely conducted in natural language. The latter point is the view advanced in Frankish (2004, p. 50), wherein it is argued that our beliefs are frequently language-involving and the type of reasoning conducted in system 2 frequently language-driven (in inner speech, we assume). We don't think this can be the case, because the sort of role these scholars are envisioning for language within system 2 is one in which inner speech would be partaking in causal mental processes, and linguistic behavior doesn't work that way (or *can't* work that way). As it happens, this is an old point about linguistic behavior, but it usually goes unmentioned, let alone considered, in this sort of discussion.

We are rather cryptically referring to the most important point Chomsky (1959) made in his review of Skinner's *Verbal Behavior* book, which Fodor has also made elsewhere (Fodor, 1965); namely, the fact that linguistic behavior is effectively stimulus independent. Paraphrasing Chomsky's more recent formulations somewhat (e.g., in Chomsky 2013), the circumstances that surround us at any given time don't as a matter of fact *compel* us to utter anything in particular, or indeed anything at all; *all* we can say about the matter is that we can be *incited* to say something, but there really is no telling what a person might say when prompted. One of the examples Chomsky (1959) used is that of a person standing in front of a painting; one might say a number of things in such circumstances—about the painting itself, its surroundings, something else entirely, or indeed nothing at all. As mentioned, inner speech is a type of linguistic behavior, and we would suggest that the general point we are making is even more pronounced in this case. (Who can control their interior monologue, after all? The author of *Stephen Hero?*).

This feature of linguistic behavior doesn't seem to apply to thought, or at least not in the same manner (thought now simply understood as the ability to entertain a concept or concepts). While there is a great deal of voluntary action involved in the tokening of an utterance, that's not quite the case in the tokening of a concept (cf. Fodor 2001). The aforementioned painting may not result in the tokening of any utterance, but that doesn't mean that its observation would proceed in a thoughtless mental vacuum; some conceptualization surely takes place; some concepts *would* be tokened—say, DUTCH PAINTING. We are not suggesting, let it be clear, that given a specific stimulus, we can be entirely certain of the precise set of concepts and thoughts that would surface, let alone the specific type of response; all we are saying is that some concepts *would* be tokened (in this case, DUTCH PAINTING), whereas no linguistic tokens need be. In this sense, thought is quasi-independent of stimuli, while linguistic behavior is fully independent.[7]

Thus, the stimulus independence of linguistic behavior makes it very unlikely for inner speech to be the vehicle in which the slow decision-making Carruthers and Frankish envision for system 2 is conducted. There is simply a great deal of uncertainty regarding what (covert) linguistic material would be elicited by a given mental representation, if any at all (and for any individual), and the general state of affairs can only be accentuated in an experimental setting aimed at examining this aspect of dual-theories of reasoning. We certainly want a causal account of reasoning, which is what such an experimental setting would in fact aim to unearth, but inner speech won't play the requisite role. (We wonder to what extent it is actually possible to construct a causal theory of reasoning, considering thought's quasi-independence of most stimuli, but we suppose this depends on what one understands by "reasoning" in the first place.)

All in all, then, much has been gathered up here against PHONs, the underlying representations of overt and covert speech, being the possible vehicles of thought, be this conscious, meditative, or else. PHONs are flat and linear, composed of morpho-phonemic primitives, and when exercised in actual externalization, in either outer or inner speech, its products are entirely independent of external stimuli. Put together, this would appear to be the wrong collection of features for the purposes of thought.

We now turn to another domain, that of spatial cognition, where PHONs will make yet another appearance, albeit its role therein will not be as central as in this section. More importantly, in the next section we will be evaluating the role of SEMs and Pietroski's (2005; 2007; 2012) semantic representations in spatial cognition and cognitive flexibility more generally, concluding that these representations too fail to account for the spatial reorientation data, and most probably do not possess the most basic of properties a system of thought must exhibit.

The Internal Uses of Language: The Case of Spatial Cognition

So much for overt speech, inner speech, and PHONs, but what other internal uses of language might there be, to use a phrase of Chomsky's (2012)? Carruthers (2002), and in later writings of his, supposes the mind to be composed of multiple domain-specific systems, with language, in one specific regimentation (that of LFs/SEMs; see *infra*), offering the inter-modular medium of communication—this view constituting what he calls the *cognitive conception* of language. Such a double-take has been defended by others (e.g., Pietroski 2007), but the ground for believing the mind to be massively modular, to focus on the first claim for now, is less than firm.

Scholars pushing that point have usually drawn from the study of animal cognition, but its import is not as convincing as these scholars believe. In particular, it has been argued, by Carruthers especially, that if animal psychology is modular, or domain-specific, this has some sort of implication as to whether the human mind itself is modular—that is, composed of domain-specific computational systems (e.g., in Carruthers 2006, p. 149). This sort of inference, however, seems to be based on an analogy between two very different "universes," and that doesn't strike us as a very persuasive argument at all. The analogy we are referring to is between the universe of animal cognition, including all the little and not so little cognitive lives in that universe, and the universe of the human mind—two very different things. More to the point, the postulation of domain-specific abilities in animal cognition, oftentimes one domain-specific ability per species in fact, does not warrant the conclusion that the human mind, the mind of *one* species, is as a result composed of various domain-specific systems, massively so or otherwise (see Wilson 2008 for a similar critique; Fodor 1983, 2000 for some precedents; and Samuels 2006 for a recent discussion of massive modularity tout court).[8]

Putting aside the question of whether the mind really is massively modular, what evidence is there for the proposition that language connects different components of the mind, whatever their number (the second claim of Carruthers's cognitive conception of language)? The one case Carruthers has consistently pointed to is that of spatial reorientation, the data on this cognitive skill mostly due to experiments conducted by Hermer-Vázquez, Spelke, and colleagues (Hermer-Vázquez, Selke, & Katsnelson, 1999; Hermer-Vázquez, Moffet, & Munkholm, 2001). It is the analysis of this phenomenon that the present section will be fully devoted to.[9]

Roughly, those papers describe a spatial re-orientation task in which subjects are placed inside a room where a specific feature stands out (e.g., a short white wall) and an object is placed within its proximity (usually hidden nearby, but always near a corner). The subjects are then disoriented while wearing a blindfold and once unmasked are subsequently asked to navigate to the location of the object, a course of action that can only be carried out in one go if the subjects use the geometrical and non-geometrical sources of information available to them (the dependent measure is the subjects' search times at each location, and obviously enough going to the correct corner of the room right away instead than searching for the hidden object elsewhere is regarded as a demonstration that subjects are able to combine the two kinds of information). The combination of the two bodies of information is argued by these authors to be only possible if language is employed for such a purpose, as exemplified in the possibility of generating or producing sentences such as *the ball is to the left of the short white wall*, where *left of* would constitute the geometrical information and *the short white wall* the non-geometrical kind.

According to the data, children under the age of 5 are unable to conjoin these two sources of information during the experiment and therefore fail to pass the task (that is, their search times are equal among the possible locations). Hermer-Vázquez and colleagues explain these particular data by arguing that children of this age are yet to acquire the relevant linguistic representations—those pertaining to what these authors call "spatial language" (viz., terms such as *left of*, etc.). The acquisition of spatial terms apparently provides the right resources to not only conceptualize the relevant sources of information, but in addition link them up in the requisite sentences. The resultant complex expression—in this case, *the ball is to the left of the short white wall*—can then be employed for further reflection and thought, and indeed to pass the task.

Adult participants also fail to adjoin geometrical and non-geometrical pieces of information if during the experiment they are asked to carry out a secondary linguistic task such as speech shadowing, but they don't seem to have much trouble if the concurrent task involves rhythm-clapping shadowing instead. Crucially, these two secondary tasks are supposed to be fairly equal in terms of the cognitive load involved, at least according to a control experiment Hermer-Vázquez et al. (1999) ran (see Note 11 *infra*, though). Put together, Hermer-Vázquez et al. (1999) argue that it is language that mediates the union of these two non-verbal representations, even if they accept that geometrical and non-geometrical information can certainly be separately entertained in a LoT. Why can't they be combined in the LoT, then?

There are various reasons to doubt many of the details of these experiments. To begin with, what is the claim here exactly? Is the actual production of spatial linguistic sentences (to ourselves or aloud) what allows us to combine geometrical and non-geometrical information on-the-fly, this ability hampered if a verbally mediated task is imposed? Or is it more a case of the acquisition of spatial language bringing about a change in mental architecture so that those two bodies of information can be combined via more abstract linguistic representations (SEMs, for instance)? The first possibility is more or less defended in Carruthers (2012), as therein he explains the data by referring to how inner speech is used to solve the tasks—no surprise "that language should have an impact upon verbally mediated tasks," Carruthers (2012, ft. 3, p. 385) says.[10] According to this view, subjects would use inner speech as a sort of executive control mechanism to solve the task (we think this is unlikely, for some of the reasons we espoused in the previous subsection, soon to be recast here). The latter possibility, on the other hand, is the position defended by Carruthers in earlier work of his (Carruthers, 2002, 2006), wherein he proposes that LF representations (for logical form; now SEMs, effectively) serve as the inter-modular form of communication.

Closer to the actual data, it turns out that languageless creatures manage to accomplish the merging of geometrical and non-geometrical information just fine (see references in Twyman & Newcombe, 2010; Varley, 2014), and 18-month-olds also demonstrate this capacity when the size of the experimental setting is big enough (Twyman & Newcombe, 2010, p. 1324; it is not clear why this is the case, though). That, at the very least, would suggest that natural language—or rather, the language of spatial relations—is not a necessary prerequisite for combining the two sources of information. Be that as it may, we would like to argue that in the case of the adult data the experiments actually point to possible processing effects due to using language in a verbally-mediated task, which ought to be explained, we submit, not in terms of how PHONs or SEMs are employed to pass the task, but in terms of what mental elements language and thought trigger, and how these interact in real-time processing.

Samuels (2002) offers such a story, but only a partial one, which we aim to complement. He rightly points out that speech shadowing is a language production task that, according to standard accounts, involves, *inter alia*, the integration of communicative intentions during the construction of the meaning of a sentence. Samuels then points out that there are in fact two sort of integrations to carry out in the speech shadowing version of the experiment: on the one hand, the integration of geometrical and non-geometrical information; and on the other, the integration of the speaker's intentions behind the linguistic message. It is plausible to claim, according to Samuels, that both integrations take place in what Fodor (1983) calls the central system (and thus supposedly in the LoT), an accumulation of factors that must cause a significant memory load; or, at least one greater than in the rhythm-clapping condition, and hence the different results.

To be more precise, speech shadowing is a *type* of linguistic production, as it doesn't incur quite the same sort of integration that one would expect in normal production. At the same time, speech shadowing also involves whatever cost comprehending the sentences to be repeated imposes, as shadowing is not an automatic parroting of the material one hears. Marslen-Wilson (1985), to cite a prominent study of speech shadowing, provides evidence that both fast and slow shadowers process both the syntax and the semantics of the input material they are exposed to before they integrate it into the sentences eventually produced. This is evidenced in the participants' error patterns and their sensitivity to the syntactic and semantic disruptions present in the input data, demonstrating that response integration and its execution are important factors in this sort of task. As a result, speech shadowing may not be a full-blown case of either language comprehension or production, but properties of both processes are operative. In short, speech shadowing involves a type of production task and a type of comprehension task. We should then expect a working-memory overload stemming from the two cognitive processes at play: perceptual integration and belief fixation, on the one hand, and language comprehension *and* production on the other.[11]

Samuels, then, offers an explanation that is based on decomposing the speech shadowing version of the task into atomic operations in order to then place these different operations in the right mental loci. This is a necessary step if we are to understand what is going on in the experiments—and in order to contrast the speech shadowing condition with the rhythm-clapping condition. Among the relevant atomic operations, we would have to include the possession of the right concepts, the integration of percepts and communicative intentions, and what we have added regarding language comprehension. As for the mental domains necessarily at play, we would argue that syntactic processing would involve the language faculty and the parser only, while (parts of) semantics and intentional content, in addition to the integration of geometrical and non-geometrical information, would instead engage the central (thought) systems. Thus, while parsing a linguistic input necessarily implicates constructing a SEM (but not a PHON), the operations carried out in the central systems would involve conceptual instead of linguistic representations (we have already mentioned that syntactic features and concepts are different things, and we will expand this point). A corollary of all this is that the spatial reorientation data tell us much about processing systems and how they relate to each other, including why the two shadowing conditions yield such different results, but they do not establish what sort of role linguistic representations play in belief fixation.

Surprisingly, the problematic aspect of employing speech shadowing in dual tasks doesn't seem to be fully appreciated, even in papers discussing the uses of this technique (e.g., Papafragou, Hulbert, & Trueswell, 2008; Varley, 2014; de Villiers, 2014). There's a tendency in the literature, in fact, to believe that employing speech shadowing "ties up" the language faculty during the resolution of a task (de Villiers's 2014 phrase, p. 106), thereby blocking the linguistic system from forming the representations needed to pass a given task. Such a take, however, assumes not only that subjects use the language faculty to form the necessary mental representations during an experiment; it furthermore presumes that subjects would be forming the *right* linguistic representations, and both assumptions are unwarranted. Take Carruthers (2012) as a case in point, where inner speech is assigned the role of a "quasi-executive function" (p. 394). In the spatial reorientation task, Carruthers argues, it is because adults and children over the age of five are able to formulate sentences such as *it is left of the red wall* that they are able to pass the task, but therein lies the troubling issue. Given the stimulus independence of linguistic behavior, we really don't know what sentences would be produced by each one of the participants, if indeed any at all (as we stressed in the previous section).[12] In any case, none of the studies we have cited take into consideration what is actually involved in speech shadowing. Carruthers (2002, p. 712), in considering and dismissing Samuels's proposal, is an exception, as he does cite Marslen-Wilson (1975) in his response to Samuels. However, and while he is aware of the claim that speech shadowers process both the syntax and the semantics of the material they repeat, an aspect of speech shadowing that would support Samuels's analysis and not his own, Carruthers points out that the data in Marslen-Wilson (1975) show that very few subjects do so. True enough of the mid 1970s, but the more wide-ranging data of Marslen-Wilson (1985) clearly backs Samuels's contention instead (and ours).

As things stand, then, we would argue that neither inner speech (PHONs) nor LFs or SEMs directly participate in the operations involved in computing spatial cognition; or at least we haven't been given any reasons to believe this is the case (the mapping between linguistic structures and LoT representations notwithstanding, naturally). There is another possibility, though. Coming from a rather different perspective, that of modern generative grammar, but reaching similar conclusions to those of Carruthers's regarding the internal uses of language, albeit with some differences worth discussing, Pietroski (2007), Boeckx (2009), Chomsky (2012), and McGilvray (in Chomsky 2012, which he edited) have also alluded to the experiments of Hermer-Vázquez and colleagues while discussing the place of language within cognition. We now turn to this, but first we must devote some space to describing what this other perspective actually entails.

According to the modern generative grammar strand just mentioned, the location of language within the mind is such that the two principal types of representations the FoL generates—recall, PHONs and SEMs—are to be regarded as "instructions" to the other mental systems with which language is embedded (for a succinct description, see e.g. Chomsky, 2007). Describing PHONs as instructions to the sensorimotor systems so that the latter can articulate and produce whatever is commanded to be articulated and produced is clear—a linear string of elements does establish how it is to be externalized—but must this also be the case regarding SEMs and what Chomsky calls the conceptual/intentional systems (or simply, the thought systems)? In particular, what systems would SEMs instruct, and how? The answer to the first question, according to McGilvray (1998), is yes; the claim that the language faculty instructs other systems "suggests, as intended," McGilvray clarifies, "that the interfaces internally constituted in the language faculty configure the other systems, not the other way around" (p. 238). Such a stand would imply a certain precedence (or prominence) of the language faculty vis-à-vis the thought systems, but the only system these scholars have unambiguously alluded to is, as stated, that of spatial cognition. Having said that, it is not easy to work out what the actual claim is meant to be here. Are we to suppose that SEMs instruct thought systems to combine geometrical and non-geometrical information, or do SEMs themselves effect the actual combination? What does it mean to say that the FoL "configures" thought systems?

Putting that aside, the linguistic instructions sent to the conceptual/intentional interface would be composed of constituents that could well prove to be incommensurable upon receipt. Lexical features, after all, are rather particular to language, a point that is magnified in the case of syntactic features. Take Stabler (2011), a work that offers a useful outline of the different syntactic features linguists have posited. According to Stabler, syntactic features divide into, at least, categorial or substantive features, such as N(oun), V(erb), A(djective), and P(reposition); selector features that require a category to be merged with (a Determiner selects a noun phrase, for instance); "goal" features that require licensing at different stages of a derivation (such as *-focus*, *-case*); and "probe" features that do this licensing (marked as + *focus*, + *case*). At first sight, then, these features appear to be exclusively specific to the syntactic engine itself, and we would venture to add that the outcome representations (SEMs, the output) could as a result prove to be incomprehensible for other mental systems.

What other options are there? Linguists certainly have a story to tell us regarding what happens at the interface between the syntactic engine and the C/I systems, the locus where the (language-internal) semantic component would operate, and perhaps SEMs are appropriately modified there before being sent to the C/I systems (so that they can instruct those thought systems). Perhaps, moreover, it is therein that the relevant *linguistic* medium of thought is to be found—or at least one that would support spatial cognition. Pietroski (2007, 2012) has put forward a proposal according to which the language faculty reformats conceptual representations at this very interface, yielding a specific type of semantic representation (which we numbered TR3 in the second section). Crucially, Pietroski argues that this whole *lexicalization* process may well explain the spatial reorientation data, and certainly belief fixation in general.[13]

As was the case with Carruthers, Pietroski (2012) also finds the argument from animal cognition to massive modularity a compelling one, and this same reasoning also suggests to him, and complementary so, that pre-linguistic concepts (his phrase) are *ipso facto* not fully integrated—i.e., that they can't freely combine with each other. The languageless mind is apparently not a connected mind, for in order to establish communication among different mental systems, Pietroski argues, a lexicalization process is needed—that is, mental integration obtains once language has made lexical items out of concepts; once, that is, the language acquisition process has resulted in a fully mature language faculty.

So what is this lexicalization process exactly? To begin with, Pietroski (2007) argues that on the whole pre-linguistic predicate concepts take more than one argument—that is, they are polyadic—but after lexicalization they are connected to "an analytically related monadic concept" (p. 343). According to this view, a relational concept such as CHASE (x, y), which takes two arguments, would after lexicalization be linked to the one-argument event CHASE (x) (Pietroski, (2007) and so on with three- and four-place concepts. Acquiring the lexical item *chase*, therefore, results in the three-way linkage of the concept CHASE (x, y), the sound for the lexical item *chase*, and the monadic concept CHASE (x), making lexicalization a much more complicated process than the simple linkage of a concept with the pronunciation of the corresponding lexical item (p. 345). Lexicalization, then, reformats polyadic concepts into monadic ones.

Semantic composition in language, Pietroski adds, consists in Davidsonian conjunctions of predicates (p. 344), and thus a conceptual representation such as FIDO CHASED FELIX, commonly represented as in (5), would after reformatting look like (6). In this sense, the computation of linguistic meaning imposes the monadic analysis in (6) on relational concepts such as CHASE (x, y), and this results in the combination of monadic concept with monadic concept—and thence the fully integrated mind (we have simplified the representations Pietroski, 2007 uses a little bit; the E in 6 stands for events).

(5) *PAST ⟨CHASE(FIDO,FELIX)⟩*

(6) $\exists E[PAST\,(E)\ \&\ CHASE\,(E)\ \&\ \exists x\,[FIDO(x)]\&\ \exists y[FELIX\,(y)]]$

We think there's some merit to Pietroski's conjunctionist theory, but the suggestion that the reformatting the FoL effects during lexicalization from relational to monadic concepts explains how the mind comes to be fully connected is not a compelling or well-composed argument. Recall, to begin with, that Pietroski (2007, p. 367) supposes animal cognition to be highly modular, and that stand goes some way into pre-empting the sort of story Pietroski wants to tell. Indeed, it is precisely because of the way Pietroski reads the literature on animal cognition that he believes pre-linguistic concepts to not be fully integrated (Pietroski, 2012, p. 129, ft. 1). What Pietroski ought to have seen, to rephrase the point we made earlier for present purposes, is that different animal species manipulate different representations for different computational problems (recall, very often one domain-specific ability to species), and while these representations are clearly incommensurable to one another, nothing at all follows regarding the human mental architecture, let alone the character of human concepts (or, indeed, the possibility of a domain-general language of thought). The point, to be as clear as possible, is that Pietroski cannot derive the postulation of the specific type of (human) pre-linguistic concepts he has in mind from animal cognition studies. That is, the fact that the representations of one species are of a different character from the representations of another species doesn't warrant the conclusion that something like that situation is mirrored in the human mind. (We also worry about the surprising fecundity of concept types Pietroski allows overall, with polyadic concepts on one plane and monadic concepts on another, but that's a different issue.)

We are not doubting that many human concepts are in fact polyadic; nor are we dismissing the conjunctionist account of monadic concepts (a semantic theory *of* language, we'd emphasize).[14] What we are doubting is the reasonableness of the two premises that seem to drive Pietroski's argument: that concept combination requires monadic concepts, and that polyadic concepts of different adicities cannot combine with each other.

Where does Pietroski find evidence for his take on things in any case? As mentioned, in the spatial reorientation data we have been discussing, with references to Carruthers's (2002) analysis of these also making an appearance here and there (Pietroski, 2005, pp. 271–274). However, we think he is clutching at straws here. Pietroski offers a rather peculiar reading of those data, one according to which the issue turns out to be the impossibility of preverbal children to form complex concepts combining geometric and non-geometric sources of information *because* the requisite polyadic concepts are yet to be reformatted into a unified, monadic code (Pietroski, 2005, p. 274). That's *not* quite what the data actually showed, though. The only effect Hermer-Vázquez and coworkers found concerned the production of sentences with spatial terms, which coupled with the fact that speech shadowing involves both comprehension and production processes, led us to believe that the data had to be explained in terms of processing effects and how different systems of the mind interact in real-time processing.

More importantly, Pietroski doesn't actually show us how (or why) geometric and non-geometric bodies of information fail to be combined into a structured conceptual representation on account of the supposed disparities in conceptual adicities these two types of representations would exhibit. He makes no attempt, that is, to show *what* polyadic concepts precisely are involved in geometric and non-geometric conceptual representations and where the disparity lies. It's hard to see how such an account would proceed, in fact, for it is not at all obvious what variety of polyadic concepts would be involved in the construction of a conceptual representation for a sentence such as *the ball is to the left of the white wall* (and what of the other possible sentences in Note 12?). But perhaps that's just as well, given that toddlers do appear to be able to form such concepts—recall, they manage to pass the task when the room is big enough—and that single result defeats the whole point Pietroski is trying to make—18-month-olds *haven't* lexicalized the relevant concepts yet.

Put together, the evidence for the centralizing role language is said to have on thought will not be found in the spatial reorientation data; something else will have to be called upon to make that claim true. Granted, Pietroski's reasoning is an appealing one, but it involves a slippery slope: any experiment purportedly showing a linguistic effect on thought can be taken as evidence in favor of Pietroskian lexicalization—or, indeed, of lack thereof—and that is something to worry about. In addition, it seems to us that Pietroski's theory is hostage to the constraints he assumes; disregard or modify those assumptions, and his account of lexicalization would be very different (and so would his take on cognitive flexibility). Namely, it is because Pietroski assumes the human mind to be largely modular that he thinks polyadic concepts cannot combine with each other; and it is because Pietroski assumes a conjunctionist composition instead of a functional application composition that he believes semantic composition to be monadic. It may well be the case that neither assumption stands on very firm ground.

CODA

The faculty of language generates four formats—PHONs, SEMs, phonological representations, and semantic representations—and we have argued that all four are inadequate for the fixation of belief, or at least for the phenomena we have discussed here. We won't repeat the main points we have put across in this chapter;

instead, we will mention another possibility altogether, which will engage us a great deal in the near future.

This possibility involves changing tack a little bit, and we only do so for the sake of the argument. Namely, we could follow McGilvray (1998) and accept his contention(s) that (a) lexical items are composed of semantic features in addition to syntactic and phonological features (Fodor would demur); (b) that these semantic features are fine-grained versions of the substantive subset of syntactic features (and from that perspective lexical items would not be very different from concepts); and, (c) that these very features partake in the construction of hierarchical objects (skeletal or rarefied versions of SEMs, perhaps) that would be of the right kind (of the right form, that is) for thought. A number of scholars defend such a position (e.g., Chomsky, 2013; Hinzen, 2013), and Fodor himself, while not quite on the same wavelength as these scholars, speculated 40 years ago that the LoT may be very similar to natural language, perhaps to the point that the resources of the former may be implicated in the resources employed in the latter (Fodor, 1975, p. 156). We think that the stronger version of this take on things faces formidable problems, and that the evidence indicates that the FoL and the LoT remain slightly different systems, designed to solve different problems of cognition—the position, we think, that Fodor eventually argues for (page 156 is a bit of an odd fish in *The Language of Thought*).

In particular, there appear to be very wide and rather varied misalignments between linguistic representations and the corresponding semantic or conceptual representations, suggesting two different systems of representation. This is not only true in the general sense that language offers both too much and too little for thought, as Collins (2007b) puts it. That is, on the one hand linguistic structure offers too much detail for the representation of thought, as Collins (2007a) points out in relation to such linguistic features as c-selection, raising/control pairs, and others, properties that would play no role in thought. On the other hand, further, linguistic representations are often inexplicit as to what thought they express, given that there are "more metaphysical distinctions. . . in the range of ideas we may intend to communicate than are marked in natural language," as Koralus (2013, p. 286) shows in a study of the different interpretations linguistic descriptions give rise to.

Rather, the point about the misalignments between language and thought applies in more dramatic fashion in the following cases: (a) ungrammatical but thinkable sentences, such as **what did Bill meet a man that made*, taken from Collins (2011), an example that suggests that the content we can ascribe to that sentence can be entertained in some logic-like system (Collins provides the following representation for such content: what X is such that Bill met the man that made X); (b) ellipsis phenomena, which in some cases at least point to different underlying representations for the elided material and the precedent, surface sentence (as in voice mismatches; Phillips & Parker 2014); and, (c) grammatical illusions, as exhibited in ungrammatical sentences that are nonetheless consistently regarded as being well-formed despite being in fact nonsense (and thus unthinkable), such as **more people have been to Russia than I have*, taken from Phillips, Wagers, and Lau (2011), a phenomenon we would want to argue indicates that some syntactic structures cannot be mapped to the corresponding thought representations that hearers put together when they attempt to process sentences such as these. To be sure, these pointers would have to be spelled out carefully and slowly, and we hope to provide the details anon, but we submit that these data *do* nonetheless show that there *has to be* a language of thought that is different from natural language in very specific ways.

AUTHORS' NOTE

We would like to thank Martin Davies and Nick Shea for various conversations and comments on some of the topics treated in this essay. We also thank the editors of this volume for their valuable feedback and support. The research presented here was partly funded by the AGAUR and the University Rovira i Virgili (2011 BP-A2 00018 and SGR2014-1444/2014 PFR-URV-B2-37).

NOTES

1. We've cited Peacocke (1986) in the text, whose concerns actually center on a Fregean understanding of propositions, but we treat this term in a different way; in particular, we do not wish to hold the view that propositions/contents are atemporal externalia (the Fregean sense), nor for that matter sets of possible worlds/situations, as some other philosophers do. Rather, here we will take the term proposition to strictly refer to the *format* properties of mental representations, a composite of various elements in the sense of Pylyshyn (1984) (and, we think, of much of Fodor's work).
2. This is precisely the feature of systematicity that Evans would have not accepted, and which distinguishes his Generality Constraint from systematicity. Evans, however, may have overlooked the fact that mental chains of inferences often come about because of the internal structure of mental representations, as we mention in the next paragraph, but this is merely a historical aside. In any case, we take it that both Evans's Generality Constraint and Fodor's systematicity are properties of thought that are in principle independent of language; in fact, we think these scholars arrived at these features because they took thought to be a sort of logic-like system rather than a derived phenomenon of natural language.
3. We are stressing this point because scholars such as Collins (2008) have argued that some of the failings of language qua thought identified by Fodor (2001)—e.g., that language is too inexplicit and ambiguous (issues we won't directly discuss here)—pertain to the external products of language (PHONs or sentences) instead than to its internal outputs, and that's not entirely correct: PHONs are also internal products of the FoL (see the next note, though).
4. Both Fodor (2001) and Gleitman & Papafragou (2005) point to the structural ambiguity of natural language, and this is taken to be a shortcoming of language qua medium of thought. This point, however, only really applies to the externalized sentences and the corresponding PHONs and not to the internal syntactic structures—one could simply (counter-)argue that we don't think with sentences

but with SEMs. The argument against SEMs qua vehicles of thought will have to be something else (see *infra*).

5. See Moro (2008, pp. 68 et seq.) on the specifier-head-complement(s) structure that all syntactic objects are said to adhere to.
6. Even more information than that is said to be obtainable thus: "we have immediate access to a particular phonological representation, together with its interpretation," Carruthers (1998, pp. 463) tells us. Surely what he means to say is that we are aware of the pronunciation of a particular sentence and that we know what such a sentence means, but the finer details of the phonology, syntax, and semantics would remain inaccessible to consciousness.
7. The (sort of) stimulus dependency of thought should not be confused with what Chomsky calls the referentialist dogma, an issue that seems to trouble him so and which he tends to bring up while discussing these very issues (e.g., in Chomsky 2012). By that moniker, he's referring to externalist theories of mental content, currently dominant among philosophers, and according to which the meaning of concepts and/or words (or rather, lexical items) is a matter of how they are causally connected to external objects. Chomsky counters that such a stand can't be right because human psychology doesn't work that way; to wit, human concepts and/or lexical items are not causally connected to externalia in the same way that the concepts/mental symbols of other species seem to be, references a-plenty to the work of Gallistel on animal cognition abounding in such discourse (e.g., Gallistel 1990, cited in Chomsky 2007, p. 10). The idea here is that the internal representations of non-human species are causally connected to features of their environment; given some environmental input, a behavioral output is subsequently observed. We can grant Chomsky's contention that human cognition doesn't quite work in the same way, but that point appears to have very little to do with externalist theories of content, as the latter are concerned with the facts that supposedly establish the meanings concepts/lexical items have, and as such these theories have rather little to say about what particular thoughts/behaviors are automatically triggered given particular stimuli. In any case, we don't think anyone has ever claimed that concepts/words are causally connected to externalia in the way other species' internal representations are said to be. Fodor (2001) does consider the connection between concepts and externalia to be causal, as opposed to the connection between externalia and language, which isn't so according to him, and he does take that reputed fact to legislate against natural language being the repository of meaning—and, therefore, of thought—but *his* causal story is very different from the one Chomsky has in mind.
8. Moreover, the specificity of these animal abilities is a rather subtle matter. The computational architecture usually postulated for these abilities is in fact a domain-general one: a rather simple read-and-write Turing Machine-like mechanism (Gallistel, 1995, 2006). The specificity comes in in the sort of data structures that this read/write mechanism manipulates in each species, and consequently in the sort of computations that the system carries out in each case.
9. The discussion of the spatial cognition data that follows is based on Lobina (2012), which it amplifies and supersedes in important respects.
10. This is also the stance we take Hermer-Vázquez and co-workers to be defending, given that language production is the one factor they found to be significant in the correlation analysis they ran.

11. As mentioned, Hermer-Vázquez et al. (1999) ran a series of controls to compare the cognitive load (or attentional demands, as these authors call it; p. 16) involved in speech shadowing and rhythm clapping. They employed a visual search task to compare them directly, and found that speech shadowing and rhythm clapping exert roughly the same amount of resources—in such a visual search task, we would like to stress (separately, they ran further controls to determine whether speech shadowing interferes with the detection and storing of either geometric or non-geometric information, concluding that it doesn't in either case). We find that running these controls was a bit of a red herring, however. Hermer-Vázquez et al. (1999, pp. 25 et seq.) consider two potential explanations for their data—either it is the case that language is used to combine geometric and non-geometric information and shadowing interferes *or* it is instead the case that these two sources can be combined elsewhere and shadowing interferes with the ability to detect/remembering non-geometric landmarks—and such a position goes some way toward biasing what sort of controls they were interested in. The actual task, however, implicates a slightly different set of concerns, as Samuels has argued: combining the two requisite sources of information *and* computing/producing language necessarily involves the same domain-general system, whatever that turns out to be, and it is in these terms that the conflict of resources would arise. As such, the visual search task can't possibly be an appropriate control to compare the cognitive load of speech shadowing and rhythm-clapping shadowing. Further, we find the conclusion in Hermer-Vázquez et al. (1999) that subjects "spontaneously overcome the effects of disorientation through the use of language" (p. 32) to be entirely unsupported, for some of the reasons we have advanced regarding inner speech and linguistic production in general.
12. After all, a sentence such as *the ball is to the left of the short white wall*, to provide a more comprehensive sentence than Carruthers's example, would be one way to codify the required information, but there are not only equivalent ways to do this (e.g., *the short white wall is to the right of the ball; the wall is short and white, the ball is right next to it; the ball is to the west of the non-black, little wall*, etc.), one can also imagine sentences that would get things wrong, and who's to tell that these sentences aren't also considered as a subject blindly circles around the experimental room?
13. We do not wish to go as far as stating that Pietroski's semantic representations constitute thought representations—we don't think Pietroski claims quite that—only that his lexicalization account yields structures that are part of thought. Whether these structures are modified versions of SEMs is not clear to us, however. We should also point out that Pietroski's lexicalization story can be applied, and in fact he does in his writings, to two phenomena: language acquisition and language evolution. We are focusing on the issue of language acquisition in this chapter.
14. As Pietroski (2012) himself discusses, though, it is rather difficult to work out the adicity of concepts, and there is a great deal of speculation in large tracts of his publications. The concept SELL takes four arguments, according to him.

REFERENCES

Boeckx, C. (2009). *Language in cognition*. Oxford, England: Wiley-Blackwell.

Carruthers, P. (1996). *Language, thought, and consciousness*. Cambridge, England: Cambridge University Press.

Carruthers, P. (1998). Conscious thinking: Language or elimination? *Mind and Language*, *13*(4), 456–476.

Carruthers, P. (2002). The cognitive functions of language. *Behavioral and Brain Sciences*, *25*, 657–726.

Carruthers, P. (2006). *The architecture of the mind*. Oxford, England: Oxford University Press.

Carruthers, P. (2012). Language in cognition. In E. Margolis, R. Samuels, & S. Stich (Eds.), *The Oxford handbook of philosophy of cognitive science* (pp. 381–401). Oxford University Press.

Chomsky, N. (1959). A review of B. F. Skinner's verbal behaviour. *Language*, *35*, 26–57.

Chomsky, N. (2005). Three factors in language design. *Linguistic Inquiry*, *36*(1), 1–22.

Chomsky, N. (2007). Biolinguistic explorations: Design, development, and evolution. *International Journal of Philosophical Studies*, *15*(1), 1–21.

Chomsky, N. (2012). *The science of language: Interviews with James McGilvray*. Cambridge, England: Cambridge University Press.

Chomsky, N. (2013). What kind of creatures are we? Lecture I: What is language? *The Journal of Philosophy*, *110*(12), 645–662.

Collins, J. (2007a). Review of *Ignorance of Language*. *Mind*, *116*, 416–423.

Collins, J. (2007b). Syntax, more or less. *Mind*, *116*, 805–850.

Collins, J. (2008). *Chomsky: A guide for the perplexed*. London, England: Continuum International Publishing Group Ltd.

Collins, J. (2011). *The unity of linguistic meaning*. Oxford, England: Oxford University Press.

Davies, M. (1986). Tacit knowledge, and the structure of thought and language. In C. Travis (Ed.), *Meaning and interpretation* (pp. 127–158). Basil Blackwell Inc.

Davies, M. (2015). Knowledge (explicit, implicit and tacit): Philosophical aspects. In J. D. Wright (Ed.), *The international encyclopedia of social and behavioral sciences* (2nd ed., pp. 74–90). Oxford, England: Elsevier Ltd.

de Villiers, J. (2014). What kind of concepts need language? *Language Sciences*, *46*, 100–114.

Evans, G. (1982). *The varieties of reference*. Oxford, England: Clarendon Press.

Evans, J. S. B. T. & Frankish, K. (Eds.). (2009). *In two minds: dual processes and beyond*. Oxford, England: Oxford University Press.

Fitch, W. T., Hauser, M. D., & Chomsky, N. (2005). The evolution of the language faculty: Clarifications and implications. *Cognition*, *97*, 179–210.

Fodor, J. A. (1965). Could meaning be an rm? *Journal of Verbal Learning and Verbal Behavior*, *4*(2), 73–81.

Fodor, J. A. (1975). *The language of thought*. Cambridge, MA: Harvard University Press.

Fodor, J. A. (1983). *The modularity of mind*. Cambridge, MA: Bradford Books/MIT Press.

Fodor, J. A. (1987). *Psychosemantics*. Cambridge, MA: MIT Press.

Fodor, J. A. (1998). *Concepts: Where cognitive science went wrong*. Oxford, England: Oxford University Press.

Fodor, J. A. (2000). *The mind doesn't work that way*. Cambridge, MA: MIT Press.

Fodor, J. A. (2001). Language, thought, and compositionality. *Mind and Language*, *16*(1), 1–15.

Fodor, J. A., Bever, T. G. & Garrett, M. F. (1974). *The psychology of language*. London, England: McGraw-Hill.

Fodor, J. A., & McLaughlin, B. (1990). Connectionism and the problem of systematicity: Why Smolensky's solution doesn't work. *Cognition, 35*, 183–204.

Fodor, J. A., & Pylyshyn, Z. (1988). Connectionism and cognitive architecture: A critical analysis. *Cognition, 28*, 3–71.

Frankish, K. (2004). *Mind and supermind*. Cambridge, England: Cambridge University Press.

Gallistel, C. R. (1990). *The organization of learning*. Cambridge, MA.: The MIT Press.

Gallistel, C. R. (1995). The replacement of general-purpose theories with adaptive specializations. In M. Gazzaniga (Ed.), *The cognitive neurosciences* (pp. 1255–1267). The MIT Press.

Gallistel, C. R. (2006). The nature of learning and the functional architecture of the brain. In Q. Jing (Ed.), *Psychological science around the world*, vol. 1 (pp. 63–71). Sussex, England: Psychology Press.

Gleitman, L. R., & Papafragou, A. (2005). Language and thought. In R. G. Morrison & K. J. Hoyoal (Eds.), *The Cambridge handbook of thinking and reasoning* (pp. 633–662). Cambridge, England: Cambridge University Press.

Hermer-Vázquez, L., Moffet, A., & Munkholm, P. (2001). Language, space, and the development of cognitive flexibility in humans: The case of two spatial memory tasks. *Cognition, 79*, 263–299.

Hermer-Vázquez, L., Spelke, E., & Katsnelson, A. S. (1999). Sources of flexibility in human cognition: Dual-task studies of space and language. *Cognitive Psychology, 39*, 3–36.

Hinzen, W. (2013). Narrow syntax and the language of thought. *Philosophical Psychology, 26*(1), 1–23.

Koralus, P. (2013). Descriptions, ambiguity, and representational theories of interpretation. *Philosophical Studies, 162*(2), 275–290.

Lobina, D. J. (2012). Conceptual structure and emergence of language: Much ado about knotting. *International Journal of Philosophical Studies, 20*(4), 519–539.

Lurz, R. W. (2007). In defense of wordless thoughts about thoughts. *Mind and Language, 22*(3), 270–296.

Marslen-Wilson, W. D. (1975). Sentence perception as an interactive parallel process. *Science, 189*, 226–228.

Marslen-Wilson, W. D. (1985). Speech shadowing and speech comprehension. *Speech Communication, 4*, 55–73.

McGilvray, J. (1998). Meanings are syntactically individuated and found in the head. *Mind and Language, 13*(2), 225–280.

McLaughlin, B. P. (2009). Systematicity redux. *Synthese, 170*, 251–274.

Miller, G. A., Galanter, E. & Pribram, K. H. (1960). *Plans and the structure of behaviour*. New York, NY: Holt, Rinehart and Winston.

Moro, A. (2008). *The boundaries of Babel*. Cambridge, MA: MIT Press.

Papafragou, A., Hulbert, J., & Trueswell, J. (2008). Does language guide event perception? Evidence from eye movements. *Cognition, 108*, 155–184.

Peacocke, C. (1986). *Thoughts: An essay on content*. Oxford, England: Basil Blackwell Publisher Ltd.

Phillips, C., & Parker, D. (2014). The psycholinguistics of ellipsis. *Lingua, 151*, 78–95.

Phillips, C., Wagers, M. W., & Lau, E. F. (2011). Grammatical illusions and selective fallibility in real-time language comprehension. In J. T. Runner (Ed.), *Experiments at the interfaces* (pp. 147–180). Bingley, England: Emerald Group Publishing Limited.

Pietroski, P. M. (2005). Meaning before truth. In G. Preyer & G. Peter (Eds.), *Contextualism in philosophy: Knowledge, meaning, and truth* (pp. 253–300). Oxford University Press.

Pietroski, P. M. (2007). Systematicity via monadicity. *Croatian Journal of Philosophy, 7*(21), 343–374.

Pietroski, P. M. (2012). Semantic monadicity with conceptual polyadicity. In M. Werning, W. Hinzen, & E. Machery (Eds.), *The Oxford handbook of compositionality* (pp. 129–148). Oxford, England and New York, NY: Oxford University Press.

Pylyshyn, Z. W. (1984). *Computation and cognition*. Cambridge, MA: MIT Press.

Samuels, R. (2002). The spatial reorientation data do *not* support the thesis that language is the medium of cross-modular thought. *Behavioral and Brain Sciences, 25*, 697–698.

Samuels, R. (2006). Is the mind massively modular? In R. Stainton (Ed.), *Contemporary debates in cognitive science* (pp. 37–56). London, England: Blackwell Publishing Ltd.

Slezak, P. (2002). Thinking about thinking: Language, thought and introspection. *Language and Communication, 22*, 353–373.

Stabler, E. (2011). Computational perspectives on minimalism. In C. Boeckx (Ed.), *The Oxford handbook of linguistic minimalism* (pp. 616–641). Oxford, England and New York, NY: Oxford University Press.

Twyman, A. D. & Newcombe, N. S. (2010). Five reasons to doubt the existence of a geometric module. *Cognitive Science, 34*(7), 1315–1357.

Varley, R. (2014). Reason without much language. *Language Sciences, 46*, 232–244.

Wilson, R. A. (2008). The drink you're having when you're not having a drink. *Mind and Language, 23*, 273–283.

12

The Neurobiological Bases for the Computational Theory of Mind

C. RANDY GALLISTEL

When we were young, Jerry Fodor and I, so, too, was the computational theory of mind, the central doctrine of cognitive science. It broke the theoretical fetters imposed by the mindless behaviorism that had dominated psychology, philosophy, and behavioral neuroscience for decades. Fodor was making major contributions to cognitive science by spelling out its implications. Some of the implications seemed to me, then and now, to be obvious, but only after Fodor had spelled them out.

One such implication was that the mind must possess (unconscious) symbols and (equally unconscious) rules for manipulating them. That is, it must have a language of thought (Fodor, 1975), just as do computing machines. Because the symbols are, on the one hand, objects of principled manipulation—they are the stuff of computation—and because, on the other hand, some of them refer to things outside the mind, it follows that the language of thought has a syntax and a semantics. When Fodor pointed this out, it seemed to me beyond reasonable dispute, although it has in fact been disputed, even unto the present day (Aydede, 1977; Laurence & Margolis, 1997: Schneider, 2009).

What I found thought provoking about Fodor's insight was just what connectionists objected to: its neuroscientific implications. If one believed in the computational theory of mind, then the symbols and the machinery for manipulating them must have a material realization in the brain.

The problem was that no one knew what it might be. If there were symbols, then they must reside in memory, because the basic function of a symbol in a computing machine is to carry information forward in time in a computationally accessible form (Gallistel & King, 2010). Most neuroscientists were and are unshakably committed to the hypothesis that memories consist of experientially altered synaptic conductances. I have been told in all sincerity that this hypothesis could not be false. Karl Popper would turn in his grave. There is, however, a problem with this hypothesis: synaptic conductances are ill suited to function as symbols (Gallistel & King, 2010). Anyone who doubts this should ask the first neuroscientist that they can corner to explain to them how the brain could write a number into a synapse, or into a set of synapses. Then, step back and watch the hands wave. In the unlikely event of an intelligible answer, ask next how the brain operates on the symbols written into synapses. How, for example, does it add the number encoded in one synapse (or set of synapses) to the number encoded in a different synapse (or set . . .) to generate yet another synapse (or set . . .) that encodes the sum?

Connectionists disliked the language of thought hypothesis because it was not readily reconcilable with what neuroscientists told them was the material realization of memory. However, as a behavioral neuroscientist, I knew how flimsy the evidence for the synaptic theory of memory was and how strongly neuroscientists' belief in it was undergirded by the associative theory of learning. The associative theory had—still has—enormous intuitive appeal: One of Lila R. Gleitman's many bon mots is that, "Empiricism is innate." Moreover, psychologists assured neuroscientists that the associative theory had dominated philosophical and psychological thinking for centuries, which is an historic truth, at least as regards Anglophone thought. So how could this theory be wrong? As a psychologist who had focused on the theory of learning since my undergraduate days in the laboratory of Tony Deutsch, I knew how profoundly flawed the theory was—and is.

For me, the synaptic theory of memory rested on a circularly reinforcing set of false beliefs: The neuroscientists' belief in the synaptic theory of memory was sustained in no small measure by the fact that it accorded with the psychologists' associative theory of learning. The psychologists' belief in the associative theory of learning was sustained in no small measure by its accord with what neuroscientists took to be the material realization of memory. Knowing this circular system of false beliefs, I was not tempted to follow where the connectionists wanted to lead, which was back to a murky, computationally hopeless associationism, motivated mostly by a misinformed assessment of what neuroscientists really knew about the material realization of memory—which was nothing.

So if symbols don't reside in altered synapses, where do they reside? I have argued against the synaptic theory of memory for decades, with no noticeable impact on the neuroscience community. My audiences always pester me with this question: If memory is not enduring changes in synaptic conductances, then what is its physical realization? I used to answer that I had no idea, but my audiences did not find that an appealing answer. Nor did I. Some years back, I began to have some ideas, but I was loath to put them in print, for fear they would further enhance my reputation for preposterous speculation. Now, however, very exciting experimental work in behavioral and systems neuroscience, which has recently appeared, provides empirical support for at least the general thrust of these ideas.

WHERE TO FIND THE SYMBOLS

We have been looking in the wrong place—for both the symbols and the machinery that operates on them. The symbols are not in the synapses, and the machinery that operates on them is not (primarily) in the neural circuits. The symbols are in molecules inside the neurons, and the machinery that operates on them is intracellular molecular machinery.

On this view, each neuron is a computational machine. It takes in information through its dendrites, processes that information with complex information processing machinery implemented at the molecular level within the neuron itself, and, at least sometimes, it then generates a signal that carries the encoded results of its processing to other neurons by means of a patterned train of nerve impulses. On other occasions, it may only update its memories and not send out any signal. Because symbolic memory is an indispensable component of any computing machine (Gallistel & King 2010), the molecular-level information processing machinery inside each neuron has, as one of its most basic constituents, molecules whose form may be altered by experience in such a way as to encode acquired information, information that has been conveyed to the cell through its synaptic inputs. These intracellular memory molecules carry the acquired information forward in a form that makes it accessible to the molecular computing machinery.

Insofar as neuroscientists are also biologists, they have known for decades where in the brain they *could* find materially realized symbols and computational machinery that operates on them. But they have assumed—without ever discussing the possibility—that what they knew about the genetic machinery in every cell was not relevant to the question of the material basis of memory. Anyone familiar with the rudiments of molecular biology knows that (most) codons[1] are symbols for amino acids. And, they know that the sequence of codons between a start codon and a stop codon is a symbol for a protein. If they have been following the evo-devo literature, they know that some proteins represent highly abstract aspects of organic form, such as *anterior, dorsal,* and *distal* (Carroll, 2005; Shubin, Tabin et al., 2009), while others represent complex organs, such as eyes. Non-biologists are sometimes startled to learn that there is such a thing as a gene for an eye; turn it on and you get an eye (Halder, Callaerts, et al. 1995; Gehring, 1998). What is more amazing, we now understand how this is possible. The old saw that genes cannot represent complex organic structures and abstract properties of a structure is simply false; they can, and they do.

The symbols strung out along the double helix carry *inherited* information, information about what worked in the ancestors of the current carrier of that information. These symbols are organized into data structures, just as are the symbols in the memory of a computer. The computational principle that makes this organization possible—the indirect addressing of stored information—is the same in the nucleus of a cell as it is in a computer (Gallistel & King, 2010). Moreover, the molecular machinery that reads the information and uses it to guide organ construction and govern cell function implements the logic gates that are the building blocks of computational machinery. In short, the DNA symbols carry information in a form that makes it accessible to computation, and there is molecular machinery that performs computational operations in the course of reading this information. Could it be that the process of evolution has found a way to make use of this machinery—or the closely related RNA machinery—to store *acquired* information and to carry out computations with it?

STORED INFORMATION IN COMPUTERS AND GENES

In computer memory, the symbols, that is, the words in memory locations, have a bipartite structure. One part digitally encodes some information; the other part, the address part, makes that information accessible to information processing operations. Genetic symbols have this same structure: every gene has a coding portion, in which the codon sequence encodes the amino acid sequence of a protein. Every gene also has one or more promoters. The promoter portion of a gene gives the rest of the cellular machinery controlled (programmed) access to the information in the coding portion.

In computer memory, the coding portion and the address portion of a symbol are both bit patterns. Thus, a copy of the bit pattern that constitutes the address portion of one symbol may be stored in the coding portion of another. The storing of addresses makes possible indirect addressing. Indirect addressing makes variable binding possible. Variable binding makes data structures possible. Data structures are the soul of a computing machine. They embody the computer's knowledge.

In genetic memory, the coding portion of a gene and the promoter portion(s) are both nucleotide sequences. The proteins called transcription factors contain segments that bind to the promoters of specific other genes. The bipartite structure of the gene and the selective binding of transcription factors to particular promoter sequences together implement indirect addressing in genetic memory. Indirect addressing in genetic memory is what makes the eye gene possible. It sits atop a genetic data structure in the cellular nucleus, just as the symbol for a document file (the name of the file) sits atop a data structure in the memory of a computer.

VARIABLE BINDING, INDIRECT ADDRESSING, AND DATA STRUCTURES

Among the first operations that the beginning student of computer programming learns is the operation of assigning a value to a variable. In most computer languages, it goes like this: $W = 135$, which translates as set the value of a variable W to 135. Conceptually, this operation creates two symbols in the memory of the computer, that is, two bit patterns stored at different locations in memory. One is the bit pattern for the variable, W. The other is the bit pattern for the current value of this variable, namely, 135. This latter bit pattern is the physical realization of the number that specifies, say, someone's weight. Each location in memory has a unique address, its own zip code, so to speak. The problem of variable binding is the problem of getting from the symbol for the variable to the symbol for its value.

Given the bipartite structure of computer symbols, it is fairly obvious how to solve this problem: Make the bit pattern for the address of the value the bit pattern that represents the variable. Then, when the machine goes to the address where the symbol for the variable is to be found, the bit pattern it finds at that address is the address of the variable's value. This bit pattern is called a pointer. To get to the value of a variable, the machine rarely goes directly to the address where the value itself is to be found; rather, it goes to an address where it finds a symbol that points to the address of the value. Or in the more complex reality, it goes to an address where it finds a number, which, after some possibly rather complex computations involving other numbers, yields the address of a variable's value. These computations are called *pointer arithmetic*. One consequence of this principle is that the contents of many words in computer memory do not refer to things outside the machine; rather they are the addresses of other locations in memory. They have a purely internal reference. That is how data structures are built up in the memory of a computer.

The genetic machinery works in the same way. The coding portion of many genes does not encode the structure of a protein that forms an element of cellular or tissue structure; rather, it encodes a protein that is a transcription factor, a genetic pointer. Transcription factors bind to the promoter regions of genes initiating the transcription (reading) of their coding portions. Thus, for example, the gene for an eye does not encode the structure of a protein found in the realized eye; rather, it encodes the structure of a transcription factor. And the genes to whose promoters that factor binds also encode transcription factors. One has to go down a fair ways in the genetic data structure to get to genes that encode proteins that form structural components of the realized eye.

The genes that encode transcription factors are symbols for variables. Indirect addressing gives the cellular machinery structured access to the data that specifies how to build an actual eye, that is, how to realize the value of the eye variable. The distinction between the symbol for the variable (the genetic symbol for an eye) and the symbols for the value of that variable in a particular case (the genetic data structure that, when appropriately read, yields a realized eye) is dramatically illustrated by the fact that the genetic symbol for an eye is homologous in the human and the fruit fly. The homology is so close that one can put the human gene into the cells of a developing fruit fly, turn it on at some location, and generate an eye at that location—the faceted dome fruit fly eye, not a human eye, with its lens and pupil (Quiring, Walldorf et al., 1994). Thus, the physical realization of the symbol for an eye is (almost) the same in the fruit fly and the human genome, but the realized eye—the value of the variable—is radically different. The physical symbol for the genetic program that makes an eye has remained the same through hundreds of millions of years, while the nature of the eye-constructing program itself, hence, the structure of the realized variable, has diverged greatly.

THE BUILDING BLOCKS OF COMPUTATION

The building blocks of physically realized computations are logic gates, simple structures that realize the logical operations AND, OR, NOT, NAND, and XOR (exclusive or). These operations are implemented at the molecular level in the reading of genetic data structures. Transcription factors often form dimers, that is, they transiently bind to one another, forming a molecular compound with functional properties its constituents lack. Transcription Factor A and Transcription Factor B may neither of them bind to the promoter of Gene X, but their dimer may do so. In that case, when either factor is present alone, Gene X is not transcribed, but when both are present, it is. This is a molecular AND gate. On the other hand, some dimers act as repressors: they bind to the promoter and thereby block promoters from so binding, but they do not initiate transcription. This is a molecular level NAND gate. (There is a proof in theoretical computer science that any computational operation may be realized with NAND gates.) Or, it may be that either A or B will bind to the promoter of Gene X, thereby activating its transcription, but when both are present, they dimerize, and the dimer no longer binds to that promoter. This implements XOR. Finally, of course repressors, whether dimers or not, implement NOT.

In short, processes operating within cells at the level of individual molecules implement the basic building blocks of computation, and they do so in close connection with the reading of stored information. The information in question is hereditary information, not experientially acquired information. Nonetheless, it is tempting to think that evolution long ago found a way to use this machinery, *or closely related machinery, or, at the very least, functionally similar molecular machinery*, to do the same with experientially acquired information.

SIZE MATTERS

The logical gates that are the building blocks of computational machinery can also be implemented by neural circuits. However, in pondering the relative plausibility of intracellular molecular implementation versus neural circuit implementation of basic computational operations, one should keep in mind the vast difference in the size of the posited machinery. One turn of the DNA helix, which contains 11 nucleotides and can encode 22 bits of information (2 bits per nucleotide), has a volume of about 1.1×10^{-26} meters (11 cubic nanometers), whereas one neuron has a volume on the order of 2×10^{-14} meters (20,000 cubic microns). Thus, machinery built at the level of molecules occupies 12 to 15 orders of magnitude less volume than machinery built at the level of neurons.

ENERGY CONSUMPTION MATTERS

Although many neural net modelers have backgrounds in physics and engineering, surprisingly little attention is paid to the huge differences in the energetic costs of executing computational operations at the circuit level versus at the molecular level. An article by Hasenstaub and others (2010) is an exception, but their analysis takes it for granted that computations are carried out at the circuit level. They consider only what ion-channel configurations would minimize the energetic costs of action-potential transmission, which they recognize are very large. ATP hydrolysis provides the energy packets required by energy-utilizing metabolic processes. Adding one nucleotide to an RNA molecule requires the hydrolysis of only 1 to 2 ATPs, in other words the bare minimum. On the other hand, the energy expended in the conduction of one action potential in unmyelinated cortical axons, together with the resulting transmitter release and reuptake, requires the hydrolysis of 7×10^8 ATPs (see Attwell & Laughlin, 2001, figure 1). Neural circuit models of computation treat action potentials as if they cost nothing. They approximate the signals sent between nodes with finely varying floating-point numbers. These floating point numbers are assumed to represent assumed population spike rates, rather than interspike intervals. In other words, interneuronal signal transmission during computation is assumed to be appropriately approximated by the massive exchange of finely varied floating-point numbers. This approximation is valid only if the numbers of presynaptic spikes carrying each such "rate" signal are very large. Neural net computations require the interchange of a great many such signals. Thus, the energetic costs of the computations imagined-to-be performed at the circuit level must be at least 10–15 orders of magnitude greater than the costs of doing the same computations at the molecular level.

It is hard to grasp how great these differences in size and energy costs are. The difference in size is roughly the difference in size between a neuron and the original Univac computer, which filled a large room. It is also very roughly the difference in size between the original Univac and a contemporary computing state-of-the art CPU. The contemporary CPU is a very much better computing machine than the original Univac, largely because it is so much smaller, faster and more energy efficient. For the same reasons, molecule-sized computing machinery inside neurons would be many orders of magnitude smaller, faster and more energy efficient than the same machinery implemented at the level of neuronal circuits, using synapses as memory elements. The message from biophysical considerations seems clear: if it can be done at the molecular level, it should be done at the molecular level. Sterling and Laughlin (2015) call attention to this succinctly: "These advantages [in biochemical computation as opposed to neural circuit computation]—compactness, energy efficiency, and ability to adapt and match—all suggest the principle compute with chemistry. It is cheaper" (p. 124).

EVIDENCE

These are the thoughts that have slowly taken form in my mind. But where is the evidence? Other than plausibility arguments for why memory and computation in nervous tissue ought to be an intracellular molecular-level process rather than an intercellular circuit-level process, is there any experimental evidence? Until very recently, I had to admit the answer was, no. However, the Hesslow laboratory in Lund have recently described work showing that the acquired information that informs the appropriately timed conditioned eyeblink response in the ferret resides within individual Purkinje cells in the cerebellar cortex (Johansson, Jirenhed et al., 2014). This same experiment shows that the cell possesses machinery capable of reading out this information into complexly structured spike trains in response to synaptic inputs, which inputs indicate simply and only the onset of a conditioned stimulus. This minimally informative input, which contains no information about the temporal relation between the conditioned stimulus and the unconditioned stimulus, produces a complex spike-train output that is informed by acquired temporal information stored within the cell.

Behavioral experiments long ago showed that a critical component of the information acquired during Pavlovian conditioning was the duration of the interstimulus interval (ISI), the interval between the onset of a predictive stimulus (the CS, which is short for conditioned stimulus) and the onset of the event it predicts (the US, short for unconditioned stimulus). Experiments showed that the timing of the acquired response to the CS varies in a systematic, functionally appropriate way with the interstimulus interval. The animal does not simply blink in response to the CS; it blinks at the right time. The latency of blink onset varies in proportion to the duration of the ISI in such a way that the eye reaches maximum closure at the moment when the CS predicts that the US will occur. The ISI-dependent timing of the conditioned response is observed in all of the simple learning preparations that are used to investigate the neurobiology of associative memory (Gallistel & Gibbon, 2000; Balsam & Gallistel, 2009; Balsam, Drew et al., 2010), so finding the mechanism that stores this temporal information is critical to a neurobiological understanding of learning and memory.

It has always been assumed that the structural change mediating an appropriately timed acquired response (a CR for conditioned response) must lurk within the mechanism of synaptic transmission. It is taken for granted in the literature on the neurobiology of learning, that learning of any kind must alter either the release of transmitter from presynaptic terminals or the mechanisms that mediate the binding of the transmitter to receptor molecules in the postsynaptic membrane, or perhaps both. However, how alterations in those synaptic processes could store the duration of an interval has always been a mystery. The neurobiologists have taken the primary challenge to be explaining the fact that the conditioned response occurs, not to explain the fact that it occurs at the right time. The latter fact has been treated as one of those mysteries that we would tackle later, once we had solved the mystery of why a conditioned response develops.

Neural net modelers (connectionists) have suggested that the information about the duration of the ISI resided in some unspecified way in alterations in the pattern of synaptic connections (synaptic weights) within a complex neural network (Martin & Morris, 2002), but these suggestions have been vague. There has never been a specification of what the synaptic code might be, nor how the stored information could be made accessible to computation. Specifying the synaptic code would enable us to understand why one pattern of synaptic connections encoded the fact that the ISI lasted 150 ms while another encoded the fact that it lasted 300 ms. Specifying how the encoded information entered into computations would enable us to understand how the first encoding produced a blink that culminated at 150 ms while the second produced a blink that culminated at 300 ms.

Other theorists have assumed that the answer lay in selective associations between neurons—or even whole networks of neurons (Matell, Meck et al., 2003; Meck, Penney, et al. 2008)—with intrinsically different temporal dynamics. It has been suggested that neurons that intrinsically (in the absence of any informative experience) respond to the CS with a firing rate that rises to a peak at 150 ms and then declines rapidly become preferentially associated with the blink response when the ISI is 150 ms, whereas neurons with a slower dynamic, peaking at 300 ms, become selectively associated with the response when the ISI is 300 ms (Grossberg & Schmajuk, 1989; Yamazaki & Tanaka, 2009). None of these models was plausible on its face and none has received empirical confirmation (Hesslow, Jirenhed et al., 2013). So the mystery of the structural change that mediates the appropriate timing of the conditioned response has remained throughout more than half a century of intensive research on the neurobiology of learning and memory.

The experiment from the Hesslow laboratory (Johansson, Jirenhed et al., 2014) builds on the progressive refinement of eye-blink preparations over the last half century. In this preparation, the CS is usually a tone whose onset warns of a perioccular shock or an air puff to the eyeball (the US). The US causes the eye to blink. If the onset of the tone CS reliably predicts the US at some fixed latency, the animal learns to blink in response to the onset of the tone. As already emphasized, the latency at which it blinks varies in proportion to the latency between tone onset and shock (the ISI).

Earlier work from the laboratories of Richard Thompson and Chris Yeo, showed that the critical circuitry was in the cerebellum (Yeo, Hardiman et al., 1984, 1985; Krupa, Thompson et al., 1993; Bao, Chen et al., 2002; Christian &Thompson, 2003). This was itself a surprise. It was also exciting, because the neuroanatomy of the cerebellum is relatively simple and extremely repetitive. For which reason, there has been a huge amount of work on the neurophysiology and neuropharmacology of cerebellar circuitry. When focus shifted to the cerebellum with its (relatively!) simple circuits, it was shown that the blink response was gated by a CS-produced pause in the endogenously generated (Cerminara & Rawson, 2004) basal firing of the large Purkinje cells (Jirenhed, Bengtsson et al., 2007). The axons of the Purkinje cells carry the output signals from the cerebellar cortex. The duration of this learned pause in Purkinje cell firing covaried with the ISI used in training (Jirenhed & Hesslow, 2011a; Jirenhed & Hesslow, 2011b). Thus, the pause in the firing of the Purkinje cells is an electrophysiological proxy for the conditioned response itself. Like that response, its timing depends on a learned duration stored in memory.

The massive dendritic tree of a Purkinje cell spreads athwart the parallel fiber system in the cerebellum (Figure 12.1). The parallel fibers coursing parallel to the folds in the cerebellar cortex are analogous to the signal bus in a computer. A Purkinje cell reads the signal pattern across a portion of this signal bus. The schematic in Figure 12.1 is grossly misleading as regards the density of the parallel fibers. The dendritic tree of a single Purkinje cell is synaptically contacted by as many as 200,000 parallel fibers (Harvey & Napper, 1991). The neural signals generated by behaviorally effective CSs reach the Purkinje cells by way of the mossy fibers that synapse on granule cells in the granular layer. The granule cells give rise to fibers that ascend almost to the cortical surface of the cerebellum, where they send branches that run along the folds in the cerebellar cortex. These branches are the parallel fiber system, the cerebellar signal bus. Signals generated by USs (predicted stimuli) reach the cerebellar Purkinje cells by way of the so-called climbing fibers, which originate from cells in the olivary nucleus

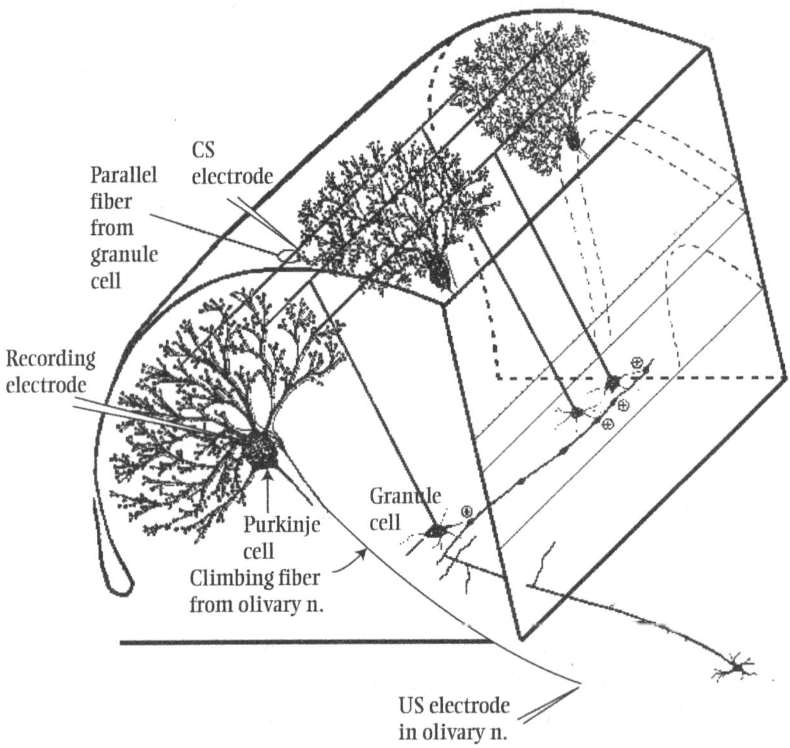

Figure 12.1 The experimental preparation in the recent experiment from the Hesslow laboratory. The artificial CS was a spike train generated in the parallel fibers by a train of stimulating pulses delivered through the CS electrode. The artificial US was a spike train in the climbing fibers generated by direct electrical stimulation of the climbing fiber. The conditioned response of the Purkinje cell was monitored via a recording electrode.

of the cerebellum. Thus, the Purkinje cell is one of several sites of convergence of CS and US signals.

A second exciting advance was the demonstration that the learned pause in the firing of the Purkinje cells was seen even when direct electrical stimulation of the parallel fibers themselves was used in place of a natural conditioned stimulus and even when stimulation of the climbing fiber from the olivary nucleus was used in place of a natural unconditioned stimulus (Jirenhed, Bengtsson et al., 2007; Hesslow, Jirenhed et al., 2013; Johansson, Jirenhed et al., 2014). This discovery radically reduces the neural circuitry that is the focus of attention (Figure 12.1). It gives experimenters unprecedented control of the inputs to the Purkinje cell.

In the most recent experiments from the Hesslow laboratory, they electrically stimulate some portion of parallel fibers while recording from a Purkinje cell that is reading some portion of the stimulated fibers(Johansson, Jirenhed et al., 2014). Thus, they directly determine and control the parallel fiber signal arriving at the dendrites of the Purkinje cell from which they record. In place of a periorbital shock for the US, they use direct stimulation of the climbing fiber from the olivary nucleus with two very short bursts of high frequency electrical pulses.

Before training, the stimulation-elicited spike train in the parallel fibers—the artificial CS—often elicits an increase in the firing of the Purkinje cell. The increased rate of firing lasts as long as the CS spike train, and ceases abruptly when that input terminates, to be followed by a profound and prolonged reduction in the basal firing rate. This is what one might expect to see, because the primary transmitter substance released by the parallel fiber synapses onto the Purkinje cell is glutamate, the brain's principal excitatory transmitter.

In their experimental protocol, they then "condition" (teach) the Purkinje cell by pairing stimulation of the parallel fibers (the artificial CS) with stimulation of the olivary nucleus (the artificial US). In different repetitions of the experiment, they use different ISIs, that is, different intervals between the onset of parallel fiber stimulation and the onset of climbing-fiber stimulation.

The training profoundly alters the Purkinje cell's response to the spike train in the parallel fibers. After training, the onset of the presynaptic spike train in the parallel fibers no longer elicits an increase in Purkinje cell firing; rather, it elicits an almost complete pause in the cell's basal firing. In other words, the training appears to convert an excitatory synapse into an inhibitory synapse, although I think that this will prove to be a misleading way of thinking about the phenomenon. This conversion in the apparent properties of the parallel-fiber-to-Purkinje cell synapses is not subtle; the training produces a huge sign-reversing change in the input-output characteristics of these synapses. (There is also a dramatic change in the duration of the pause in Purkinje cell that follows the offset of the CS.) The Purkinje cell's posttraining response to the artificial CS signal is both complex and radically different from its pretraining response.

Most importantly, the duration of the pause in Purkinje cell firing varies in proportion to the training ISI. And most important, the duration of this learned pause depends only on the training ISI, not on the duration of the presynaptic spike train that initiates it. The duration of the Purkinje cell's firing pause does not vary in response to large (posttraining) variations in the interspike intervals within the presynaptic spike train, nor to large changes in the overall duration of this spike train. The learned, well-timed posttraining response of the Purkinje cell is the same when the presynaptic spike train is produced by a stimulus train lasting only 17.5 ms and containing 8 pulses (hence, with an interpulse interval of slightly more than 2 ms) as when the it lasts 800 ms and contains 81 pulses (hence, with an interpulse interval of 10 ms). In short, radically different synaptic input patterns produce the same learned, timed output from the Purkinje cell.

These results show that the temporal information acquired during the training experience—the remembered duration of the ISI—is expressed in the time course of the Purkinje cell's response to the onset of a presynaptic spike train, under circumstances where it is almost inconceivable that this temporal information is in the activating input (the presynaptic spike train) or in the synaptic conductances between the parallel fiber input and the postsynaptic Purkinje cell. All that the input appears to do is trigger the output; the temporal characteristics of the resulting learned output are quite unrelated to the temporal characteristics of the input.

Traditionally, the only function of a synapse is to amplify or deamplify the effect of the input signal on the postsynaptic neuron. That is why, notional synapses are represented by scalars (weights) in contemporary neural net modeling. The information about the duration of the ISI is not present in the artificially produced and controlled presynaptic signals, but it is manifest at the recording electrode on the postsynaptic cell. The information must have entered the signal somewhere between the stimulating electrode on the presynaptic fibers and the recording electrode on the Purkinje cell. The only structures in that path are the synapse itself, which, as traditionally conceived cannot produce the observed appropriately timed pause, and the complex molecular structures inside the Purkinje cell itself or embedded in its membrane. Thus, the results would seem to imply that the acquired temporal information is encoded by a change in some molecular-level structure intrinsic to the Purkinje cell itself.

In the posttraining Purkinje cell, the onset of a presynaptic spike train causes the information in this cell-intrinsic structure to be read out into a spike train whose temporal complexity depends not at all on the temporal structure of the presynaptic spike train that activates the read out, but rather on: (1) a cell-intrinsic mechanism that has stored the temporal information acquired from experience; and (2) on intracellular machinery capable of converting the stored information into a complex output signal when activated by a simple input signal. The output conveys the stored information to the neurons in the deep nuclei on which the output axon of the Purkinje cell synapses. The signal from the Purkinje cells has been shown to control the timing of the conditioned response (Heiney, Kim, et al., 2014). Thus, the experiment appears to show that the acquired information about the ISI duration that is expressed in the timing of the conditioned response resides inside the Purkinje cell.

This result is incomprehensible on the basis of the simple properties that neural net theorists imagine neurons to possess, which are those of a leaky integrator with a threshold on its output. This result is perfectly intelligible, however, if one imagines that the physical basis of memory is not in the synapse qua conductor but rather in information-storing changes in molecules intrinsic to the neuron and if one further imagines that the neuron also contains the molecular level machinery necessary to read that stored information out into a complexly patterned spike train. This spike train is informed almost entirely by acquired information that has been stored by a cell-intrinsic symbolic storage medium rather than by the information conveyed to the neuron through its synaptic inputs or by the intrinsic dynamics of the neuron itself.

Given results this revolutionary in their implications, it is natural and appropriate to ask whether some other interpretation is possible. How sure can we be that it is the parallel fiber input that it critical to both the pretraining response and the radically different posttraining responses of the Purkinje cell? In the top layer of the cerebellar cortex, one finds not only the dense parallel fiber system but also two other kinds of interneurons, stellate cells and basket cells. Both of these make inhibitory synapses on the Purkinje cell. It is natural to wonder whether these inputs might somehow explain the appropriately timed pause in the posttraining firing of the Purkinje cell, because it is possible, perhaps even likely, that the electrical stimulation of the parallel fibers stimulates some of these neurons as well.

To address this question, Johansson et al. (2014) turned to another phenomenon observable in the same preparation: When one stimulates parallel fibers that are "off beam," that is, that do not synapse on the Purkinje cell from which one is recording, one observes a profound inhibition of the basal firing in the cell from which one is recording. There is reason to believe that this inhibition is mediated by either the stellate cells or the basket cells, both of which are known to make inhibitory synapses on the near dendrites and cell body of the Purkinje cell. When Johansson and his colleagues inject a drug that blocks the action of the inhibitory transmitter, the inhibitory effect of off-beam stimulation on the basal firing of the Purkinje cell is eliminated. However, this drug injection has no effect on the cell's learned, well-timed response to the artificial CS. This is strong evidence against a role for these inhibitory cells in explaining the timing of the learned pause in the Purkinje cell's response.

MORE EVIDENCE: ABRUPT CHANGES IN HIPPOCAMPAL FRAMES OF REFERENCE

The evidence from the Hesslow lab is the most direct evidence that machinery for storing acquired information resides in a cell-intrinsic information-storing medium rather than in the synaptic conductances, and so does the machinery for reading out that information into a spike train. Less direct evidence comes from at least three other sources: (1) the learned signaling characteristics of the neurons in the hippocampus and associated structures and (2) learned alterations in presynaptic transmitter release from olfactory neurons; (3) experiments showing that eliminating learning-induced changes in synapses does not eliminate the memory.

The firing of neurons in the hippocampus and in other closely connected regions of the medial temporal lobe is dramatically dependent on previously acquired spatial and temporal information (see for recent review Gallistel & Matzel, 2013). The firing of these neurons is not determined by what if anything the rat currently sees or hears or smells or feels. Rather, it is determined by the animal's location and orientation on its cognitive map, as computed by its brain from a variety of past sensory inputs (Gallistel & Matzel, 2013).[2]

A location and an orientation are represented by systems of coordinates. But coordinates represent a location or orientation in an experienced environment only when they have been anchored to an experienced frame of reference. This anchoring is what endows a system of coordinates with a semantics, that is, with a specific spatial reference. The location and orientation specified by a set of coordinates depends on the learned frame of reference to which they refer. What is innate in the brain's system for representing the experienced geometry of its environment are systems of coordinates, the machinery that implements vector spaces. A vector space is a symbolic system that can in principle represent the geometry of an environment. For a vector space to represent an actually experienced space, it must be anchored to a frame of reference within the experienced environment. In the course of constructing its cognitive maps, the brain anchors systems of coordinates to many different frames of reference. In one frame, location may be specified by reference to a prominent white card on an otherwise black wall. In another frame, the same location may be signaled by reference to the geometry of the enclosure or by the geometry of the large space that contains the enclosure (Keinath, Julian, et al., 2017).

The firing of the head-direction cells, place cells, and grid cells that signal the animal's current location and orientation is anchored to different frames of reference, even for one and the same neuron. There are object-based frames of reference, enclosure-based frames of reference, and large scale (extra-enclosure) based frames of reference (Gallistel & Matzel, 2013). The same grid or place cell or the same head-direction cell may signal location or direction within one of these frames of reference at one moment and a small fraction of a second later signal location or direction within a different frame of reference (Gothard, Skaggs et al., 1996; Gothard, Skaggs et al., 1996; Frank, Brown et al., 2000; Redish, Rosenzweig et al., 2000; Rivard et al., 2004; Diba & Buzsáki, 2008; Derdikman, Whitlock et al. 2009). The astonishingly abrupt changes in the frame of reference within which the cell's firing specifies a location or direction is difficult to explain if one assumes that the acquired information about the geometry of the experienced environment is encoded in complex patterns of synaptic strength spread throughout an extensive neuronal network (a so-called distributed representation). Just how difficult it is to explain these abrupt transitions on such a basis is hard to say, because the antecedent question of how patterns of synaptic strengths might encode environmental geometry has not been addressed. There are no theoretical proposals about how to embed a vector space in a set of synapses, only hand waves. An extremely abrupt (< 80ms) change in a frame of reference is much easier to explain if one assumes that the acquired information about the geometry of the animal's environment is stored within the neurons themselves rather than in complex neural circuits feeding signals to them.

I digress here to emphasize the following point: whereas there are no theories about how geometric information might be stored in a pattern of synaptic conductances, we know very well how information of any kind might be stored in DNA. The structure of DNA permits the storage of information at 2 bits per nucleotide, because any of the 4 nucleotides may follow any other in the sequence of nucleotides in a DNA or RNA molecule. A single nucleotide is approximately 1/3 of a nanometer in length. Therefore, DNA stores information at a linear density of 6 bits per nanometer. (To return for a moment to the consideration of size, the width of a synaptic cleft is about 20 nm; the diameter of the presynaptic vessicles that package neurotransmitters for release from presynaptic terminals is 35 nm.) A basic truth of computer science is that a medium suited to the storage of one kind of information is suited to the storage of any kind of information. When it comes to information storage and transmission, information is information; it's all just bits. That is why even poems can be stored in bacterial DNA (Gardiner, 2010). That is why there are laboratories actively exploring the use of DNA as the memory component in a future computing machine (Team, 2010, Goldman; Bertone et al. 2013). If and when DNA becomes the memory component of a computing machine, geometric information will be stored in it in essentially the same way it is now stored in the memories of the servers that you access when you use a map application. In short, there is no mystery about how to store geometric information in the structure of DNA-like molecules or in RNA-like molecules, whereas there is a profound mystery about how to store information in synapses.

STILL MORE EVIDENCE: LEARNED, SELECTIVE IMMEDIATE ENHANCEMENT OF NEUROTRANSMITTER RELEASE FROM FIRST-ORDER OLFACTORY NEURONS

When we grasp the fact that acquired information may be stored in a complex molecular computing machine intrinsic to individual neurons, there is no reason not to assume that this occurs in every kind of neuron, including sensory neurons and motor neurons. Sensory neurons may use acquired information to help them interpret the information picked up by their transducer elements. That is, the process of interpreting current sensory input in the light of previous experience may begin within the first-order sensory neurons themselves. Recent quite astonishing findings from the laboratory of my colleague, John McGann, suggest just that.

The McGann laboratory brings state-of-the-art neurobiological visualization methodology to bear on the question of how the brain represents olfactory input. We do not experience smells as a meaningless sensations; rather, they are freighted with learned significance: We smell bacon or the sea or manure or eucalyptus or the odor of a loved one. (One is reminded of Napolean's famous epistolary admonition to Josephine: "Coming home in three days; don't bathe.") It seems likely that the same is true for non-human animals, perhaps even more so than for us, as odor plays a larger role in the sensory/perceptual life of many animals than it does in ours.

However, until recently, the study of olfactory perception was a neurobiological and psychophysical backwater. This changed with the advances in the understanding of olfactory neuroanatomy consequent upon the discovery of the molecular biology of olfactory transduction (Mombaerts, Wang et al., 1996; Su, Menuz et al., 2009).

From a functional/computational standpoint, a basic property of any sensory system is the number of distinct channels that are operative. Each functionally distinct channel filters the stimulus in a different way and adds a degree of freedom (a dimension in a vector space) to the brain's representation of that stimulus. The scotopic visual system, which operates in dim light, has only one channel; the photopic system, which operates in brighter light, has three; the auditory system has thousands. It turns out that the olfactory system has hundreds. Each functional olfactory channel is composed of neurons that express one and only one of the hundreds of olfactory receptor molecules in the receptor end of the sensory neuron in the olfactory mucosa.

Remarkably, all of the neurons that express the same receptor in their mucosal transducer portion project their signal-carrying axons to one or two glomeruli in the brain's olfactory bulb. Glomeruli are small spherical synapse-rich structures in the olfactory bulb. Each glomerulus receives projections from only one odor channel. Thus, the functional unit—the olfactory channel—maps to an anatomical unit—the glomerulus. Every different odorant creates a different pattern of activation of the olfactory glomeruli. For any given odorant, most glomeruli are inactive, but a few show a pattern of activation in which there is odorant-specific variation in the relative strengths of the activation.

McGann's laboratory visualizes these activation patterns in mice both before and after they have been trained with different odorants as discriminative stimuli. One of the odorants the mouse sniffs during training predicts shock; the other odors do not. McGann and his students find that this training selectively increases neurotransmitter release from the presynaptic endings of the first-order olfactory neurons synapsing on the glomeruli encoding the shock-predicting odor. In other words, information gained from an experienced predictive relationship between that odor and a fear-inducing shock finds its way to the presynaptic endings of the first stage olfactory neurons. This acquired information selectively alters their signaling at the point where they pass on to the rest of the brain the information they have gleaned from the odorant molecules currently binding to their receptors in the olfactory mucosa.

Remarkably, when the spectrum of glomeruli activated by the predictive odorant overlaps to some extent with the spectrum activated by a non-predictive odorant, the enhanced neurotransmitter release in the glomeruli in the intersect is specific to the predictive odorant. The release of transmitter caused by the non-predictive odors in those glomeruli is not enhanced; only the release produced there by the predictive odor is enhanced.

As in most sensory systems, there are extensive efferent projections from higher levels of the brain to the synaptic endings of these sensory neurons. Thus, there is no neuroanatomical mystery as to how the information acquired from the experience of the predictive relation between a given odor and shock may reach these presynaptic endings. There are, however, two quite different stories that one may imagine about how information conveyed by these efferents comes to inform the release of neurotransmitter from those endings. On one hypothesis, the information about the predictive relation between the one odorant and shock is not stored in the presynaptic endings themselves; on the other hypothesis, it is.

On the one hand, one may imagine that the acquired information about the predictive relation between odor and shock is stored more centrally in the brain. A connectionist would imagine that the acquired information is stored in some distributed pattern of synaptic conductances in some complex circuit, perhaps located in the amygdala, which is known to play an important, but ill-defined role in fear conditioning, or perhaps in the neocortex. On this story, each time an odorant evokes a spike train in the first-order sensory neurons, the first few spikes in this train cause postsynaptic activity that propagates to the complex central circuit in which the information acquired from the training experience is stored. These initial afferent signals activate the complex central circuit in such a way as to cause it to generate an efferent signal that propagates back to the endings of the first-order sensory neurons. This efferent recognition signal enhances the release of neurotransmitter by later portions of the sensory spike train. On this story, the predictive significance of the sensory signal is recognized centrally—as has always been assumed.

A different possibility—until recently, almost unthinkable—is that when computations on the temporal map of past experience (Balsam & Gallistel, 2009; Balsam, Drew et al., 2010) reveal the predictive relation between a specific odor and shock, this information is relayed to the presynaptic endings of the first-order neurons to be stored there. Then, as in the cerebellar circuit studied in the Hesslow lab, this intracellularly stored acquired information alters the release of neuro-transmitter by the odorant-induced spike train. In this way, the signal passed on from the first-order sensory neurons to the postsynaptic circuit is already partially interpreted in the light of the predictive relation revealed by previous experience.

On the first hypothesis, which almost any neuroscientist would judge to be far more plausible, the selective enhancement of neurotransmitter release from the presynaptic endings of the first-order olfactory neurons can occur only some while after the onset of the odorant-evoked spike train in the first-order sensory neurons, because it depends on real-time feedback from the central circuits where the recognition of the signal's predictive significance occurs. On the second hypothesis, by contrast, the enhancement of neurotransmitter release can occur at signal onset, because it does not depend on real-time feedback. It depends instead on locally stored information conveyed to the presynaptic endings by earlier "off-line" feedback. This earlier off-line feedback came from the more central structures that computed the predictive relation from a time-stamped record of past events (Balsam & Gallistel, 2009).

In fact, the enhancement that McGann's lab observes is present throughout the signal. As best they can determine, it is already there at signal onset. If the evidence for the immediate enhancement of transmitter release holds up, it strongly favors the second hypothesis, the local, intracellular storage of acquired information. It will be interesting to see just how much information is stored locally at that earliest possible stage of sensory signal processing, and at what level of abstraction.

In short there is now evidence that acquired information relevant to the interpretation of sensory signals may be stored within the sensory neurons themselves. One wonders whether the evidence for learning at the spinal level (Windhorst, 2007; Wolpaw, 2007) will lead to the discovery that acquired information relevant to the regulation of muscle activation and joint control is stored within the motor neurons themselves.

As regards the third line of experimental evidence favoring the intracellular molecular basis of memory, we have recent reports from two labs, one working with Aplysia (Chen, Cai et al., 2014), one with mice, showing that the synaptic changes produced by a conditioning experience may be abolished without abolishing the memory itself (Ryan, Roy, et al., 2015).

BACK TO JERRY FODOR

What I lay at Fodor's door is an insight that—in the fullness of time—may transform neuroscientists conceptual framework in ways as profound as the transformation in biochemists' conceptual framework wrought by the identification of the molecular structure of the gene. Fodor realized that there must be symbols in the brain, just as Mendel realized that there must be physically mysterious "particles" in seeds, particles that carried heritable information from generation to generation, quite independently of whether the information they carried was expressed in the observable structure of the organisms produced in any one generation. The physical realization of the symbols that carry acquired information is at this time as mysterious as was the physical realization of Mendels' particles. Like Mendel's particles, the information carried by these symbols is often not expressed in behavior. Fodor also realized that there must be computational machinery that operates on those symbols, the machinery that embodies the syntax. He realized, in other words, that the brain must have a language in exactly the sense in which a computing machine has a language. This was a truly profound insight, which is, of course, why it has also generated so much debate. The old ways of thinking die hard, very hard. To paraphrase Planck, science progresses one funeral at a time.

If, as I expect, Fodor's insight comes to inform the foundations of neuroscientific thinking, there will be great ironies. Fodor is conspicuous among cognitive scientists for his indifference to the question how the language of thought might be implemented in the brain. He commented on neurobiologically inspired theories of cognition only so far as to point out that they lacked the productivity, systematicity, and compositionality that are seen in a machine that has a language. He and Zenon Pylyshyn rightly argued that these properties were such salient properties of thought that any model of thought that denied these properties, either explicitly or implicitly, as neurobiologically inspired cognitive theories generally did, was clearly untenable in the face of the behavioral evidence (Fodor & Pylyshyn, 1988). Fodor was sublimely indifferent to the protests from some philosophers and many psychologists and cognitive psychologists that there was no neurobiological foundation for the language of thought. Like the classical geneticists who were unperturbed by the biochemists' claims that the gene was biochemically incomprehensible, Fodor believed in the implications of the data he knew. He was unperturbed by the neuroscientists who took absence of neurobiological evidence to be evidence of neurobiological absence. What an irony it will be if the language of thought hypothesis becomes the key to understanding the neurobiology of cognition.

I am immensely excited by the prospect that Fodor's insight may finally begin to influence neuroscientific thinking. Until the recent discoveries that I have described, I thought there was no prospect that we would know the physical identity of the brain's symbols in my lifetime. I thought there was even less prospect that we would know the machinery that implemented its computational operations. I suspected that the answers were to be found at the molecular level within neurons, rather than at the circuit level, where neuroscientists have assumed they must lie and where, therefore, they have looked for them to little avail throughout my career. Until these recent discoveries, there was no neurobiological evidence in favor of the hypothesis that acquired information is stored in a cell-intrinsic molecular medium, where it is operated on by molecular level computational machinery. Now that there is at least some neurobiological evidence pointing in that direction, my hope is that the molecular biologists will jump in and begin a serious quest for the intracellular molecular biology of neural memory and computation.

NOTES

1. A codon is a reading-frame triplet of nucleotides. There are four nucleotides. During the transcription of a gene, the double helix is read in the sense direction along the sense strand (as opposed to the antisense strand) in nucleotide triplets (three-letter words, written in the four-letter alphabet of nucleotides). The reading frame is determined by the nucleotide from which transcription starts. In

addition to the codons (words) that code for amino acids, there are punctuation codons that indicate the beginning and end of a codon sequence that constitutes one gene.

2. Recent work from Eichenbaum's laboratory (MacDonald et al., 2011; Eichenbaum, 2013) shows that these cells also signal temporal location, that is, the current temporal distance from recent events that function as temporal landmarks in that they occur at a fixed (temporal) distance from other events of interest. Thus, these cells appear to signal the animal's spatio-temporal location in a spatio-temporal cognitive map.

REFERENCES

Attwell, D., & Laughlin, S. B. (2001). An energy budget for signaling in the grey matter of the brain. *Journal of Cerebral Blood Flow and Metabolism, 21*, 1133–1145.

Aydede, M. (1977). Language of thought: The connectionist contribution. *Minds & Machines, 7*, 57–101.

Balsam, P. D., Drew, M. R., & Gallistel, C. R. (2010). Time and associative learning. *Comparative Cognition & Behavior Reviews, 5*, 1–22.

Balsam, P. D., & Gallistel, C. R. (2009). Temporal maps and informativeness in associative learning. *Trends in Neurosciences, 32*(2), 73–78.

Bao, S., Chen, L., Kim, J., & R. Thompson, J. (2002). Cerebellar cortical inhibition and classical eyeblink conditioning. *Proceedings of the National Academy of Sciences, 99*, 1592–1597.

Carroll, S. B. (2005). *Endless forms most beautiful: The new science of Evo Devo and the making of the animal kingdom*. New York, NY: Norton.

Cerminara, N. L., & Rawson, J. A. (2004). Evidence that climbing fibers control an intrinsic spike generator in cerebellar Purkinje cells. *Journal of Neuroscience, 24*, 4510–4517.

Chen, S., Cai, D. Pearce, K. Sun, P. Y.-W. Roberts A. C., & Glanzman, D. L. (2014). Reinstatement of long-term memory following erasure of its behavioral and synaptic expression in Aplysia. *eLife*, e03896.

Christian, K., & Thompson, R. (2003). Neural substrates of eyeblink conditioning: Acquisition and retention. *Learning and Memory, 10*(6), 427–455.

Derdikman, D., Whitlock, J. R., Tsao, A., Fyhn, M., Hafting, T., Moser, M.-B., & Moser, E. I. (2009). Fragmentation of grid cell maps in a multicompartment environment. *Nature Neuroscience, 12*, 1325–1332.

Diba, K., & Buzsáki, G. (2008). Hippocampal network dynamics constrain the time lag between pyramidal cells across modified environments. *Journal of Neuroscience, 28*(50), 13448–13456.

Eichenbaum, H. (2013). Memory on time. *Trends in Cognitive Science, 17*(2), 81–88.

Fodor, J. A. (1975). *The language of thought*. Trowbridge, Wiltshire, England: Crowell Press.

Fodor, J. A., & Pylyshyn, Z. (1988). Connectionism and cognitive architecture: A critical analysis. *Cognition, 28*, 3–71.

Frank, L. M., Brown E. N. & Wilson, M. (2000). Trajectory encoding in the hippocampus and entorhinal cortex. *Neuron, 27*, 169–178.

Gallistel, C. R., & Gibbon, J. (2000). Time, rate, and conditioning. *Psychological Review 107*(2), 289–344.

Gallistel, C. R., & King, A. P. (2010). *Memory and the computaitonal brain: Why cognitive science will transform neuroscience.* New York, NY: Wiley/Blackwell.

Gallistel, C. R., & Matzel, L. D. (2013). The neuroscience of learning: Beyond the Hebbian synapse. *Annual Review of Psychology, 64*, 169–200.

Gardiner, B. (2010). Recombinant rhymer encodes poetry in DNA. *Wired* (April).

Gehring, W. J. (1998). *Master control genes in development and evolution: The homeobox story.* New Haven, CT: Yale University Press.

Goldman, N., Bertone, P., Chen, S., Dessimoz, C., LeProust, E. M., Sipos, B., & Birney E. (2013). Towards practical, high-capacity, information storage in synthesized DNA. *Nature, 494*, 77–80.

Gothard, K. M., Skaggs, W. E. & McNaughton, B. L. (1996). Dynamics of mismatch correction in the hippocamapal ensemble code for space: Interaction between path integration and environmental cues. *Journal of Neuroscience, 16*(24), 8027–8040.

Gothard, K. M., Skaggs, W. E., Moore, K. M., & McNaughton, B. L. (1996). Binding of hippocampal CA1 neural activity to multiple reference frames in a landmark-based navigation task. *Journal of Neuroscience, 16*(2), 823–835.

Grossberg, S., & Schmajuk, N. A. (1989). Neural dynamics of adaptive timing and temporal discrimination during associative learning. *Neural Networks, 2*, 79–102.

Halder, G., Callaerts, P., & Gehring, W. J. (1995). Induction of ectopic eyes by targeted expression of the *eyeless* gene in *Drosophila*. *Science, 267*, 1788–1792.

Harvey, R. J., & Napper, R. M. A. (1991). Quantitative studies on the mammalian cerebellum. *Progress in Neurobiology, 36*, 437–463.

Hasenstaub, A., Otte, S., Callaway, E., & Sejnowski T. J. (2010). Metabolic cost as a unifying principle governing neuronal biophysics. *Proceedings of the National Academy of Science, 107*(27), 12329–12334.

Heiney, S. A., Kim, J., Augustine, G. J., & Medina, J. F. (2014). Precise control of movement kinematics by optogenetic inhibition of Purkinje cell activity. *The Journal Neuroscience, 34*, 2321–2330.

Hesslow, G., Jirenhed, D.-A., Rasmussen, A., & Johansson, F. (2013). Classical conditioning of motor responses: What is the learning mechanism? *Neural Networks, 47*, 81–87.

Jirenhed, D. A., Bengtsson, F., & Hesslow, G. (2007). Acquisition, extinction, and reacquisition of a cerebellar cortical memory trace. *Journal of Neuroscience, 27*(10), 2493–2502.

Jirenhed, D. A., & Hesslow, G. (2011a). Learning stimulus intervals: Adaptive timing of conditioned Purkinje cell responses. *The Cerebellum, 10*, 523–535.

Jirenhed, D. A., & Hesslow, G. (2011b). Time course of classically conditioned Purkinje cell response is determined by initial part of conditioned stimulus. *Journal of Neuroscience, 31*, 9070–9074.

Johansson, F., Jirenhed, D.-A., Rasmussen, A., Zucc, R., & Hesslow, G. (2014). Memory trace and timing mechanism localized to cerebellar Purkinje cells. *Proceedings of the National Academy of Science, 111*, 14930–14934.

Keinath, A. T., Julian, J. B., Epstein, R. A., & Muzzio, I. (2017). Environmental geometry aligns the hippocampal map during spatial reorientation. *Current Biology, 27*, 309–317.

Krupa, D. J., Thompson, J. K., & Thompson, R. F. (1993). Localization of a memory trace in the mammalian brain. *Science, 260*, 989–991.

Laurence, S., & Margolis, (1997). Regress arguments against the language of thought. *Analysis, 57*(1).

MacDonald, C. J., Lepage, K. Q., Eden, U. T., & Eichenbaum, H. (2011). Hippocampal "time cells" bridge the gap in memory for discontiguous events. *Neuron, 71*(4), 737–749.

Martin, S. J., & Morris, R. G. M. (2002). New life in an old idea: The synaptic plasticity and memory hypothesis revisited. *Hippocampus, 12*, 609–636.

Matell, M. S., Meck, W. H., & Nicolelis, M. A. (2003). Interval timing and the encoding of signal duration by ensembles of cortical and striatal neurons. *Behavioral Neuroscience, 117*(4), 760–773.

Meck, W. H., Penney, T. B., & Pouthas, V. (2008). Cortico-striatal representation of time in animals and humans. *Current Opinion in Neurobiology, 18*, 145–152.

Mombaerts, P., Wang, F., Dulac, C., Vassar, R., Chao, S., Nemes, A., Mendelsohn, M., . . . Axel, R. (1996). The molecular biology of olfactory perception. *Cold Spring Harbor Symposia on Quantitative Biology, 61*, 135–145.

Quiring, R., Walldorf, U., Kloter, U., & Gehring, W. J. (1994). Homology of the *eyeless* gene of the *Drosophila* to the *small eye* gene in mice and *aniridia* in humans. *Science, 265*, 785–789.

Redish, A. D., Rosenzweig, E. S., Bohanick, J. D., McNaughton, B. L., & Barnes, C. A. (2000). Dynamics of hippocampal ensemble activity realignment: Time versus space. *Journal of Neuroscience, 20*, 9298–9309.

Rivard, B., et al. (2004). Representation of objects in space by two classes of hippocampal pyramidal cells. *Journal of General Physiology, 124*, 9–25.

Ryan, T., Roy, D., Pignatelle, M., Arons, A., & Tonegawa, S. (2015). Engram cells retain memory under retrograde amnesia. *Science, 348*, 1007–1013.

Schneider, S. (2009). LOT, CTM, and the elephant in the room. *Synthese, 170*, 235–250.

Shubin, N., Tabin, C., & Carroll, S. (2009). Deep homology and the origins of evolutionary novelty. *Nature, 457*, 818–823.

Sterling, P., & Laughlin, S. B. (2015). *Principles of neural design*. Cambridge, MA: MIT Press.

Su, C.-Y., Menuz, K., & Carlson, J. R. (2009). Olfactory perception: Receptors, cells, and circuits. *Cell, 139*(1), 45–59.

Team, H. K.-C. (2010). Bacterial based storage and encryption device. Chinese University of Hong Kong iGEM.

Windhorst, U. (2007). Muscle proprioceptive feedback and spinal networks. *Brain Research Bulletion, 73*, 155–202.

Wolpaw, J. R. (2007). Spinal cord plasticity in acquisition and maintenance of motor skills. *Acta Physiolologica, 189*, 155–169.

Yamazaki, T., & Tanaka, S. (2009). Computational models of timing mechanisms in the cerebellar granular layer. *The Cerebellum, 8*, 423–432.

Yeo, C. H., Hardiman, M. J., & Glickstein, M. (1984). Discrete lesions of the cerebellar cortex abolish the classically conditioned nictitating membrane response of the rabbit. *Behavioural Brain Research, 13*, 261–266.

Yeo, C. H., Hardiman, M. J., & Glickstein, M. (1985). Classical conditioning of the nictitating membrane response of the rabbit. III. Connections of cerebellar lobule HVI. *Experimental Brain Research, 60*, 114–126.

INDEX

Page references for figures are indicated by *f* and for boxes by *b*.

abstract lexical representations, 49, 50*f*
acoustic input, 88–89
acquisition
  language, 25–26, 88–89 (*see also specific topics*)
    computational problem, 88
  learning as, 220
  profiles, language production, 57–58
addressing, indirect, 279
agrammatic disorders, 56–57
Ajemian, R., 27
Aksu-Koç, A. A., 172, 173, 183, 185
algorithms, 116, 134n2, 135n4
Altmann, G., 72, 73
Alzheimer patients, verb-thematic hierarchy, 118
analysis-by-synthesis, 64, 79, 98
analyticity, 12
analytic/synthetic distinction, 12, 14, 15, 124–125, 128, 130, 135n7
Arai, M., 70
argument structure effects, 44–45
associationism, Putnam's critique of, 227–228
associative and derivate processes, alternation between, 97
atomism, 14–15
  conceptual representation, 8, 9
  lexical, 13
auxiliary inversion, 37n10
availability, parsing, 42
background information, non-linguistic, sentence processing and, 42
backward inferences, unconscious comprehensive processes with, 92–98
Balsam, P. D., 291–292
Beach, C. M., 70
begin-entity, 125–126. *see also* indeterminate sentence
begin-event, 125
belief fixation, 7, 34, 113
  perception in, 8–9
Bever, T. G., 2–3, 43, 44–45, 78, 89, 102, 139–140
  The Psychology of Language: An Introduction to Psycholinguistics and Generative Grammar, 3
bias
  contextually driven interpretive, 44
  plausibility (priors), 64–65
biasing effect variance, 45–46
biasing foil, indeterminate sentence, 132–133, 133*f*
Bicknell, K., 93
binary distinction, 192
biolinguistic framework, 25, 27
Bizzi, Emilio, 27
Bock, J. K., 51, 55, 245
Bock, K., 51
Boeckx, C., 263
Bornkessel, I., 119

boundaries, linguistic modularity, 117–119
Broadbent, D., 89
Broca's aphasia, 56
Brown, C., 49–50
Brown, M., 93–94
Bruner, J., 202, 213, 213*f*

Caramazza, A., 56, 57
Carnap, R.
  Carnaptron, 226
  meaning postulates, 16
  Putnam's critique of, 218
Carruthers, P., 255–257, 259–262
Cartesianism, 38n21
Cartesian Psycholinguistics, 3
Cattell, James, 3
causative verb
  CAUSE, 13
  CAUSE TO DIE, 14
  cause to die, kill from, 13–14, 15–16
center-embedded sentences, 139–166
  doubly center-embedded relative clause constructions, 139–141, 141*b*
  elicited prosody experiments, 146–163 (*see also* prosody experiments, elicited)
    experiment 1 (rating task with familiarization), 147–155
    experiment 2 (in search of "missing VP illusion"), 155–163
central access, limited, 65
central system, language, 26
central-system modularity, 26–32
  Merge, 30–32
  minimal computation, 29–32 (*see also* minimal computation (MC))
  Move, 31
  vs. parsing system, 32–34
  principles, 28
  Rigidity Rule, 28
  Rule of Structure-dependence, 28, 29–30
  vision and language, 28
cerebellum, 283–286, 284*f*
Chang, F., 97
children's language learning, 87–106
  ambiguous word meanings, lexical decision task, 92
  associative and derivate processes, alternation between, 97
  comprehensive processes, unconscious, with backward inferences, 92–98
  computational fractal, 99
  conscious experience, implications for, 99–101
  immediacy assumption, 98–99
  language acquisition
    acoustic input problem, 88–89
    computational problem, 88
  language processing
    feed forward, 90, 91
    feed forward and feed backward processing, 91
  language processing units, 88–92
    child discovery, 88
    location, 89–90
    unity, and experience of language, 90–92
  late assignment of syntax theory, 97, 98
  phase, 96–97
  poverty of the stimulus, 87–88
    real, 101–104
  reanalysis processes, 92–93
  research, implications for, 104–105
  sentence like miniature opera, 87–89
  serial encoding, 91
  structure dependence theories, 88, 102
Chomsky, N., 1, 2, 27, 140, 161, 220–222, 257–259, 263
Chomskyan linguistics, 1, 3
Christianson, K., 92–93
chronostasis, 99
C/I interface, 250, 251
Clark, A., 93–94, 98
Clark, H. H., 46
clauses, meaning comprehension time, 242–243
cleaving, 192
click-mislocation, 89
Clifton, C., 71–72, 73, 76–77
"close shadowing" performances, 54
codon, 277, 278, 293n1
coercion, 123
  with interpolation, 123–125

cognition
as holistic, context-sensitive, 6
tout court, 16
Townsend, 43, 78, 97
cognitive architecture, 115–117, 197–205
connectionist, 10–11, 18, 63, 78, 90, 97, 134–135n3, 275–276, 282–283, 291
constants, variables, parameters, and explanatory power, 197–198
physical-mental gap, present view consequences, 198–201
symbolic, 5, 6, 10–11, 16, 17, 19n7, 115–121, 135n5, 205n4
visual perception modularity, 201–205, 203*f*
cognitive conception, 259
cognitively penetrable perception, 9, 197
cognitive revolution, 1
internal representations, 6–7
cognitive science
as anarchic, 7
concepts in, 7
connectionism, 10
Fodorian, 1–18 (*see also* Fodorian cognitive science)
productivity, 11
cognitivist/functionalist explanations, 5
Collins, J., 267
Combinatory Categorial Grammar, Steedman's, 72
competence *vs.* performance, distinguishing, 34
complex representations, 11–12
compositionality, 11–12, 116, 264
as "nonnegotiable assumption," 14
propositional content, 121
Compositionality Papers, The (Fodor), 12
comprehension
language production
as comprehension filter, 52–58 (*see also* language production, as comprehension filter)
profiles predict "comprehension" performance, 54–55
unconscious processes, with backward inferences, 92–98
computation, building blocks, 280
computational fractal, 99
computational mental processes, 116
computational problem, language acquisition, 88
computational relation, 6
computational-representational paradigm, 251–252
Computational/Representational Theory of Mind (C/RTM), 6, 10, 14. *see also* Representational/Computational Theory of Mind
computational theory of mind, neurobiological basis, 275–294
computation building blocks, 280
energy consumption, 280–281
evidence, 292–292
cerebellum and Purkinje cells, 283–286, 284*f*
Hesslow laboratory, 282, 283–286
hippocampal frames of reference, abrupt changes, 287–289
neural net modelers, 282–283, 287
neurons/neuron networks selective associations, 283
neurotransmitter release, learned, selective immediate enhancement from first-order olfactory neurons, 289–292
Fodor and, 292–293
size, 280
stored information, computers and genes, 278
symbols
brain, 277–278
manipulation rules, 275
synaptic theory of memory, 276
variable binding, indirect addressing, and data structures, 278–279
computers, stored information in, 278
concepts
learning, Bruner and Piaget on, 7
nature of, 7
processing, immediacy, 239–246 (*see also* immediacy, conceptual processing)
representation, atomism, 8, 9 (*see also* atomism)

Concepts: Where Cognitive Science Went Wrong, 13
conceptual nativism, 7
conceptual processing, immediacy, 239–246
  conceptual short-term memory, 245
  dual coding theory, 240–241
  perception of scenes, temporal limit, 245–246
  Potter's research question and protocol, 239–240
  sentence comprehension studies, 240–245
    lexical ambiguity, resolution, 244
    perceptual experiences as mundane symbols, 241–242
    phrases and clauses, meaning comprehension time, 242–243
    recall, long sentence *vs.* unrelated word, 244–245
    second language, representation, 242
    two-clause sentences, spoken, 243
    two-stage modular interactive model, word processing, 244
    verbal and perceptual information linked to abstract concepts, 241
    word understanding and sentence meaning processing, at 12 words per second, 243–244
  syntactic priming, 245
conceptual short-term memory (CSTM), 245
conjunctionist theory, Pietroski's, 264–265
conjunction problem, 205
connectedness
  Euler's theorem, 225
  Seymour Papert on, 223–224
connectionism, 10. *see also* cognitive architecture
  cognitive architecture, 10–11, 18, 63, 78, 90, 97, 134–135n3, 275–276, 282–283, 291
  on language of thought hypothesis, 276
  neural net modelers, 282–283, 287
Connine, C. M., 93
conscious experience, 99–101
constraint, natural, 200
construal, sentence processing, 74
constructing, Piagetian, 218
content, mental, 216
  internal structure, 252
  thought, 252
context
  contextually driven interpretive bias, 44
  language production, 55
  parsing, constraint, 42, 44
  sensitivity, 130–133, 133*f*
  sentence processing, 71–73
  utterance interpretation, 43
correspondence problem, 193, 194*f*
Crain, S., 44, 55, 71–72
cross-induction, 229–230
CTM. *see* Representational/Computational Theory of Mind
Cudworth, R., 29
Cutler, A., 77
Cutting, J. C., 50

Dalrymple-Alford, E. C., 89
data structures, 279
de Almeida, R. G., 19nn17–18, 118, 126, 128–129, 129*f*, 131
decompositional views, 14
definitions, on concept acquisition, 217
deictic gestures, 192
Dell, G. S., 97
DeLoache, J. S., 241
derivate and associative processes, alternation between, 97
Descartes, R., 26, 27, 28–29
Desmet, T., 76–77
Dilley, L. C., 93
discourse model, 50*f*
  indeterminate sentence, 121, 122, 130, 132, 133, 135n8
  language production, 55
  vs. real world model, 121, 135n8
distinction(s), 192
  drawing, 192
  process-architecture, 198
DNA, symbols in, 277–278
domain-specific modules, 66
doubly center-embedded relative clause constructions (2CE-RC), 139–141, 141*b*
  syntactic structure, 141–142, 142*f*

dual coding theory, 240–241
Dwivedi, V. D., 128–129, 129*f*

early vision, 197, 198–199, 202
Eberhard, K. M., 74
elicited prosody experiments, 146–163. *see also* prosody experiments, elicited
Elliot, Bernard, 241
Elm and the Expert: Mentalese and Its Semantics, The (Fodor), 13
embedding, 34
empirical data, theoretical advances, 14
empiricism, 8
encapsulation
  information, 66, 80, 113, 114, 117–118
  perceptual analysis, 9, 113
  psycholinguistics, 10
enduring individuality, 192
energy consumption, computational theory of mind, 280–281
epistemic boundedness, 38n21
Euler's theorem, 224–225, 224*f*
Evans, Gareth, 252–253
Evans, K. K., 245
exceptionalism, Fodorian, scientific theories, 191–206
  blindness to object properties, 193–194, 194*f*
  cognitive architecture, 197–205
    constants, variables, parameters, and explanatory power, 197–198
    physical-mental gap, present view consequences, 198–201
    visual perception modularity, 201–205, 203*f*
  cognitively penetrable perception, 9, 197
  correspondence problem, 193, 194*f*
  distinctions, drawing, 192
  exceptionalism, 205n1
  FINST, 192–193, 198–199, 205–206nn4–6
  indexing (FINSTing), 192, 193, 201
  indexing *vs.* selecting-by-property, 192
  Multiple Object Tracking, 193–196, 195*f*
  Phi illusion, 193
  picking-out *vs.* locating, 192
  process *vs.* processing architecture, 192
  Q and *not-Q*, 191
  seeing *vs.* seeing as, 197
explanation, 4–5
expressions, production, 26
externalization, sensory modality and, 26
external Merge, 31

facilitative prosodic phrasing, 141–146, 142*f*, 143*b*, 146*b*
faculty of language (FoL), 19n11, 27, 34, 36, 115, 205n2, 250, 251, 254, 255, 262–265, 267
faddish science, 10
Farmer, T. A., 93–94
Faulconer, B., 240–241
feature-placing language, 199
feed-backward processing, 91
feed-forward perception, 90
feed-forward processing, 91
Fernández, E. M., 164
Ferreira, F., 4, 69, 71–72, 73, 76–79, 92, 94
Ferreira, V. S., 50
Feyerabend, P., 10
field linguists, 28, 37n7
filler-gap problems, 32, 33
Finger of INSTantiation (FINST), 192–193, 198–199, 205–206nn4–6
fings, 193
FINGs, 193, 199–200, 202
FINSTing (indexing), 192, 193, 201
  vs. selecting-by-property, 192
  Spencer-Brown on, 192
  visual perception modularity, 202, 203*f*
Firestone, C., 204
"First Law of the Non-existence of Cognitive Science" (Fodor), 27
Fitch, W. T., 257
fMRI (functional magnetic resonance imaging), sentence indeterminacy, 127, 128–129
Fodor, J. A., 253. *see also specific topics and works*
  Compositionality Papers, The, 12
  Concepts: Where Cognitive Science Went Wrong, 13
  In Critical Condition: Polemical Essays on Cognitive Science and the Philosophy of Mind, 15

Fodor, J. A. (*cont.*)
  Elm and the Expert: Mentalese and Its Semantics, The, 13
  "First Law of the Non-existence of Cognitive Science," 27
  on functionalism, 4
  Holism: A Shopper's Guide, 12
  Hume Variations, 15
  Language of Thought, The, 6–7, 249
  Language of Thought Revisited, The, 15
  Mind Doesn't Work That Way, The, 16
  Minds without Meanings: An Essay on the Content of Concepts, 16–17, 232n5
  Modularity of Mind, The, 8–9, 114
  Psychological Explanation, 4
  Psychology of Language: An Introduction to Psycholinguistics and Generative Grammar, The, 3
  Psychosemantics: The Problem of Meaning in Philosophy of Mind, 12–13
  Representations: Philosophical Essays on the Foundations of Cognitive Science, 5, 7, 13, 15
  Theory of Content and Other Essays, A, 12–13
Fodor, J. D., 4, 14
Fodorian cognitive science, 1–18
  on analyticity, 12
  belief fixation, 7, 34
    perception, 8–9
  complex representations, 11
  compositionality, 11–12
    nonnegotiable assumption, 14
  Computational/ Representational Theory of Mind, 6, 10, 14
  concepts, 7
  conceptual representation, atomism, 8–9, 14
  connectionism, 10
  on decompositional views, 14
  empiricism, 8
  functionalism, 4, 5
  history, 2–3
  on holism, 12, 14–15
  impact, 17–18
  internal representations, theory of, 6–7
  kill/cause-to-die, 13–14, 15–16
  lexical atomism, 13
  lexical semantics and semantic markers, 13
  linguistic wars, 13–14
  meaning as myth, 16–17
  mental content, 7
  modularity
    of mind, 8–10, 65–66
    of perception, 9
  modules, 3, 8–9 (*see also* modularity *(On Modularity)*)
  nativism, 7–8
  PET FISH problem, 14
  referents and symbolic representations, 16–17
  on theories, 14
Foldi, N., 46–47
form
  logical, 120
  morphological, grammatical meaning and, 170
  one-to-one mapping between meaning and, 172–173
format (thought), 254–266
  internal uses of language, spatial cognition, 259–266
  PHONs and inner speech, 254–259
  representational vehicle, 252
Forster, K. I., 43–44, 55, 56, 239
Fraisse, P., 100
Frankish, K., 257, 258
Frazier, L., 4, 43, 67
Frege, G., 16
functionalism
  in cognitive science, 4
  as materialism, 5
  in philosophy of mind, 5

Galanter, E., 255
Galileo Galilei, 205n2
Gallistel, C. R., 27, 288, 291–292
Garrett, Merrill, 2–3, 14, 25, 78–79, 89, 92, 139–140
  The Psychology of Language: An Introduction to Psycholinguistics and Generative Grammar, 3

Generality Constraint, 252–253
generalizations, 103
generative enterprise, 25, 27
generative semantics, 13
genes, stored information in, 278
Gibbs, R., 46
Gibson, E., 139, 156–157, 221
Gillon, B. A., 128–129, 129*f*
Gleitman, Lila R., 201, 232n4, 276
glomeruli, 290
good-enough processing, 64, 74–75, 93
Goodman, Nelson, 213–214
grabbing mechanism, 201
gradual learning graphs, 232n4
grain, 99
grammar, phrase structure, 4
grammatical meaning, morphological form and, 170
grammatical morpheme, Turkish
  acquisition, 184–186
  children's initial use, 169
  lexical root frequency and, child speech, 176–183, 178*t*, 179*t*, 181*t*
grammatical transformations, communicative efficiency and, 33
Gricean Maxim of Quantity, 71

Hagoort, P., 49–50
Halle, M., 161
Hasentaub, A., 281
Hauser, M. D., 257
Hemholtzian unconscious inference, 99–100
Hempl, Carl G., 213
Henderson, J. M., 69
Hermer-Vázquez, L., 259–260, 266
Hesslow, G., recent research, 282, 283–286
heuristic rules, 116, 134n2
hierarchically structured predictions, ongoing confirmation with error corrections, 93–94
hippocampal frames of reference, abrupt changes, 287–289
Hirose, Y., 164
Hoffman, Donald, 28
holism, 12, 14–15
Holism: A Shopper's Guide (Fodor and Lepore), 12
Holmes, V. M., 68, 69
horizontal faculties, 117
Huang, Y., 47
Hudson, R., 140
Hume Variations (Fodor), 15
Husserl, E., 100
hypothesis testing, 212–214, 213*f*
  hypothesis, source, 214
hypothesis formation and, 214

I-language, 26–27, 33, 37n2, 222
  internalization, 35
  unbound nesting, 34
imagery (mental), 204–205
  contemporary theories, 191
  Paivio's model, 240
imagining it, 256
immediacy assumption, 97–99
In Critical Condition: Polemical Essays on Cognitive Science and the Philosophy of Mind (Fodor), 5
indeterminate sentence, 115, 123–130
  begin-entity, 125–126
  biasing discourse, 132–133, 133*f*
  as challenge to modularity, 130
  coercion with interpolation, 123–127
  definition and example, 123
  fMRI studies, 127, 128–129, 129*f*
  MEG studies, 126, 127–128
  type shifting, 124
  underdetermining, 123
indexing (FINSTing), 192, 193, 201
  vs. selecting-by-property, 192
indication, 192
indirect addressing, 279
indirection, 46–47
individuality, enduring, 192
individuating, 200
induction, 215
  Carnap's theory, Putnam's critique of, 218
inductron, 226
inference, scalar, 47–48
inferior frontal gyrus (IFG), 128

information. *see also specific types*
  encapsulation, 66, 80, 113, 114, 117–118
  language-internal sources, 68–71
  non-linguistic background, sentence processing and, 42
  stored, computers and genes, 278
Inhelder, B., 221
innateness, all (basic) concepts, 7–8, 13, 211–233
  conclusion, author's, 230–231
  Fodor's argument, 212, 214
  Fodor's suggestion, Royaumont debate, 218–219
  learning, 212–214, 213*f*
  learning paradigms, three most popular
    1, 215
    2, 215
    sub-option 1, 215–216
    sub-option 2, 216–217
    suggestion "doomed to fail," 217–218
  meaning *vs.* sorting, 219–220
  objections and counters, 223–230
    Euler's theorem, 224–225, 224*f*
    Fodor's counter, 225–226
    Fodor strikes back, 228–230
    Papert on connectedness, 223–224
    Putnam's critique of Fodor, 226–227
    Putnam's own critique of associationism, 227–228
    Putnam's rejoinder, 230
  triggers, 216, 220–223, 233n15
inner speech, 254–258
  behaviorists on, 255–256
  PHONs and, 254–258
  in thought, 256–257
intellectual revolutions, 1
interaction
  sentence processing and non-linguistic background information effects, 42
  weak, 72–73
interactive perspective, *vs.* modularity, 41
internal Merge, 31
internal representations, theory, 6–7
internal uses of language, spatial cognition, 259–266
interpolation, coercion with, 123–127
interpretive semantics, 13
interstimulus interval, 282
intra-modular mental processes, 116
intra-modular semantic representations, 116

Jackendoff, Ray, 14
Jarema, G., 128–129, 129*f*
Jespersen, Otto, 26
Johansson, F., 287

Katz, Jerrold, 13–14
Kayne, Richard, 37–38n11
Kempen, G., 53
Ketrez, F., 173, 183, 185
kill/cause-to-die asymmetry, 13–14, 15–16
knowing X, 35
knowledge-based process, symbolic/ algorithm level, 5
Konopka, A. E., 51
Koralus, P., 267
Kornfilt, J., 172
Kortes third "law" of apparent motion, 193
Kroll, J. F., 242

Ladefoged, P., 89
language acquisition, 25–26, 88–89
  acoustic input problem, 88–89
  computational problem, 88
  as reflex, 25–26
language and thought, 249–270
  coda, 266–268
  computational-representational paradigm, 251–252
  content or proposition, 252
  format, 254–266
    internal uses of language, spatial cognition, 259–266
    PHONs and inner speech, 254–259
  Generality Constraint, Evans's, 252–253
  overview, 249–250
  PHONs and SEMs, 251
  representational vehicle of thought representation, 252
  representation types, 250–251
  SM and C/I systems, 251
  speech shadowing, 260, 261–263
  systematicity of thought, 10–12, 219, 253, 268n2, 293

language disorders, language production, 56–57
language experience, language processing unit and, 90–92
language generation *vs.* language production, 26–27
language-internal sources, of information, 68–71
language learning. *see also specific topics*
  children's, 87–106 (*see also* children's language learning)
  module for, 27–28
  as reflex, 25
language of thought (LoT), 6–8, 103, 249–270. *see also* language and thought
  connectionists on, 276
Language of Thought, The (Fodor), 6–7, 249
Language of Thought Revisited, The (Fodor), 15
language processing
  feed forward, 90, 91
  feed forward and feed backward processing, 91
  time course, 42
language processing units, 88–92
  child discovery, 88
  location, 89–90
  unity of, and experience of language, 90–92
language production. *see also specific aspects*
  expressions, 26
  vs. generation, 26–27
  lexically based routines, 53
  speed of lexical recognition and structure projection, 54
  study of, 48–52, 50*f*
language production, as comprehension filter, 52–58
  acquisition profiles, 57–58
  context, discourse, and plausibility effects, 55–56
  fundamentals, 52–53
  language disorders, 56–57
  lexically based production routines, 53
  profiles predict "comprehension" performance, 54–55
  speed of lexical recognition and structure projection, 54
language universals, 37n9
late assignment of syntax theory (LAST), 64, 74, 78, 97, 98
late closure, 67
Lau, E. F, 267–268
learning
  as acquisition, 220
  bona fide, 212
  concepts, Bruner and Piaget on, 7, 13, 211–233 (*see also* innateness, all (basic) concepts)
  constraints on, 213–214
  gradual learning graphs, 232n4
  innateness of all concepts and, 212–214, 213*f*
  language, 25, 27–28 (*see also specific topics*)
  multipurpose strategies, 230
  Piattelli-Palmarini on, 212–214, 213*f*
  serial learning models, 102
  single stimulus learning, 223
learning function, Papert's, 224
lemmas, 49, 50*f*
Lennenger, Eric, 27
Lepore, Ernie, 12, 125, 130
Levelt, W. J. M., 49
lexical ambiguity, resolution of, 244
lexical (conceptual) atomism, 8, 9, 13, 14–15
lexical bias–plausibility interaction, 45
lexicalization process, 264–265
lexically based production routines, 53
lexical preferences, 53
lexical recognition, speed, 54
lexical representations, abstract, 49, 50*f*
lexical retrieval, for syntactic integration, 49–51, 50*f*
lexical roots. *see* roots, lexical
lexical semantics, 13
limited central access, 65
Linear Correspondence Axiom, Kayne's, 37–38n11
Linebarger, M., 56
linguistic universals, 37n9
linguistic wars, 13–14

linking hypothesis, 73–74
literal meaning, sentence, 41
Locke, John, 1, 212
  program failure, 214, 216–217
Loebell, H., 51
logical form, 120
long sentence recall, 244–245
Lovrić, N., 164
Lucy, P., 46
Lungu, O., 128–129, 129*f*
Lurz, R. W., 256

Macdonald, M. C., 97
magnetoencephalography (MEG), sentence indeterminacy, 126, 127–128
mandatory operation, 66
Manouilidou, C., 128–129, 129*f*
Marr, David, 5
Marslen-Wilson, W., 43–44, 54, 55, 261, 262–263
Matzel, L. D., 288
Mazuka, R., 70
McDonald, M. C., 54
McElree, B., 126
McGann, John, 289–292
McGilvray, J., 263, 267
McLaughlin, B., 253
meaning *vs.* sorting, 219–220
memory
  charge of original representation, on later interpretation, 77
  computer, 278–279
  conceptual short-term, 239, 245
  false, 133
  genetic, 278
  long-term, stimulus activation, 245
  overload, 164, 262
  symbolic, 275–276, 277
  synaptic theory of, 276, 281
  word representation, 73
merge, 30–32, 250
metaphor, 46
Miller, C. A., 55
Miller, G. A., 2, 3, 34n18, 140, 255
mind, modularity of, 8–10, 65–66
Mind Doesn't Work That Way, The (Fodor), 16
Minds without Meanings: An Essay on the Content of Concepts (Fodor and Pylyshyn), 16–17, 232n5
minimal attachment, 67, 71
minimal computation (MC), 29–32
  Merge, 30–32
  Move, 31
Minimalism, 88, 102
Mitchell, D. C., 68, 69
Mode of Presentation (MOP), 217
modularity. *see also specific topics*
  case for, 114–115
  central-system, 26–32 (*see also* central-system modularity)
  definition, 113
  degrees, 66
  vs. interactive perspective, 41
  of mind, 8–10, 65–66
  parsing and, 42–46 (*see also* parsing)
  of perception, 9
  proprietary databases, 68
  research programs, 113
  of sentence processing, 63–81 (*see also* sentence processing, modularity)
  of visual perception, 201–205, 203*f*
modularity, limits of, 41–58
  experimental pragmatics and intimations of modularity, 46–48
  fundamentals, 41
  language production
    as comprehension filter, 52–58 (*see also* language production, as comprehension filter)
    study of, 48–52, 50*f*
  parsing, 42–46 (*see also* parsing)
modularity, two notions, 25–38
  biolinguistic framework and generative enterprise, 25, 27
  central-system, 26–32 (*see also* central system)
    vs. parsing system, 32–34
  embedding, 34, 38n19
  expressions, language production, 26
  grammatical transformations and communicative efficiency, 33
  I-language, 26–27, 33, 37n2
  input modules, 26

inspiration for, 25
internal cognitive system, 27
knowing X, 35
language learning
module for, 27–28
as reflex, 25
neocartesianism, 34
neural mechanism development, 36
parsing as reflex, 25–26
parsing system, 27, 32
performance *vs.* competence, distinguishing, 34
production *vs.* generation, language, 26–27
Quinean and isotropic internal systems, 9, 34, 35–36
semantic role assignment, order-independent, 30
modules, 3, 9. *see also* modularity *(On Modularity); specific types*
automatic and quick operation, 66
biological properties, 65
domain-specific, 66
emergence during development, 65
information encapsulation, 66
input, 26
limited central access, 65
neural specialization, 65
selectively impaired, 65
semantics for, 113–134 (*see also* semantics, for module)
shallow outputs, 65, 119–123
superficiality, 65
Monod, J., 221
morphemes, grammatical
children's initial use, 169
English *vs.* Turkish, 170
lexical roots *vs.*, child identification, 170
Turkish, 170
acquisition of, 184–186
lexical roots and, child speech, 176–183, 178*t*, 179*t*, 181*t*
morphological form, grammatical meaning and, 170
morphologically complex words, in adult Turkish speech to children, 174–176, 175*t*
morphology acquisition, Turkish, 169–188. *see also* Turkish morphology acquisition
morphosyntax, Turkish, 172–173
Moses illusion, 75
motherese, child-directed, *vs.* normal conversation, 101, 104
Multiple Object Tracking (MOT), 16, 193–196, 195*f*
multipurpose learning strategies, 230

Nakamura, C., 70
NAND gates, 280
nativism. *see* innateness, all (basic) concepts, 7–8
natural constraint, 200
neocartesianism, 34
neural mechanism development, 36
neural net modelers, 282–283, 287
neural specialization, modules, 65
neurobiological basis, computational theory of mind, 275–294. *see also* computational theory of mind, neurobiological basis
neuroscientists, cognitive scientist, 5
neurotransmitter release, learned, selective immediate enhancement from first-order olfactory neurons, 289–292
Newcombe, N. S., 261
Newell, Alan, 5
New Look in Perception, 196
Nicol, J., 55
noisy channel/rational communication, 64, 76, 79

olfactory neurons, first-order, learned, selective immediate enhancement of neurotransmitter release, 289–292
Optimality Theory, 142
order-independent semantic role assignment, 30
Osgood, C., 2–3, 103
output, language model, shallow proposition, 65, 119–123

Paivio, A., 240–241
Papert, S., 214
  on connectedness, 223–224
paradoxical syntactic capacity, 56
paradoxic comprehenders, 57
Parker, D., 267
parsing
  definition, 26
  forward, 90
  meaning representations and, 121
  models, early, 4
  as reflex, 25–26
parsing, modularity and, 42–46
  argument structure effects, 44–45
  availability, 42
  biasing effect variance, 45–46
  cognitive function compromise, 42
  contextual constraint, 42, 44
  contextually driven interpretive bias, 44
  lexical bias–plausibility interaction, 45
  meaning variables lacking, 43
  prepositional phrase attachment, 45
  semantic constraint, 42
  structural processing ambiguities, 42
  subcategorization of trace following intransitive verb, 45
  syntactic effects and plausibility, 43
  time course for processing, 42
  two stage proposals, 42
  verb phrase interpretation as relative clause modifier, discourse setting on, 44
parsing system
  vs. central-system modularity, 32–34
  embedding, 34, 38n19
  filler-gap problems, 32
  nesting on, 34
  performance *vs.* competence, 34
Pavlovian conditioning, 282
perception
  autonomy, 9
  cognitively penetrable, 9, 197
  experiences of, as mundane symbols, 241–242
  feed forward models, 90
  modularity, 9
  of scenes, temporal limit, 245–246
perceptron
  Euler's theorem, 224–225, 224*f*
  Papert on, 223–224
perceptual analysis, constrained, 9
perceptual circle, 16–17
perceptual information, linked to abstract concepts, 241
performance *vs.* competence, distinguishing, 34
PET FISH problem, 14
phase, 96–97
Phi illusion, 193
Phillips, C., 98, 267–268
phonological cues, children's language acquisition, 170
PHONs, 251
  inner speech and, 254–258
  representations, 250, 254
phrase
  listener-imposed structure, click-mislocation, 89
  meaning comprehension time, 242–243
  predicting final words, 90
  psychological reality, 90
  surface breaks, relative depth, 89
phrase structure grammar, 4
physical-mental gap, present view consequences, 198–201
Piaget, J., 8, 214, 221
Pickering, M. J., 126
picking-out *vs.* locating, 192
Pietroski, P. M., 135n10, 251, 258–259, 263, 264–266, 270nn13–14
pitfalls, 204
plausibility
  bias towards, 64–65
  language production, 56
  lexical bias–plausibility interaction, 45
  sentence processing, 71–73
  syntactic effects, 43
pointer arithmetic, 279
Popper, Karl, 276
positive time displacement, 100
"possible candidates" problem, 201
poverty of the stimulus, 87–88
  real, 101–104

pragmatics
  definition, 46
  experimental, intimations of modularity and, 46–48
  experimental findings, conflict, 43
  indirection, 46–47
  metaphor, 46
  processing, 115, 127
  vs. psycholinguistics, rationalization of conflicting findings, 43
  right and left hemisphere damage on, 46–47
  scalar inference, 47–48
prepositional phrase attachment, 45
preposition effect, 99
Pribram, K. H., 18–19n2, 255
priming
  lexical and structural, 51
  syntactic, 51
principles and parameters, 220–221
prior entry effect, 99, 100
probabilistic language of thought (pLoT), 103–104
process-architecture distinction, 198
process *vs.* processing architecture, 192
productivity, 11
proposition, 252
propositional content, 121
prosodic information, on sentence processing, 69–70
Prosodic Phrase Processor, 145
prosodic phrasing, facilitative, 141–146
  balanced, 143
  constraints, 142, 143*b*
  difficulty, intuitive judgments, 146, 146*b*
  syntactic tree structure, 141–142, 142*f*
prosody experiments, elicited, 146–163
  experiment 1 (rating task with familiarization), 146, 147–155
    materials, 147–148, 147*b*
    participants and procedure, 148–149
    predictions, 150
    results, comprehensibility judgments, 152–153, 152*f*
    results, produced prosodic phrasing, 153–155, 154*f*–155*f*
    results, pronounceability judgments and produced prosody evaluation, 150–152, 150*f*–151*f*
  experiment 2 (in search of "missing VP illusion"), 146, 155–160
    discussion, 160
    materials, 156–157, 157*b*
    participants and procedure, 157
    purpose, 155–156
    results, 157–160, 158*t*–160*t*, 159*f*
  explanations, 161–162, 162*f*, 163*f*
  fundamentals, 146
psycholinguistics, 103
  Cartesian, 3
  encapsulation, 10
  experimental findings, conflict, 42–43
  vs. pragmatics, rationalization of conflicting findings, 43
Psychological Explanation (Fodor), 4
psychology of language, 3
Psychology of Language: An Introduction to Psycholinguistics and Generative Grammar, The (Fodor, Bever, and Garrett), 3
Psychosemantics: The Problem of Meaning in Philosophy of Mind (Fodor), 12–13
Purkinje cells, 283–286, 284*f*
Pursuit of Truth (Quine), 36
Putnam, Hilary, 33, 213, 214, 218
  critique of Fodor, 223, 226–227
  "doing OK with," 222
  Fodor on Putnam's critique, 230
  Fodor on Putnam's critique, Putnam's rejoinder to, 230
  own critique of associationism, 227–228
Pylyshyn, Zenon, 5, 9, 10–11, 134n1, 293
  Minds without Meanings: An Essay on the Content of Concepts, 16–17, 232n5

qua explanatory level, 5
Quine, W. V. O., 12, 35–36, 199, 200, 214
Quinean and isotropic internal systems, 9, 34, 35–36, 117
Quine's Gavagai problem, 201

rapid serial visual presentation (RSVP), 239
reading, self-paced, 69

reading frame, 293–294n1
real world model, *vs.* discourse model, 121, 133, 135n8
reanalysis processes, 92–93
Referential Theory, 74
referents, symbolic representations and, 16–17
referring, 199
reflex
 language acquisition as, 25–26
 parsing as, 25–26
replicability, 126
Representational/Computational Theory of Mind, 116, 134–135n3. *see also* Computational/Representational Theory of Mind (C/RTM)
 size, 280
representational vehicle (format), of thought representation, 252
representations
 algorithm level, 5
 complex, 11
 complex, compositionality and, 11–12
 conceptual, 253
 thought, 252–253
Representations: Philosophical Essays on the Foundations of Cognitive Science (Fodor), 5, 7, 13, 15
Rigidity Rule, 28
Riven, L., 128–129, 129*f*
Rizzi, Luigi, 220
roots, lexical, 170
 frequency, in adult Turkish speech to children, 174–176, 175*t*
 vs. grammatical morphemes, child identification, 170
 Turkish, 183–184
  grammatical morphemes and, child speech, 176–183, 178*t*, 179*t*, 181*t*
  morphology acquisition, 183–184
Royaumont Debate, 214
 Fodor's suggestions, 218–219
RTM. *see also* Representational/ Computational Theory of Mind
Rule of Structure-dependence, 28, 29
rules, innate, 216
Ryle, G., 4
Sachs, J. S., 132
Saffran, E., 56
same-different matching test, 44
Samuels, R., 261, 262–263
Sanford, A. J., 77
Sanz, M., 43
"sausage machine," 4, 43, 145
scalar inference, 47–48
Schlesewsky, M., 119
Scholl, B. J., 204
Schwartz, M., 56
search space, 4
second language, representation of, 242
Sedivy, J., 74
seeing, 196–197
seeing as, 197
selectively impaired modules, 65
self-embedding, 34, 38n19
self-paced reading, 69
Selkirk, E., 170
semantic composition. *see* compositionality
semantic constraint, 42
semantics
 control, 49
 features, 12
 generative, 13
 interpretive, 13
 for Katz and Fodor, 13
 lexical, 13
 markers, 13
 role assignment, order-independent, 30
semantics, for module, 113–134
 case for modularity, 114–115
 cognitive architecture and, 115–117
 context sensitivity, 130–133, 133*f*
 linguistic, boundaries, 117–119
 output, shallow proposition, 65, 119–123
 sentence indeterminacy, 115, 123–130, 129*f*
SEM representation, 253
SEMs, 251
sentence comprehension studies, Potter's, 240–245
 perceptual information linked to abstract concepts, 241

phrases and clauses, meaning comprehension time, 242–243
two-clause sentences, spoken, 243
two-stage modular interactive model, word processing, 244
verbal and perceptual information linked to abstract concepts, 241
verbal information linked to abstract concepts, 241
word understanding and sentence meaning processing, at 12 words per second, 243–244
sentence indeterminacy, 115, 123–130, 129*f.* *see also* indeterminate sentence
sentence processing
computations, *vs.* sequences of syntactic transformations, 3–4
construal, 74
literal meaning, 41
non-linguistic background information, 42
nonmodular view, 63–64, 70
prosodic information, 69–70
verb information, initial processing, 69
verb-specific syntactic information, 69
sentence processing, modularity, 63–81. *see also specific topics*
context and plausibility information, 71–73
history, 63
language-internal sources of information, 68–71
mind modularity, 8–10, 65–66
nonmodular view *vs.*, 63–64
original proposal, 64
sentence comprehension shallowness, 64–65
shallow processing, 76–79
two-stage model, 63, 67–68
visual world paradigm, 64, 73–76
serial encoding, 91
serial learning models, 102
shallow outputs, modules, 65, 119–123
shallow processing, 64
shallow processing, sentence
comprehension shallowness, 64–65
modularity, 76–79
short-term memory, conceptual, 245
single stimulus learning, 223
Skinner, B. F., 255–256
Slezak, P., 256
Slobin, D. I., 170, 184, 185–186
SM interface, 250–251
Snedeker, J., 47
sorting, meaning *vs.*, 219–220
spatial cognition, internal uses of language, 259–266
spatial language, 260–261
speech, inner, 254–258
speech shadowing, 260, 261–263
speed
language processing, 42
lexical recognition and structure projection, 54
meaning comprehension, of clauses, 242–243
Spelke, E., 259
Spencer-Brown, George, 192
Spivey, M. J., 74
Stabler, E., 263–264
Steedman, M., 44, 55, 71–72
Steedman's Combinatory Categorial Grammar, 72
Strawson, P. F., 199, 200
structural processing, ambiguities, 42
structure-dependence, 88, 102
Structure-dependence, Rule of, 28, 29–30
structure dependence theories, 88, 102
structure projection, speed, 54
Sturt, P., 77
superficiality, modules, 65
Swets, B., 76–77
Swinney, D., 92, 244
symbolic/algorithmic level, 5
symbolic architecture, 116
symbolic cognitive architecture, 5, 6, 10–11, 16, 17, 19n7, 115–121, 135n5, 205n4
symbolic-computational view, 115–117
symbolic representations, 115–116
referents and, 16–17
symbols
brain, 277–278
manipulation rules, xxx

symbols (*cont.*)
  mundane, perceptual experiences as, 241–242
  and rules for manipulating them, 275
  synapses and, 276
  types, 117
synapses
  in memory, 276
  symbols and, 276
synaptic theory of memory, 276
syntactic capacity, paradoxical, 56
syntactic effects, plausibility and, 43
syntactic integration, lexical retrieval for, 49–51, 50*f*
syntactic priming, 51, 245
syntactic processing effects, 41
syntax, complexity, 12
System 1, 78
System 2, 78
systematicity of thought, 10–12, 219, 253, 268n2, 293

Tanenhaus, M., 45, 74–75, 93–94
theoretical advances. *see also specific types*
  empirical data in, 14
theories, Fodor on, 14
Theory of Content and Other Essays, A (Fodor), 12–13
theory of mind
  computational, 275 (*see also* computational theory of mind, neurobiological basis)
  modules, 3, 9 (*see also* modularity *(On Modularity)*)
Thomas, J., 139, 156–157
Thompson, Richard, 283
thought
  content or proposition of, 252
  Generality Constraint, 252–253
  inner speech, 256–257
  language and (*see* language and thought)
  representations (format), 252–253
  systematicity, 10–12, 219, 253, 268n2, 293
time course, language processing, 42
tout court cognition, 16
Townsend, D., 43, 78, 97
tracking moving objects, 16, 193–196, 195*f*
transcription factors, 278–280
Traxler, M. J., 131
Triesman, A., 245
triggers, 220–223, 233n15
  result and, 216
Truckenbrodt's Wrap constraint, 143*b*, 161
Trueswell, J., 45
Turing, Alan, 1, 3, 16, 19n9, 26, 34–35, 269n8
Turkish morphology acquisition, 169–188
  data to date, 173
  discussion, general
    grammatical morpheme acquisition, 184–186
    lexical roots, 183–184
  form and meaning, one-to-one mapping between, 172–173
  lexical root frequency and morphologically complex words, in adult speech to children, 174–176, 175*t*
  morphemes
    English *vs.* Turkish, 170
    grammatical use, children's initial, 169
    Turkish, 170
  morphological form and grammatical meaning, 170
  morphosyntax, 172–173
  Turkish language linguistics, 169–170
  vowel harmony, 170, 171–172, 181–182, 184, 185, 186
two-clause sentences, spoken, 243
two-stage model, sentence processing, 63, 67–68
two-stage modular interactive model, word processing, 244
Twyman, A. D., 261
Tyler, L., 43–44, 55
type shifting, 124
types of representations (TRs), 250–251

unbound nesting, 34
unconscious comprehensive processes, with backward inferences, 92–98

unconscious inference, Hemholtzian, 99–100
Uniformity/Balance principle, 143
unity of consciousness and consciousness of unity, 87–106. *see also* children's language learning
Universal Grammar (UG), 28, 102
language universals and, 37n9
universals, language and linguistic, 37n9
unrelated word recall, 244–245
utterance interpretation, context on, 43

Van de Ven, M. A. M., 95
Van Turennout, M., 49–50
variable binding, 278–279
verb
information
initial processing, 69
linked to abstract concepts, 241
sentence processing, initial, 69
intransitive, subcategorization of trace following, 45
types, 117
verb phrase interpretation as relative clause modifier, discourse setting on, 44
verb-specific syntactic information, sentence processing, 69
verb-thematic hierarchy, Alzheimer patients, 118
Veres, C., 55
vertical faculties, 117
visual perception. *see also* exceptionalism, Fodorian, scientific theories
early vision, 197, 198–199, 202
modularity of, 201–205, 203*f*
visual world paradigm (VWD), 64, 73
modularity, 73–76
von Humboldt, Wilhelm, 26
vowel harmony, 170, 171–172, 181–182, 184, 185, 186

Wagers, M. W., 267–268
Watson, J. B., 255–256
weak interaction, 72–73
Wexler, K., 221
wh-island constraint, 38n19
Whitney, William Dwight, 26
words
final in phrase, predicting, 90
meanings, ambiguous, on child's lexical decision task, 92
morphologically complex, in adult Turkish speech to children, 174–176, 175*t*
understanding, at 12 words per second, 243–244
unrelated, recall, 244–245
world-mind gap, 205
Wundt, W., 3, 99

Yeo, C. H., 283

Zurif, E., 56, 57
Zwitserlood, P., 54

Printed in the USA/Agawam, MA
October 12, 2020

762622.008